Plant Genetic Resources and Traditional Knowledge for Food Security

Romesh Kumar Salgotra
Bharat Bhushan Gupta
Editors

Plant Genetic Resources and Traditional Knowledge for Food Security

Springer

Editors
Romesh Kumar Salgotra
School of Biotechnology
Sher-e-Kashmir University
of Agricultural Sciences
and Technology of Jammu
Jammu, Jammu and Kashmir, India

Bharat Bhushan Gupta
Division of Plant Breeding and Genetics
Sher-e-Kashmir University
of Agricultural Sciences
and Technology of Jammu
Jammu, Jammu and Kashmir, India

ISBN 978-981-10-0058-4 ISBN 978-981-10-0060-7 (eBook)
DOI 10.1007/978-981-10-0060-7

Library of Congress Control Number: 2015959595

Springer Singapore Heidelberg New York Dordrecht London

Printed on acid-free paper

Springer Science+Business Media Singapore Pte Ltd. is part of Springer Science+Business Media (www.springer.com)

Foreword

Growing population, declining productivity, shrinking resource base, changing climate and ethnic concerns are the challenges towards food security globally. Plant Genetic Resources (PGRs) and Traditional Knowledge (TK) constitute our invaluable assets in this direction. They play an increasing role in world food security and economic development. They are crucial for sustainable agricultural growth and offer livelihood security to agrarian society. Therefore, conservation and effective utilization of PGRs and TK, which constitute a unique global heritage, is of immediate concern.

The United Nation Department of Economic & Social Affairs Population Division has estimated world population around 7.4 billion (July, 2015). It is expected to stabilize around 9 billion by 2050. Even today, millions of people go hungry every day, and the scenario may be more serious and disturbing in future. The agriculture sector today is facing challenges like shrinking cultivable land, dwindling water resources, changing climate, energy crisis, labour issues and increasing cost of inputs. The agriculture has to grow against all these odds with efficient utilization of all inputs and management skills.

Crop improvement is one way of achieving the desired goals, which requires a very strong bio-diversity basis. The IPR, protection and promotion of farmers' rights, access and benefit sharing to genetic resources and relevant knowledge are key issues for the developing and developed countries. These will provide steady and fairly legal basis for relevant stakeholders' parties. Regarding challenges of food security in future, a common outlook in management and use of plant resources genetic, indigenous/traditional knowledge and intellectual property should be discussed, shared and released.

In addition, various international conventions have addressed the issue of genetic erosion and declining use of agro-biodiversity in modern intensive agriculture. Concern about the future vulnerability of agricultural production, food security and environmental stability has moved the conservation and sustainable use of plant genetic resources to the top of the international development agenda. An enhanced use of plant genetic diversity is essential to address these and other future challenges.

The book, *Plant Genetic Resources and Traditional Knowledge for Food Security*, is designed to provide a comprehensive picture of the global situation and trends regarding the conservation and use of PGRs and TK. It describes the current status and the significant improvements made through PGRs and TK and brings out major gaps between erosion of PGRs and TK to that of their preservation to set future priorities for the conservation and sustainable utilization of plant genetic resources for agriculture. The book emphasizes on the importance of an integrated approach for the management of PGRs and TK. It points out the need to secure broad diversity of crop plants, including their wild relatives and underutilized species in the accessible conservation systems, and to increase capacities for plant breeding and seed delivery worldwide.

One of the important issues explored in this publication is the need for a joint effort between farmer innovators and researchers to further improve modern and traditional farm practices. Equally important to the application of TK is the monitoring and assessment of its effects in combating desertification. Many of the issues surrounding the promotion, protection and use of plant genetic resources and traditional knowledge are being addressed through international collaborative efforts. It has been emphasized to realize the significance and benefit of such works at the local level.

It is hoped that this publication will serve to sensitize the scientific community, local partners and the various specialized agencies for their willingness and commitment in countering the problem of desertification. I hope and believe that the information contained in this book will be useful for policy planners, researchers, teachers, students and farming community.

Chatha
Jammu, Jammu and Kashmir, India

(Pradeep K. Sharma)
Vice Chancellor
SKUAST of Jammu (J&K)

Preface

The preservation, protection, and promotion of the Plant Genetic Resources (PGR), Traditional Knowledge (TK), innovations, and practices of local and indigenous communities are of key importance for the development of a country. The rich endowment of PGR and TK plays a critical role in health care, food security, culture, religion, identity, environment, sustainable development, and trade. But these valuable assets are at risk in many parts of the world, and here are concerns that PGR and TK are being used and patented by third parties, with few or none of the benefits being shared with the original traditional knowledge holders and without their prior informed consent. While such concerns have pushed PGR and TK to the forefront of the international agenda, the best ways of addressing the range of issues related to its preservation, protection, further development, and sustainable use are not yet clear.

PGR and TK represent the basic building blocks for the sustainable development of the agro-based industries in a country. Empowerment of local communities is a prerequisite for the integration of PGR and TK into the development process. The integration of PGR and TK into development programs has already proved to contribute to efficiency, effectiveness, and sustainable development. Supporting local and regional networks of traditional practitioners and community exchanges can help to disseminate useful and relevant indigenous knowledge and to enable communities to participate more actively in the development process. However, experience has shown that this cannot be done by one institution acting alone. Therefore, partnerships are needed to support this process at all levels.

A great part of the modern pharmaceutical industry has been developed on the basis of medicinal plants discovered by indigenous peoples and local communities; nevertheless, the economic benefits that these medicines give remain in the companies, without any type of recognition or compensation for those who created these valuable resources. Supporting small-scale farming systems and indigenous communities who are having indigenous knowledge, improving their agricultural productivity and livelihood base, is crucial in this context.

The growing global demand for food, feed, fiber, and bio-based renewable materials, such as biofuels, is changing the conditions for genetic resources development and bio-resource production worldwide. TK in the fields of medicine, healing, and biodiversity conservation are well known, and the need for protection of this TK is a crosscutting issue at the moment involved in discussions from different institutions, with different approaches. Only 1 % of TK is known by scientists and accepted for commercial purposes. Part of the modern pharmaceutical industry is developed on the basis of PGR discovered and use by indigenous peoples and local communities, even though the economic benefits are not equitably shared.

International conventions and treaties together with the rapid bioscience development have led to new conditions for the access of genetic resource and knowledge. The Convention on Biological Diversity, 1992 (CBD), mandates the contracting parties to preserve and maintain knowledge, innovations, and practices of indigenous peoples and local communities embodying traditional lifestyles relevant for the conservation and sustainable use of biological diversity and promote their wider application with the approval and involvement of the holders of such knowledge, innovations, and practices and encourage the equitable sharing of benefits arising from their utilization. The Nagoya Protocol further regulates the conditions for accession to genetic resources and the fair and equitable division of monetary and nonmonetary benefits raised from their use. This protocol regulates the rights and obligations of entities that use genetic resources for research, development, and commercialization or that use traditional knowledge associated with genetic resources.

Recent developments related to Intellectual Property Rights (IPRs) for PGR and TK have created both threats and opportunities for developing countries. With the advent of genetic engineering and novel multilateral IPR agreements, a new chapter in the history of gene hunting has started while genetic resources have been important raw materials in agriculture and in the natural products industry, biotechnology, and the information technology, and novel IPR regimes are opening new frontiers. The demand for PGR and TK is likely to grow rapidly with the improved techniques for rapid genetic characterization and screening of valuable genetic traits and the possibilities to improve, modify, and add value to genetic resources. Thus, the era of bioinformatics and genetic engineering will have a major impact on the trade of PGR and TK and the development of natural products industries worldwide.

Farmers' Rights are essential for maintaining plant genetic diversity, which is the basis of all food and agricultural production in the world. This book shows the necessity of realizing Farmers' Rights for poverty alleviation and food security, the practical possibilities of doing so, and the potential gains for development and society at large. It provides decision-makers and practitioners with a conceptual framework for understanding Farmers' Rights and success stories showing how each of the elements of Farmers' Rights can be realized in practice. The success stories have brought substantial achievements as regards one or more of the four elements of Farmers' Rights: the rights of farmers to save, use, exchange, and sell farm-saved seed; the protection of TK; benefit-sharing; and participation in decision-making.

This book provides an opportunity for various authors from widely differing backgrounds to explore some of the issues raised by conserving PGR and TK. A wide range of expertise and experiences will be needed to develop realistic approaches for sustainable utilization of genetic resources. Genetic, ecological, agricultural, social, economic, and legal concerns all have to be considered and integrated in developing practical work plans at international, national, and local levels. In this book, authors with experience in different fields explore some of the problems and possibilities from their perspective.

This book, *Plant Genetic Resources and Traditional Knowledge for Food Security*, is designed to provide in-depth information of PGR and TK, their protection, conservation, access, and commercial use on benefit-sharing basis in a balanced and comprehensive fashion to researchers, professionals, and students. The book is well illustrated to help students, researchers, and professionals to better understand the principles and concepts about the PGR and TK. International treaties for sustainable utilization of PGR and TK are well discussed. The style of presentation throughout the book is easy to follow and comprehend. Professionals, researchers, and students are constantly reminded of previous topics of relevance to current topics being discussed. This book is not only an excellent teaching tool, but it is also a suitable reference source for professionals.

Jammu, Jammu and Kashmir, India Romesh Kumar Salgotra
Jammu, Jammu and Kashmir, India Bharat Bhushan Gupta

Contents

Contributors

Hasan M. Abdullah Department of Agroforestry and Environment, Bangabandhu Sheikh Mujibur Rahman Agricultural University, Gazipur, Bangladesh

Vijai Pal Bhadana Crop Improvement Section, ICAR, Indian Institute of Rice Research (IIRR), Hyderabad, Andhra Pradesh, India

Javaid Akhter Bhat Division of Plant Breeding and Genetics, Sher-e-Kashmir University of Agricultural Sciences and Technology of Jammu, Chatha, Jammu, Jammu and Kashmir, India

Dharminder Bhatia Punjab Agricultural University, Ludhiana, Punjab, India

Shiv Datt Indian Council of Agriculture Research (ICAR), IP & TM Unit, Krishi Anusandhan Bhavan-I, Pusa, New Delhi, India

Bharat Bhushan Gupta Division of Plant Breeding and Genetics, Sher-e-Kashmir University of Agricultural Sciences and Technology of Jammu, Chatha, Jammu, Jammu and Kashmir, India

Sulpha Gupta Plant Biotechnology Division, Council for Scientific and Industrial Research (CSIR), Indian Institute of Integrated Medicine (IIIM), Jammu, Jammu and Kashmir, India

Veena Gupta Division of Germplasm Conservation, National Bureau of Plant Genetic Resources, New Delhi, India

Mehfuz Hasan Department of Genetics and Plant Breeding, Bangabandhu Sheikh Mujibur Rahman Agricultural University, Gazipur, Bangladesh

Nguyen Thi Ngoc Hue Plant Resources Center, Hanoi, Vietnam

Awsi Jan Division of Food Science and Technology, Sher-e-Kashmir University of Agricultural Sciences and Technology of Jammu, Chatha, Jammu, Jammu and Kashmir, India

Tripta Jhang Department of Genetic Resource Management, Genetics and Plant Breeding Division, CSIR-Central Institute of Medicinal and Aromatic Plants, Lucknow, Uttar Pradesh, India

Zoran Jovovic Biotechnical Faculty Podgorica, University of Montenegro, Podgorica, Montenegro

Gopal Katna Department of Crop Improvement, CSK Himachal Pradesh Krishi Vishvavidyalaya, Palampur, Himachal Pradesh, India

Rajinder Parshad Kaushik Rice Research Station, CSK Himachal Pradesh Krishi Vishwavidayala, Palampur, Himachal Pradesh, India

Suzana Kratovalieva Ministry of Agriculture Forestry and Water Economy, State Phytosanitary Laboratory, Former Yugoslav Republic of Macedonia

Roohi Mushtaq Department of Biotechnology and Bioinformatics, S.P. College, Srinagar, Jammu and Kashmir, India

Muslima Nazir Department of Botany, University of Kashmir, Srinagar, Jammu and Kashmir, India

Pankaj Pandotra School of Biotechnology, Sher-e-Kashmir University of Agricultural Sciences and Technology of Jammu, Chatha, Jammu, Jammu and Kashmir, India

Deepak Rajpurohit Section of Biotechnology, Department of Post Harvest Engineering and Technology, AICRP on PHET, CTAE, Maharana Pratap University of Agriculture and Technology, Udaipur, Rajasthan, India

Romesh Kumar Salgotra School of Biotechnology, Sher-e-Kashmir University of Agricultural Sciences and Technology of Jammu, Jammu, Jammu and Kashmir, India

Gulzar S. Sanghera Regional Research Station, Punjab Agricultural University, Kapurthala, Punjab, India

Manmohan Sharma School of Biotechnology, Sher-e-Kashmir University of Agricultural Sciences and Technology of Jammu, Chatha, Jammu, Jammu and Kashmir, India

Pawan Kumar Sharma ICAR-National Bureau of Agriculturally Important Microorganisms, Mau, Uttar Pradesh, India

Anju M. Singh Division of Genetics, ICAR-Indian Agricultural Research Institute, New Delhi, India

Sumer Pal Singh Division of Genetics, ICAR-Indian Agricultural Research Institute, New Delhi, India

Monika Sood Division of Food Science and Technology, Sher-e-Kashmir University of Agricultural Sciences and Technology of Jammu, Chatha, Jammu, Jammu and Kashmir, India

Vinod Kumar Sood Department of Crop Improvement, CSK Himachal Pradesh Krishi Vishvavidyalaya, Palampur, Himachal Pradesh, India

Steve Taiwo Federal Ministry of Agriculture and Rural Development, Garki, Abuja, Nigeria

K.S. Thind Punjab Agricultural University, Ludhiana, Punjab, India

Nguyen Van Kien Plant Resources Center, Hanoi, Vietnam

Aijaz Ahmad Wani Department of Botany, University of Kashmir, Srinagar, Jammu and Kashmir, India

Sajad Majeed Zargar Centre for Plant Biotechnology, Division of Biotechnology, SKUAST-K, Srinagar, Jammu and Kashmir, India

Showkat Ahmad Zargar Department of Botany, University of Kashmir, Srinagar, Jammu and Kashmir, India

Abbreviations

AAA	American Anthropological Association
ABS	Access and Benefit-Sharing
AFLP	Amplified Fragment Length Polymorphism
AgBD	Agricultural Biodiversity
AKT	Agroecological Knowledge Toolkit
AVRDC	Asian Vegetable Research and Development Center (World Vegetable Center)
AYUSH	Ayurveda, Yoga, Unani, Siddha, and Homeopathy Systems
BDA	Biological Diversity Act
BSI	Botanical Survey of India
CAPS	Cleaved Amplified Polymorphic Sequences
CBD	Convention on Biological Diversity
CBDCOP	Convention on Biological Diversity Conference of the Parties
CBM	Community Biodiversity Management
CBO	Community-Based Organization
CBR	Community-Based Register
CGIAR	Consultative Group on International Agricultural Research
CGRFA	Commission on Genetic Resources for Food and Agriculture
CIAT	Center International de Agricultural Tropical
CIMMYT	Centre International de-Mejoramients de Maize Trigo
CIP	Centro Internacional de la Papa
CITES	Convention on International Trade in Endangered Species
COP	Conference of the Parties
CPGR	Commission on Plant Genetic Resources
CWR	Crop Wild Relatives
DNA	Deoxyribonucleic Acid
EIAs	Environmental Impact Assessments
ELISA	Enzyme-Linked Immune Sorbent Assay
EPO	European Patent Office
EU	European Union
FAO	Food and Agriculture Organization
GATT	General Agreement on Tariffs and Trade
GBIF	Global Biodiversity Information Facility
GCDT	Global Crop Diversity Trust
GDP	Gross Domestic Product
GEAC	Genetic Engineering Approval Committee
GEBMAP	Genebank for Medicinal and Aromatic Plants

GI	Geographical Indicators
GIAHS	Globally Important Agricultural Heritage Systems
GIFTS	Germplasm Information Funds Technologies and Farming Systems
GIPB	Global Initiative Plant Breeding
GIS	Geographic Information Systems
GMO	Genetically Modified Organism
GNARI	Global Network of Agricultural Research Institutions
GNP	Gross National Product
GPA	Global Plan Action
GREEN	Genetic Resource Ecology Energy Nutrition Foundation
GTZ	German Agency for Technical Cooperation
HKH	Hindu Kush Himalayan
IBPGR	International Board of Plant Genetic Resources
ICAQC	Indian Craftsmen and Artists of Quebec
ICESCR	International Covenant on Economic, Social and Cultural Rights
ICGTK	Intergovernmental Committee on Intellectual Property and Genetic Resources, Traditional Knowledge and Folklore
ICMBAs	Instituto de Ciências Biomédicas Abel Salazar
ICRISAT	International Crops Research Institute for Semi-Arid Tropics
IES	Indian Environment Society
IGC	Inter-governmental Committee
IHCS	International Herbal Cross Society
IIFM	Indian Institute of Forest Management
IITA	International Institute of Tropical Agriculture
IK	Indigenous Knowledge
ILC	Indigenous and Local Communities
ILO	International Labour Organization
IP	Intellectual Property
IPBES	Inter-governmental Platform on Biodiversity and Ecosystem Services
IPC	International Patent Classification
IPCC	Inter-governmental Panel on Climate Change
IPGRI	International Plant Genetic Resources Institute
IPR	Intellectual Property Rights
IRRI	International Rice Research Institute
ITPGRFA	International Treaty on Plant Genetic Resources for Food and Agriculture
IUCN	International Union for Conservation of Nature and Natural Resources
IUPGR	International Undertaking on Plant Genetic Resources for Food and Agriculture
KEPDA	Kenya Economic Pastoralist Development Association
KIPO	Korean Intellectual Property Office
KNS	Kopatgiri Nandiveerimath Seva
KTKP	Korean Traditional Knowledge Portal
LK	Local Knowledge

LN	Liquid Nitrogen
MAB	Marker-Assisted Breeding
MABC	Marker-Assisted Backcross
MAPs	Management Action Plans
MAS	Marker-Assisted Selection
MoEF	Ministry of Environment and Forests
MPA	Marine Protected Area
MPCA	Medicinal Plant Conservation Areas
MSBP	Millennium Seed Bank Project
MSSRF	M.S. Swaminathan Research Foundation, India
MTA	Material Transfer Agreement
NAGS	National Active Germplasm Sites
NBA	National Biodiversity Authority (of India)
NBPGR	National Bureau of Plant Genetic Resources (India)
NCCAM	National Center for Complimentary and Alternative Medicine
NCPMH	National Center for the Preservation of Medicinal Herbs
NGO	Non-governmental organization
NTFP	Non-timber Forest Product
NWFP	Non-wood Forest Product
OAU	Organization of African Unity
OECD	Organisation for Economic Co-operation and Development
PBR	Plant Breeders' Rights
PCT	Peoples Clinic Trust
PGR	Plant Genetic Resources
PGRFA	Plant Genetic Resources for Food and Agriculture
PIC	Prior Informed Consent
PIS	Plant Introduction Stations (US)
PPB	Participatory Plant Breeding
PPP	Public-Private Partnership
PVP	Plant Variety Protection
PVP&FRA	Plant Varieties Protection and Farmers' Rights Act
PVS	Participatory Variety Selection
QTL	Quantitative Trait Loci
R&D	Research and Development
RAPD	Random Amplified Polymorphic DNA
RFLP	Restriction Fragment Length Polymorphism
SARTHE	Society for Agriculture Research Training and Health Education
SCAR	Sequence-Characterized Amplified Region
SDC	Swiss Agency for Development and Cooperation
SGSV	Svalbard Global Seed Vault
SHER	Society for Himalayan Environmental Research
SID	Seed Information Database
SMTA	Standard Material Transfer Agreement
SPS	Sanitary and Phytosanitary
SSR	Simple Sequence Repeat
TBGRI	Tropical Botanical Garden Research Institute
TCK	Traditional Crop Knowledge
TIK	Traditional and Indigenous Knowledge

TIP Rights	Traditional Intellectual Property Rights
TK	Traditional Knowledge
TKDL	Traditional Knowledge Digital Library
TKRC	Traditional Knowledge Resource Classification
TMP	Traditional Medical Practitioners
TRIPS	Agreement on Trade-Related Aspects of Intellectual Property Rights
UDHR	Universal Declaration of Human Rights
UN	United Nations
UNCED	United Nations Conference on Environment and Development
UNCEF	United Nations International Children's Emergency Fund
UNCTAD	United Nations Conference on Trade and Development
UNDP	United Nations Development Programme
UNEP	United Nations Environment Programme
UNESCO	United Nations Educational, Scientific and Cultural Organization
UNFCCC	United Nations Framework Convention on Climate Change
UPOV	The International Union for the Protection of New Varieties of Plants
US	United States
USDA	US Department of Agriculture
USPTO	US Patent and Trademark Office
WARDA	West African Rice Development Association
WB	World Bank
WHO	World Health Organization
WIPO	World Intellectual Property Organization
WTO	World Trade Organization

Introduction

A primary goal of agricultural research is to develop technology that will enable the world's farmers to produce enough food in a manner that is sustainable and economically viable for generations. The conservation and promotion of the genetic resources and traditional knowledge innovations and practices of local and indigenous communities are of key importance for developing countries. The rich endowment of these resources and associated knowledge play an important role in health care, food security, culture, religion, identity, environment, and sustainable development. The conservation and sustainable utilization of plant genetic resources and traditional knowledge, as well as the sharing of the benefits arising out of that, are very imperative to secure the food security in the future.

This book meticulously documents the progress of initiatives undertaken to shape the conservation and sustainable utilization of genetic resources and associated indigenous knowledge. Each chapter is well documented and gives in-depth information about indigenous practices developed by the local and indigenous communities and their legal protection and recognition. This book is indispensable for decision-makers and practitioners dealing with agriculture and food production. The potentials and utilization of plant genetic resources and traditional knowledge for crop improvement have been well explained in the first two chapters of the book. Chapter 8 explains how ex situ and in situ or on farm conservation of plant genetic resources and traditional knowledge help to safeguard food, health, and nutritional security in the future. This book provides a wealth of information for all concerned about and engaged in strengthening Farmers' Rights to secure their livelihood. Chapter 5 illustrates how Farmers' Rights can and need to be implemented in order to ensure the continuous conservation and development of agro-biodiversity in the future. Chapter 14 explains how the role of women is important for conserving genetic resources and traditional knowledge and their utilization in daily needs to secure food security.

This book also clarifies about intellectual property rights of native and indigenous local people. The protection of indigenous peoples' right over the genetic resources and traditional practices at the national and international forums has been well discussed. The success stories in the book have brought substantial achievements. The concept of conservation and utilization of local traditional practices is partly an effort by the scientific community to honor its debt to the legacy of farming peoples who created the biological basis of crop production.

Plant Genetic Resources and Traditional/Indigenous Knowledge: Potentials and Challenges

1

Romesh Kumar Salgotra
and Bharat Bhushan Gupta

Abstract

Genetic resources and indigenous/traditional knowledge are the major resources on which human being has relied for their very livelihood and their demand will increase in the future due to increase in global population. The relation between plant genetic resources and traditional knowledge is of recent origin. Plant genetic resources are the heritable materials contained within and among plant species of present and potential value. On the other hand, traditional/indigenous knowledge is the outcome of intellectual practice in a traditional perspective. The genetic resources and indigenous knowledge protect biological resilience that exists in a natural way. The problems of food security are of global significance and are further compounded by precedential increase in world population resulting in over-exploitation of genetic resources and diversity. Nevertheless, these resources are lost at alarming rates due to anthropogenic effects such as climate change, pollution, genetic erosion, gross mismanagement of these resources and population growth. A vast amount of genetic resources are threatened and endangered, and some have even gone extinct due to genetic erosion and environmental changes. In order to meet current global challenges, it is obligatory for all nations and institutions to discover, collect and conserve potentially valuable plant genetic resource and traditional knowledge and utilise them sustainably.

R.K. Salgotra, Ph.D. (✉)
School of Biotechnology, Sher-e-Kashmir University of Agricultural Sciences and Technology of Jammu, Jammu 180009, Jammu and Kashmir, India
e-mail: rks_2959@rediffmail.com

B.B. Gupta, Ph.D.
Division of Plant Breeding and Genetics, Sher-e-Kashmir University of Agricultural Sciences and Technology of Jammu, Chatha, Jammu 180009, Jammu and Kashmir, India
e-mail: dr_bbgupta@ymail.com

R.K. Salgotra, B.B. Gupta (eds.), *Plant Genetic Resources and Traditional Knowledge for Food Security*, DOI 10.1007/978-981-10-0060-7_1

1.1 Introduction

The food security of the world could be endangered if we fail to conserve the wild relative species which are genetically related to crop species. The increased world population from 2.5 billion to 6.3 billion in the last five decades and the population explosion threatens that another 3 billion people will inhabit the planet by 2100. Currently, about 854 million people are chronically malnourished (FAO 2012). More than half of the world population suffers from malnutrition. The low crop productivity remains the main cause of poverty and malnutrition in the world especially in Africa and Asia. In these developing and underdeveloped countries, a number of diseases are the results of malnutrition. It is estimated that more than 60 % of malaria deaths are caused by malnutrition (Sanchez and Swaminathan 2005). Today about 30 crops are providing 95 % of food energy in the human diet. Out of the 30 main crops, wheat, rice, maize and potatoes are the main crops providing more than 60 % of energy. Keeping in view the importance of a these crops in maintaining the world food security, it is imperative to maintain and conserve the genetic diversity within these crop species (FAO 1996). As a small number of plant species are supplying the world's protein and energy in human diet, it is immensely important to maintain the diversity within these species.

The ever-increasing demand for food and other basic needs has necessitated the sustainable utilisation and conservation of Plant Genetic Resources (PGR) and traditional knowledge (TK). It is immensely important to conserve the diversity particularly in agricultural sector in terms of addressing the accompanied challenges associated with increasing crop productivity. The utilisation of PGR in agriculture has not only brought about profound changes in the crop productivity and quality but has also opened up newer and unforeseen potential vistas including improvement of novel traits by domestication, manipulating plant architecture and molecular farming. Challenges like biotic and abiotic stresses, gradually reducing crop productivity and environmental safety, and demand for newer products have been constantly addressed successfully in terms of phytoneutraceuticals, exotic plant types, healthy planting materials, postharvest losses, export quality plant products and designer agricultural crops using genetic resources. A major challenge of sustainable livelihood in the developing and underdeveloped nations can be met with the judicious and rational use of PGR along with other equally powerful crop production systems. PGR and indigenous knowledge constitute our invaluable assets to meet up the growing demand of the population, particularly in developing countries. PGR are the plant genetic material possessing the potential values which determine their characteristics and ability to adapt and survive (Hammer 2003). These genetic resources start to establish as a result of growing worries about the erosion of biodiversity, its conservation and sustainable utilisation (Gepts 2006). No doubt numerous international programmes have been working to control the population growth rate, but global food security is a major challenge for the future generations (Hoisington et al. 1999). To conserve and secure PGR and TK for future human kind, it has become a priority for most of the countries. Total gene pool of crop species includes crop varieties, landraces, wild relatives, advanced breeding material, inbred lines, genetic stock, modern varieties, obsolete varieties, etc. (Varaprasad and Sivaraj 2010). These genetic resources constitute the basic material for crop improvement programmes to secure the food security for the future.

PGR and TK constitute a unique global heritage, and their conservation and sustainable utilisation are of immediate concern to increase crop production. The available genetic information on PGR and the loss of genetic diversity of agricultural crops have resulted in great effort for the collection of these genetic resources (Briggs 2000). On the other hand, TK is dynamic in nature which modifies its character when the requirements of the people change. Moreover, TK is deep-rooted in indigenous people's lives which is very difficult to separate from them. It is advocated that TK and traditional practices are contributed to enhancement and preservation of biological diversity (Greaves 1996). The indigenous and native people's lifestyles and traditional practices are devised in such a way so that the environment as well as biodiversity can be preserved.

1.2 What Are PGR?

PGR are the building blocks for the improvement of agricultural and industrial crops and the agro-processing sector. PGR are the pillars upon which world food security depends especially with expanding global population (Ogwu et al. 2014). PGR include materials considered of systematic importance and applicable in cytogenetic, phylogenetic, evolutionary biology, physiological, biochemical, pathological and ecological research and breeding. They encompass all cultivated crops and those of little to no agricultural value as well as their weedy and wild relatives (Ulukan 2011). The growing global demand for food and bio-based renewable materials, such as biofuels, is changing the conditions for PGR development and bio-resource production worldwide. Most of the developing countries' Gross Domestic Product (GDP) depends on agriculture is an agrarian type. To feed the ever-increasing population is a great confrontation in the development of many parts of the world, but the problems are worst in sub-Saharan Africa, where there has been stagnation and even decline in the agricultural productivity of small-scale farmers.

The conservation research, improvement and sustainable use of PGR are the components of a complex system in a dynamic interaction which is based on relationships among different types of representative with specific functions within a system called the "plant genetic resources system". The indigenous farmers and communities, research institutions, breeder, private seed companies, curators, collectors and farmers are the different representatives in the PGR system which performs functions within a framework of legal rules (Fig. 1.1).

Development and utilisation of the PGR base and the connected knowledge are central in such efforts. International conventions and treaties together with the rapid bioscience development have led to new conditions for the access of PGR and TK. The Convention of Biological Diversity (CBD) and the requirements under World Trade Organization (WTO)/Agreement on Trade-Related Aspects of Intellectual Property Rights (TRIPS) have led to stricter mechanisms for access to PGR for protection of biological diversity. To access the PGR through the CBD is directly under the control of sovereign states. These corporations are forcing to bargain with the developing countries rich in genetic diversity for direct access to these resources.

The survey, collection, preservation and sustainable utilisation of genetic resources in an organised way are the utmost responsibility of all nations. To best conserve and sustain utilisation of existing biodiversity for the benefit of society in a country, all the policymakers, planners and scientists are engaged to get the best out of it (Tanksley and McCouch 1997). With the introduction of newly developed crop varieties and domestication of crop species, the genetic diversity among the genotypes is declining (Hyten et al. 2009). In order to prevent vulnerability of crop species and genetic erosion, there is a need to preserve and sustainably use these valuable collected germplasm (Frankel 1984). To have the maximum plant genetic diversity of the entire collected germplasm, the concept of developing core sets was introduced (Brown 1989). The evaluation and characterisation of germplasm on the basis of morphological characters does not reflect the maximum genetic diversity; however, the use of molecular markers gave the maximum diversity at the genic levels (van Hintum 1999).

In India, particularly Northwest Himalaya is well known for their rich biodiversity of PGR among field and vegetable crops (Wani et al. 2003). Among the field crops there exists a wide range of scented/aromatic rice (*Oryza sativa* L.) – Mushkbudji, Safed budji, Qadir ganai, black rice, Kamad and Tila zag – which are prominently grown in southern Kashmir particularly in specific areas of Kokernag, Saagam, Larnoo and Khudwani, and thus these places can be developed as hot spots for seed production by farmers, thus helping in the in situ conservation of the

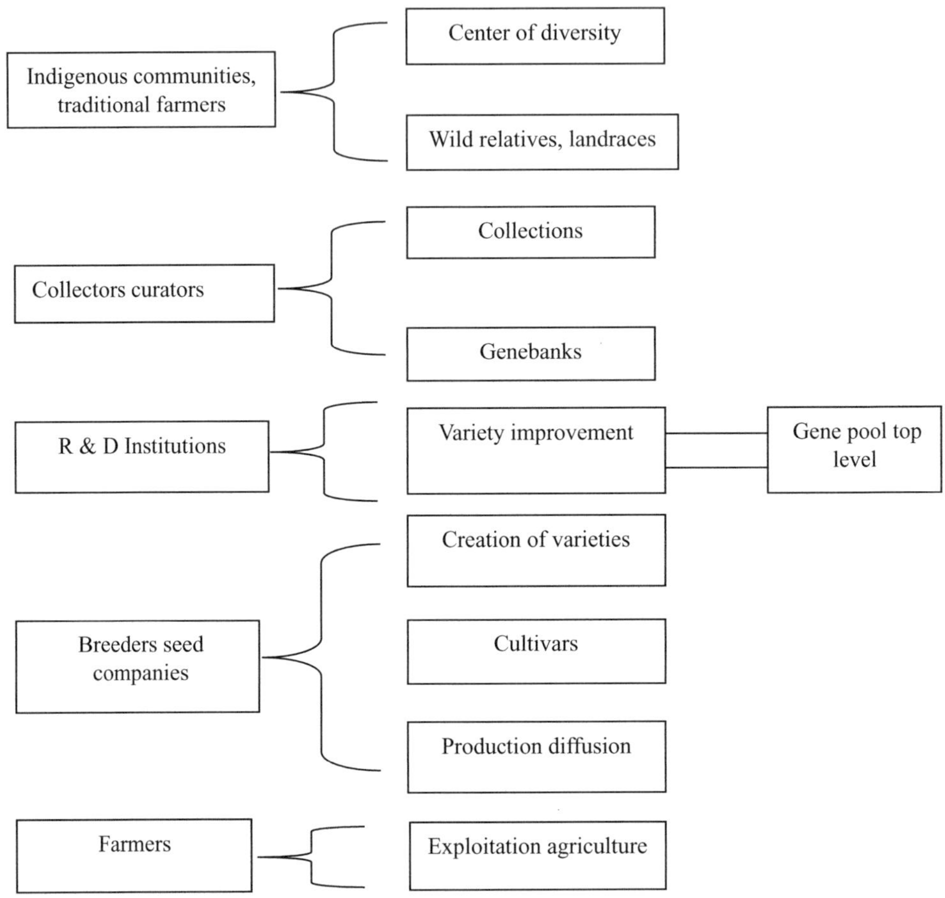

Fig. 1.1 Plant genetic resources system

valuable PGR. Basmati rice grown in Jammu region of Jammu and Kashmir is world famous for its palatability and aroma (Salgotra et al. 2015). Non-aromatic rice – Mehwan (green), Laer Beoul and red rice – are also in great demand because the staple food of the local inhabitants is rice (Gupta et al. 2009). There also exists great genetic diversity among maize local races in Gurez and parts of southern Kashmir possessing tolerance against biotic and abiotic stresses. Some of them include Anantnag Safed, Tangwin safed and Aru wuzg. Herbal vegetables grown in the state are wild garlic (*Allium ursinum*), top onion (*Allium cepa* L.), *Taraxacum* (*Taraxacum officinale*) (Haand), mallow (Sochel), *Rumex* (*Rumex crispus*) (Obuj) and red gourd (*Cucurbita pepo*) (Parim). Potatoes (*Solanum tuberosum*) grown in Hurpur (Shopian), rajmash (*Phaseolus vulgaris* L.) mixed with maize (*Zea mays* L.), local moong (*Vigna radiata*) and broad bean (*Vicia faba*) are potential pulse crops grown in pockets of Shopian, Uri, Bhaderwah, Mandi, Poonch, Budhal and Gurez (Salgotra and Gupta 2005). Underwater vegetable lotus stem (Nadroo) has a rich potential due to great local demand. Kashmiri red chillies (*Capsicum annuum* L.) due to their bright red colour and moderate pungency are very famous in industrial and processed products and are important export commodities, grown over an area of more than 1700 ha, and if exploited could form one of the most potential high-value commodities for export.

Underutilised cereal crops like medicinal buckwheat and sweet buckwheat/duck wheat which are grown for human purposes also serve as the feed for livestock and poultry as well as

Fig. 1.2 Natural genetic diversity in plant genetic resources

green manure. The buckwheat honey has a very high potential in difficult hilly terrains of Leh, Zanskar and Gurez and adjoining areas to meet the livelihood and food security of the people (Wani et al. 2003). There is a vital need to pay attention for conservation and utilisation of these PGR which will further lead to improvement of food, nutritional and livelihood security of diverse inhabitants of Jammu and Kashmir states (Fig. 1.2).

1.3 Traditional Knowledge (TK)

TK is the mature, long-standing traditional practices, knowledge and wisdom of local communities and indigenous people of a region. To sustain the indigenous people and local communities and their culture, TK is used to maintain the PGR for their survival. "TK and associated genetic resources" encompass all forms of technology-based knowledge, beliefs, know-how skills and traditional practices which come overtime and generations in a particular environment (Dowd 2009). TK is the mental inventories of PGR which include landraces, wild relatives and other indicator plant species which show soil salinity and flower at the beginning of the rains. Traditional production techniques of agriculture such as seed treatment, crop planting, intercropping, introducing new plant species into farming system, harvesting and different seed/grain storage methods are all included in TK. TK also includes belief systems, and indigenous peoples depend upon it for their livelihood, health (introducing a healer test of a medicine plant) and protecting the environment (Dowd 2009).

TK is the indigenous knowledge of local communities which is based on traditions and does not entail that this knowledge is old and non-technical in nature. This is a "traditional" because the knowledge is developed on the reflections of traditional practices and knowledge of the indigenous communities and does not develop by nature itself. TK is a collective effort of the local communities and is the property of entire indigenous people who have created and developed it over generations. TK is maintained by the local cultures and indigenous communities and transmitted through demonstrations, orally, elders, specialists and few people within a community (Hirwade and Hirwade 2012).

TK has got a lot of attention nowadays due to its utility all over the world and has become a focus in international forums. The CBD has paid due attention for protection of TK under Intellectual Property Rights (IPRs) of indigenous knowledge. The local and indigenous people also have their traditional festivals, local songs, practices, stories, etc. as a method of transmitting TK to other people and young generations. TK is also preserved and transmitted from father to next generations on some specific artefacts. TK can be found in a multitude of fields such as food, nutrition, agriculture and fisheries, human health, veterinary care, handicrafts, performing arts, folk songs, religion and astrology and many other day-to-day customs and practices (Hirwade and Hirwade 2012). Therefore, the protection of this knowledge by indigenous communities is very imperative for most of the developing countries.

Indigenous knowledge plays a significant role in boosting the economic growth of the countries, particularly the developing countries. The utilisation of this TK is to attain developmental goals of the developing countries such as increased agriculture production, health and preservation of biological diversity. The countries are the owners of genetic resources and have their sovereign right for access to potentially exploitation of their natural resources under the CBD. Under the CBD, all the countries are responsible for accessing the genetic resources and equal sharing of benefits arising out of their utilisation. To access these resources prior information content (PIC) is needed as per mutual agreement conditions (Gupta 2006). The major objective and goal of the CBD is to preserve the genetic resources in their territories, while the goal of TRIPS is to stimulate technological advancement, giving individual rights to the inventor through IPRs. The World Intellectual Property Organization (WIPO) and Intergovernmental Committee on Intellectual Property and Genetic Resources, Traditional Knowledge and Folklore (ICGTK) are functioning on the issue of contractual practices. The WIPO Member States have complied indigenous knowledge databases and registries with sui generis possible system for its protection. The Traditional Chinese Medicine (TCM) Patent Database contains more than 12,024 deeply indexed records of China. The Korean Intellectual Property Office (KIPO) has planned to devise information of a database of TK. The Traditional Knowledge Digital Library (TKDL) in India has documented and digitised the TK available in the public domain and translated into five international languages, i.e. English, German, Japanese, French and Spanish.

India is the country which has been nurturing a tradition of civilisation over about 5,000 years, and ancient scriptures consist of four Vedas, 108 Upanishads, two epics, Bhagavad-Gita, Brahma sutras, 18 Puranas, Manusmriti, Kautilya shastra and smritis. The rich heritage of India has already documented over 47,000 plant species besides more than 350 wild-related species and landraces of agricultural crop species. The vast repository of diverse genetic resources of TK in India further helped to enrich the eighth Vavilovian centre of origin. The diverse biological resources and multitude of natural wealth have created a renewed interest in the traditional medicinal system, which includes the Ayurveda, Yoga, Unani, Siddha and Homeopathy (AYUSH) systems. Ayurveda is the oldest and most effective systems of medicine in developing and least developing countries. The ancient scriptures of the Ayurveda are full of instances where medicinal properties of herbs were used not only for curative purposes but for increasing physical and mental efficiency (Hirwade and Hirwade 2012).

1.4 PGR and TK for Food Security

The PGR diversity is the major component for the improvement of crop varieties to secure the food security. The conservation and sustainable utilisation of these resources are vital and important to secure food security for future generations. For the continuous improvement of these crop varieties, these resources should be easily available to the plant breeders and the farmers. On the other hand, TK is completely based on the traditional know-how of the indigenous and local people in a particular environment which also helps in the preservation of PGR and their sustainable utilisation. In most of developing and underdeveloped countries, farmers are using their own saved seeds for sowing purposes to secure their livelihood and food security. The Leipzig Declaration (1996) emphasised to save the seed and planting material to avoid the genetic vulnerability and shortage of food under adverse conditions.

The importance of PGR is reflected in every facet of human endeavour as it provides the gene pool from which resistant and improved varieties can be developed. The studies conducted by

Alston et al. (2000) and Evenson and Gollin (2003) have been well documented to know the economic values of increase in agriculture productivity through the introduction of improved and newly developed crop. Similarly, Koo et al. (2004) and Smale and Koo (2003) have also estimated the cost-benefits of preserved PGR in gene banks which are commercially for use by farmers. Thus, genetic resources including TK can play roles in ensuring food and income security especially for developing and underdeveloped countries where majority of livelihood is hinged upon these biological resources.

The plant species diversity loss has been significantly increased in the last four decades due to climate change and mainly due to the activities of the earth's dominant species. Land clearing for agriculture and infrastructure development, wars and deforestation all impacted negatively to destroy natural habitats and the diversity contained therein. Sodhi and Erhlich (2010) emphasised that the earth dominant species have destroyed, degraded and polluted earth's natural habitat which is a key in life support. More so, serious illnesses and ecological destruction can be attributed to the drilling of oil which has caused widespread destruction of rainforest (Hvalkoff 2001). The deforestation not only narrow down the genetic diversity of plant species but also sometimes completely diminished in some crop species due to genetic vulnerability and over-exploitation of species by the human being as well as by adverse conditions of the climate.

PGR and TK have great scopes for economic as well as social growth of the developing countries. These will also offer a sustainable food security in the future. To realise the maximum benefits from exiting diversity in genetic resources in most of the countries, particularly in in situ conservation, building close relationship between the development and preservation and sustainable utilisation of these resources are needed (FAO 2001). To achieve this challenge, there is a need that all stakeholders including plant breeders, farmers, government policymakers, seed banks and other agencies work in close for systematic conservation, development, access to information, PGR diversity and new technologies. To promote this linkage among the different stakeholders, the Food and Agriculture Organization (FAO) is providing technical know-how and regulatory mechanisms to the entire world. This will definitely open new opportunities for the global communities to develop linkage for efficient use and management of these resources (FAO 2001). To establish and strengthen the system of accessing and sharing of benefits arising from these resources, the stakeholders and policymakers must frame the national regulations and laws through the International Treaty on Plant Genetic Resources for Food and Agriculture (ITPGRFA).

For proper preservation and sustainable utilisation of PGR, the Global Plan of Action (GPA) helps the utilisation of per se capacity of the local communities to save their farm seeds and share and market their Plant Genetic Resources for Food and Agriculture (PGRFA) to secure the food security. This emphasises need for diversified agriculture system with maximum utilisation of traditionally local landraces and underutilised crop species. The FAO also emphasised that the TK of local communities should be recognised for further conservation of diversity to enhance food security in the future. A harmony was developed in 1992 between the International Undertaking on Plant Genetic Resources (IUPGR 1983) and the CBD for the conservation of PGR under the umbrella of the Earth Summit of the United Nations Conference on Environment and Development (UNCED).

1.5 Potentials of PGR

The diverse PGR are the main resources used for the improvement of crop varieties to meet the ever-increasing demand for food in the developing countries. These plant species have the ability to survive and adapt in a diverse environment (Varaprasad and Sivaraj 2010). PGR profile includes landraces, wild relatives and improved varieties of crops which are conserved by indigenous people for food and health. The biological resource conservation without use has little point and importance; moreover, using these resources without conservation leads to completely vanishing these resources, and there will be no PGR for the future (Hammer and Teklu 2008). Genetic resources have great potential to strengthen food security and make agricultural systems more productive and resilient to climate change. Using PGR for crop improvement can increase yields and nutritional value and make crops more resilient to pests, diseases, drought and flooding, thus reducing the region's dependence on food imports.

Although there are many genetic resources of plant species that meet smaller and often local need, the plant species that moved into greatest world prominence are those that met the greatest number of food needs in the widest range of world food systems. As a result of global movement of favoured crop species with travellers and immigrants around the world and, in more recent times, because of global trade in plant-derived commodities, only 50 species now account for most of the cultivated world crop acreage. There are four major crops such as wheat, rice, maize and potatoes which contribute more than 60 % of food energy in the world. There is a very uneven distribution of these crop species, whether they are cereal grains, pulse crops, vegetables, fleshy fruits or root crops. This contribution is even greater in developing and least developing countries, where meat products are consumed at a lower level because of cultural custom, high cost or limited availability. The much higher dependency on plants as a source of food and essential protein in developing compared to developed countries is also very evident in these statistics (Hammer and Teklu 2008). This difference also leads to a greater primary dependency on biological resources in developing countries, although this trend is known to change as developing countries become more affluent and diversified in their food demands towards greater use of meat products (Table 1.1).

Table 1.1 Distribution of PGR to meet the nutritional needs in the developed or developing countries

	Digestible energy (%)			Protein		
Source	World	Developed	Developing	World	Developed	Developing
Plant products						
Cereals	50.2	26.4	57.7			
Pulses and nuts	3.9	2.4	5.6			
Roots and tubers	6.9	3.7	6.2			
Sugars	9.1	13.0	9.7			
Vegetable oils/fats	4.9	4.8	4.4			
Stimulants/alcohol	3.9	6.7	1.5			
Subtotal	83.7	68.3	91.4	65	44	79
Animal products						
Meat	7.6	15.3	3.1			
Milk and cheese	4.4	8.5	3.1			
Animal oils/fats	2.5	4.7	1.1			
Eggs	0.9	1.6	0.4			
Fish	0.9	1.6	0.7			
Subtotal	16.3	31.7	8.6	35	56	21
Total kcal (per person per day)	2,630	3,390	2,350			
Total (M) (per person per day)	11.0	14.2	9.8			
Total protein (g person per day)				68	99	57

Source: Crop evolution, adaptation and yield, Cambridge, UK

1.5.1 PGR: A Source of Germplasm in National Depository

The use of exotic and wild relatives of crop species in crop improvement programmes to increase the yields has been significantly enhanced for the last three decades due to systematic development of PGR information in all Member States of the CBD (Bonham et al. 2010). For effective use of PGR in crop improvement programmes, there is a need to further evaluate and characterise these PGR on the basis of various morphological characters. For proper documentation of genetic resources in the gene banks, characterisation on the basis of molecular markers is essential. Networking is another important way of widening the use of PGR which helps in bringing all stakeholders to work together and share their tasks and priorities related to PGR. This will also help in conservation of genetic resources in a collective manner. Today more than 150 countries have PGR networking which is easily available on the worldwide networks which are being shared among these nations for crop improvement programmes (Hammer and Teklu 2008). Due to this all the nations shared responsibilities for conservation of PGR, development of new technology and establishment of common strategies for efficient utilisation of these resources based on their interests (Hammer and Teklu 2008).

1.5.2 Role of PGR in Agriculture

PGR play a significant role in the agriculture, particularly in crop variety development and improvement programmes. PGR diversity provides the major sources of important genes such as diseases and insect pest resistance and yield and quality improvement.

1.5.2.1 Core Collections and Utilisation of Germplasm for Food Security

"Core collections" are the core sets that represent genetic diversity of entire germplasm which is otherwise difficult to handle. However, these core sets are prepared on the basis of morphological characters and passport data of these germplasm. The newly developed core sets will become more useful to the plant breeders as well as the farmers in crop improvement programmes. The scientists are not using these resources extensively because of lack of detailed information available about the important traits. The scientists are now evaluating these core collection sets at multilocation trials for evaluation instead of evaluating the entire collected germplasm.

1.5.2.2 New Sources of Germplasm

Newly developed core sets of collected germplasm will be evaluated at multilocations to identify the diverse parent with particular character of interest. Their use in breeding programmes would result in broad-based cultivars. To strengthen crop improvement programmes, core sets of chickpea, pigeon pea and groundnut have been developed and evaluated at multilocations at International Crops Research Institute for the Semi-Arid Tropics (ICRISAT) for screening of diverse genotypes for specific trait of interest. For identification of specific trait, these genetic resources should be characterised using molecular markers, such as Random Amplified Polymorphic DNA (RAPD), Restriction Fragment Length Polymorphism (RFLP) and Simple Sequence Repeat (SSR) markers. The available techniques and information will be used for population structure analysis which helps to know the association between the specific trait of interest and the molecular markers.

1.5.2.3 Landrace Varieties

The traditional varieties developed and selected by the indigenous communities are termed as landrace varieties. These landrace varieties possess a number of desirable traits to be used in crop improvement programmes. These landrace varieties have numerous important characters like resistance to disease, insect pests, heat stress, salinity, etc. No doubt that the landraces are adapted to a specific environment and region, but these are the reservoir of various desirable traits to be used in plant breeding programmes.

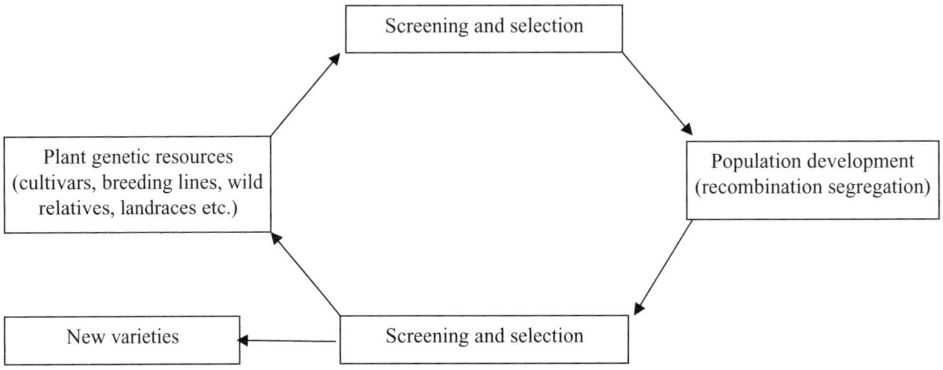

Fig. 1.3 Plant genetic resource for development of new cultivars

1.5.2.4 Wild Relatives

The wild relatives of plant species contribute a major portion in PGR. Like landraces, wild relatives of crop species also possess desirable characters for crop improvement. Compared to landrace varieties, wild species are difficult to cross with cultivated varieties due to some crossability barriers. To overcome these barriers, techniques like embryo rescue, protoplast fusion and genetic engineering are being used to transfer the desirable characters from the wild relatives to cultivated varieties.

1.5.2.5 Use of PGR as Varieties

PGR contain diverse genotypes which enable the plant breeders and the farmers to make selection among the diverse genotypes with specific characters. The potential and promising PGR can be used directly as cultivar. The diverse PGR genotypes are adapted to specific environment with potential yield and can be released as a promising variety. These released varieties can be used in breeding programmes for the transferring of the desirable traits to other well-adapted varieties. The use of PGR as a direct cultivar is more desirable if trait of interest is easily identifiable among the germplasm (Fig. 1.3).

1.5.2.6 Source of Resistance to Plant Diseases and Insect Pests

The varietal resistance is the most effective method for developing resistance varieties. The diversity in PGR is a rich source of resistance genes for plant diseases and insect pests. These resources can be effectively utilised in plant breeding programmes for the improvement of crop varieties. Modern plant breeding methods, particularly the molecular breeding, involve the transfer of resistance genes in a widely cultivated variety through Marker-Assisted Selection (MAS) and Marker-Assisted Backcross Breeding (MAB) methods. The pyramiding of different resistance genes has been successfully introduced in a number of crop varieties (Singh et al. 2001; Xu et al. 2006; Sundaram et al. 2010).

1.6 Challenges in PGR

The impact of humans upon biological plant biodiversity has gradually increased with growing industrialisation, technology, population, production and consumption rates. Food security and accessibility along with gross mismanagement are contending issues which impact on PGR. Since the inception of green revolution, the landraces and the wild relatives of crop species are being eroded which narrow down the genetic diversity of PGR. This is further compounded by issues arising from patent rights. Most of the developing countries are continuously facing the problem of climate change, population growth and biotic and abiotic stresses in agriculture which directly cause instability in food production.

1.6.1 Habitat Loss and Modification

The over-exploitation of PGR to meet the daily needs of people affects the habitat to a great extent. The utilisation of PGR without conservation vanishes these resources completely. For example, the extraction of oils with deforestation leads to air pollution and degradation of environment (Olubisi and Oluduro 2012). Therefore, intensive measures should be undertaken for conserving these valuable resources and sustainable utilisation to secure the food security for the next generations.

1.6.2 Climate Change

A significant negative impact of climate change has been observed on the PGR and related resources. The climate change affects the environment and often leads to perturbations such as drought, flood and disease. Sometimes extreme weather conditions also likely to affect the crop production and diminish crop productivity in many areas. The rise in sea level, causing loss of coastal land and saline water intrusion, also leads to PGR depletion (Pisupati and Warner 2003). This will impact on the distribution of PGR and most likely alter their physiognomy.

1.6.3 Insect Pest and Diseases

The effects of human activities have introduced certain levels of stress to natural resources including PGR. This stress will overtime weaken the immunity of the affected population. More so, PGR are now susceptible to different new diseases and insect pests absent in the original population. Reduction in gene pool increases vulnerability to various diseases and insect pests. It has been observed that varietal resistance is the most effective method of crop protection against the diseases and insect pests (Bhullar et al. 2012).

1.6.4 Alien Species

The plant species which are introduced in a new area of their origin are termed as the alien species. These species are invasive in nature and outcompete the native and indigenous plant species which sometimes completely vanish the habitat of that area. The nonindigenous plant species become noxious and the native crop completely lost, threatening the genetic diversity of PGR which is the second most important threat after habitat loss (CBD 2005). The genetic diversity loss of plant species is now comparable to habitat loss if it occurs on an island (Baillie et al. 2004). With the invasion of alien genotypes, the ecosystem of that region is completely distorted by changing the composition of different plant species existing in that area (McNeely et al. 2001). Not all alien plant species are invasive, but policies should be framed to check and avoid the invasive behaviour of these species to prevent the loss of biodiversity.

1.6.5 Patent Rights for the Protection of PGR

The protection of PGR in developing countries is important to avoid the over-exploitation of genetic resources by the developed nations. Moreover, there is a huge difference in Intellectual Property Rights (IPRs) given to the farmers or indigenous people and the plant breeders. The farmers and indigenous people are the main stakeholders who are preserving and sustainably using these resources. The indigenous communities are facing the challenge of complete loss of these genetic resources by over-exploitation. The PGR protection policies should be made in order to avoid the over-exploitation of these important resources by the developed countries. All these are challenges because they consider plant species not for what they are but for the value that can be derived thereof.

1.6.6 Genetic Erosion

The genetic erosion of PGR is of great concern for most of the developing countries. Most of the genetic resources are being extinct from their native area due to various biotic and abiotic factors, climatic factors and the development of modern varieties. The replacement of landrace varieties with modern varieties is one of the main causes of genetic erosion of PGR (Rosendal 1995). Although seed banks and gene banks play an important role in conserving and maintaining these varieties, but these resources are being lost even in the gene banks due to improper handling and management of these resources (FAO 1998).

Creating and maintaining crop genetic diversity and their varieties in production systems can help to reduce vulnerability and can be said to impact on ecosystem stability. In current plant breeding methods, we are mostly favouring the selection of few desirable traits controlled by single gene/allele. This will narrow down genetic diversity in PGR (Prada 2009). This process reduces the levels of genetic diversity. Moreover, most of the modern cultivars are descended from landraces which significantly reduced the diversity in plant species (Bhullar et al. 2012).

1.6.7 Conservation of PGR

Conservation of PGR is the process that actively retains the diversity of the gene pool with a view of actual or potential utilisation. Utilisation is the human exploitation of that genetic diversity. The main aim of genetic resources conservation is to explore, collect and preserve adaptive gene complexes for present or future use (Hammer and Teklu 2008). The conservation of PGR is very important for sustainable utilisation of these resources to secure the food security for the future. The farmers have domesticated these plant species including wild relatives, landraces and economically important medicinal plants. They preserved these materials for future planting to secure their livelihood. At that time indigenous people are unaware about the importance of these resources for commercial use. Later on the plant breeders used these diverse materials in breeding programmes for the development of new crop varieties (Hammer and Teklu 2008).

1.7 Potentials of TK

TK is the experience obtained over years through traditional practices and is the common property of the local and indigenous communities. Each member in the community contributes towards this knowledge over time. Over the last two and a half decades, the knowledge of these resources has been significantly increased in the developed as well as developing countries (Sinclair and Walker 1999). TK may be different from the Western knowledge which is based on global experiences, scientific discoveries and other cultures (Greaves 1996). The indigenous people are linked with these traditional practices for various social, religious, economic and cultural activities (Agrawal and Gibson 1999).

1.7.1 TK in Agricultural Practices

Traditional and local technical knowledge are often used by the local and indigenous people in their agricultural practices. Traditional practices used by the indigenous communities in farming system help to maintain the genetic diversity in genetic resources also. TK practices in agriculture have been ignored with introduction of modern scientific methods of farming system in agriculture. With the inception of green revolution, most of the traditional practices are wiped out from the agriculture mainly in Asian countries. But TK practices are being recognised again in the developed and developing nations. These are being preserved and utilised in agricultural crop productions in most of the countries. The growing knowledge and information on vast potentials of TK help the people to use these resources on sustainable basis. According to Agrawal and Gibson (1999), scientists, policymakers, nongovernmental organisations (NGOs) and community-based organisations (CBOs) have collected the information on traditional practices being used in agricultural practices. But this information is very meagre, and we have to collect more information on TK practices being used by the farming communities in agriculture.

Indigenous knowledge developed through traditional practices over generations and the information are being transferred orally as well as through practices. These traditional practices may be agricultural crop production techniques or related to health (Anuradha 2007). TK has multiple roles in agriculture practices and in curing human health. The traditional practices in agriculture provide the food security to local and indigenous people on sustainable basis without disturbing the genetic diversity.

1.7.2 Prevention of Soil Erosion

It is estimated that the frequency of soil erosion has been significantly increased with the inception of modern agricultural practices. Compared to modern practices of crop production, indigenous practices have less soil erosion. This may be due to cultivation of various landrace crop varieties in different combinations, i.e. deep-rooted crop followed by shallow crops, growing of different crops in the same area to avoid the disease and insect pest outbreaks, traditional methods of conserving soil moisture, etc. Keeping the crop residues after harvest prevents the soil from direct exposure to wind and water. Soil and water conservation is also promoted by residue of remains on the surface of soil. To avoid erosion of soil, the farmers are planting deep-rooted ornamental crops at the base (Fig. 1.4). The indigenous people also have diversified system of farming. The indigenous farmers are growing different agricultural and horticultural crops in different combinations to prevent the soil erosion and conservation of soil moisture. These practices are very helpful for conserving soil and water and nutrient management and reduce soil erosion and genetic diversity.

1.7.3 Intercropping

In this system different crops are grown in the same season and in the same field. Despite the strong campaign for promoting the modern system of crop production such as monoculture, intensive agriculture and high-yielding varieties (HYVs), the indigenous farmers are still using traditional practices for the cultivation of different crops. They are using their own techniques for the production of different crops in a diversified system and grow the crops in different combinations. The indigenous people are generally growing the deep-rooted crop with shallow-rooted crops,

Fig. 1.4 Soil conservation by planting ornamental crops in the slope

Fig. 1.5 Soil and water conservation by planting lemon grass on the bunds of paddy field

pulses with cereals to conserve the nitrogen in the soil and tall varieties with small varieties to act as windbreak. Farmers are planting lemon grass at the bunds of paddy field to conserve soil as well as water in the field and to get additional income from lemon grass (Fig. 1.5). The lemon grass protects the soil from erosion and conserves the water inside the paddy field. These practices no doubt conserve soil and water but also provide extra income to the farmers by extracting oil from the lemon grass. The intercropping practices are very important where erratic rainfall occurs. In this case farmers can get some income from at least one crop even if its companion crop fails.

This cropping system also provides continuous source of income to the farming communities. The growing of different crop species on the same field also conserves the biodiversity of that region. It is estimated that the diversity in plant species is more in developing countries compared to developed nations. The livelihood of indigenous people in most of the developing countries depends on this type of farming system which provides continuous source of income and food security.

1.7.4 Indigenous Knowledge of Crop Plants

The indigenous communities have the knowledge of importance of each valuable crop plant. They cultivate these crop plants depending upon their need. The farmers know the value of crop plants, production techniques, nutrient value, taxonomy, market value, etc. The indigenous farmers are well aware about biological diversity and the dietary values of crop plants for food security. It is estimated that about 614 racial groups in Africa depended on wood from forest for their livelihood (Makombe 1993). They recognise crop varieties suited to different ecological conditions, i.e. in alkaline soils, sodic soils, wet lands, etc. The indigenous farmers are very conscious to use different crop plants and varieties for particular seasons, tolerant to drought, heat and others stress tolerance factors. They know that the newly developed modern crop varieties are less resistant to various biotic and abiotic stresses. Farmers have a strong commitment to use particular variety in different situations to avoid the complete crop failure.

1.7.5 TK Methods for Storage of Seeds

The indigenous farmers have various traditional methods of storing their farm produce and seeds for the subsequent seasons. These methods are well adapted to various ecological conditions. They generally mixed ash with grains of different cereal and leguminous crops kept in bins. The maize crop seeds are stored in the cobs itself and placed or hanged on the trees. In some cases the farmers mix their grain with leaves of neem tree for long storage and to save their produce from the attack of any insect pests and diseases. Some farmers used sand for storage of leguminous crops. The sand is mostly used in humid conditions to keep the storage bins dry and to avoid any attack from insect pests.

1.7.6 TK in Treatment of Seeds before Sowing

The seed treatment is generally given before sowing the crops to eliminate the infested seeds from diseases or insect pests. This technique also promotes good and fast germination of seeds. Generally farmers soak their seeds overnight and sow the seeds next day for fast germination. The farmers also mix ash in water before soaking the seeds. Some farmers mix the ash with the wet seeds after soaking. Sometimes the leaves of neem trees are mixed with water before soaking the seeds. This will disinfect the seeds and reduce the infestation of any diseases before sowing.

1.7.7 Traditional Methods of Foodstuff Storage

Traditional technologies are being used by the indigenous people for storing the processed foodstuffs in the form of dried fruits and vegetables, butter converted into cheese or ghee and other fermented food items during the lean period to avoid any bacterial or fungal attacks. The foods stored by using these traditional methods are more nutritional compared to modern methods, as in traditional methods of storage of food items, there were no chemicals used to increase the longevity of food items. For fermentation and processing of other material, traditional biotechnology is more effective even today. Keeping in view the nutritional importance of storage food by traditional methods, the Western nations are interested to use these techniques.

1.7.8 TK to Control Pests and Diseases

The insect pests in crop production areas are causing a huge loss to crops. Indigenous people are having the deep knowledge of breeding cycles of various insect pests and diseases in a season. The farmers are also aware of the congenial environment for the multiplication of these insect pests. The farmers clear their land immediately after the harvest of crops to avoid the breeding cycle of the insect pests and diseases. In the cropping seasons, the farmers keep sanitation in the fields by clearing the bunds. Other methods to reduce the incidence of insect pests and diseases are to use well-rotten farmyard manures in the field before sowing. This will completely destroy the pests and avoid the multiplication of diseases, as fresh farmyard manure is the alternate host for some insects. The indigenous people are rarely using the insecticides, fungicides and fertilisers for the production of crops which adversely affect the ecosystem. To control the insect pests and diseases, farmers are using traditional practices such as timely sowing of suitable varieties, proper tillage of soil, burning of crop residue after harvest, zero tillage to capture the soil moisture and botanical control of diseases and insect pests. To avoid the vulnerability the indigenous people are using diversified system in agriculture.

1.7.9 TK in the Development of Healthcare Products

Most of today's medicines are originated from the ancient system of medicine production in which traditional practices have been used. These traditional practices are likely to be used in the future also (Ameenah 2011). Presently, the natural products or their derivatives are contributing a major portion in the medicine system. Today drug development and discovery prepared by using natural resources are in great demand because of numerous drawbacks in developing new synthetic molecules from active components or pure chemicals. In the traditional method of medicine production, most of the drugs are developed on the basis of pharmacokinetics. In the modern system of medicine production, drugs are developed like a bullet which strike at the particular target place (Gangadharan 2005). In the future, we will be able to redesign systems by modification or drugs to have completely new systems properties (Hood 2003). All of modern drugs are created within the confines of a chemical paradigm of medicine and drug therapy (Gangadharan 2005). In Ayurveda, a drug brings the equilibrium of the body and corrects the damages brought by imbalance of the *doshas*. For example, when Triphala (*Terminalia bellirica*) is taken as purgative, it removes the unwanted materials from the system, but does not in any way disturb the intestinal flora in the process. According to Patwardhan et al. (2003), indigenous medicine knowledge and drug development can reduce the toxicity effect of medicine compared to allopathic medicines.

1.8 Challenges in TK

The commercialisation of TK starts with the rise in industry. Indigenous people were unaware about the importance of their genetic resources. After the inception of modern techniques, the concept of protection of indigenous peoples' knowledge on traditional practices as well as resources has risen. IPR was enforced to protect the interest of these people. IPR is a legal concept that deals with creations of human ingenuity. After the inception of IPR protection norms, farmers are getting their equal shares from the benefits arisen out of TK. Modern technologies and TK are not just about incompatible values, attitudes and practices but also relate to rights within a legal framework. Moreover, whenever we talk about innovation, we fail to recognise the innovation done by artisans, farmers, tribes or other grass-roots innovators; therefore, these informal innovators have nurtured, refined and generated a rich system of the TK (Chandra 2010).

The indigenous communities are not aware about IPRs and protection of their traditional resources. Rather than protecting TK of indigenous people, IPR has facilitated the taking and commercialisation of this TK by individuals or entities without giving any royalties or compensation to the communities who have generated such knowledge. This use or taking of this valuable knowledge without consent or compensation has been characterised as "biopiracy" or "misappropriation". No doubt indigenous knowledge has been easily available to everyone for commercialisation, but now IPRs play a significant role to safeguard the interests of indigenous communities. TK of indigenous people can be protected under IPRs system and some benefits arisen out of it can be utilized for conservation and sustainable utilization of these resources. To provide the complete benefits to indigenous people, TRIPS Agreement was introduced. The main objective of this Agreement is to provide the claim and rights to holders on the TK. In TRIPS, indigenous knowledge is completely associated with social and cultural norms. However, some of the indigenous people are not aware of TRIPS and are getting full benefits of sharing their TK.

Despite TK is having rich information on utilisation and economic values, there are some barriers which prevent its extensive utility. Moreover, the indigenous knowledge is known in local languages, and their widespread is very less compared to modern practices. The international debates are going on in most of the countries for its globalisation, protection and sustainable utilisation (Chandra 2010). TK is having its own importance, and it is being acknowledged by the world communities. Now, most of the developed and developing countries are emphasising for its protection. The current IP rights regimes fail to provide any rewards to the public domain foundations also known as "traditional knowledge", on which the innovations may be based. Thus, it becomes a matter of great concern when the IP regime is extended to the biodiversity and TK domain (Mandal 2011).

Indigenous peoples' social and cultural activities are totally based on TK, and most of them depend for their food security (Varkey 2011). These people have deep knowledge of TK for creations and innovations, and they develop all these things in an environment where they dwell. According to the World Health Organization (WHO), it is estimated that most of the world population relies on traditional medicine (Erstling 2011).

1.9 Key Common Factors Eroding PGR and TK

There are numerous factors responsible for degradations of genetic resources, but some of the important key factors responsible for erosion of PGR and TK are listed below:

1. Encouragement of newly released crop varieties and other subsidiaries provided by the government from time to time.
2. Change in demand of consumers for newly developed products by the modern technologies and R&D institutions all over the world. With the popularity of these products, the demand for traditional products decreased subsequently.
3. The increase in population growth also leads to the fragmentation of landholding of farming communities. The farming communities have changed their profession and used this land for other purposes which directly affect the over-exploitation of PGR.
4. New policies and treaty have been framed for the protection of newly developed high-yielding varieties (HYVs) and new products. Recently, the newly developed varieties have been protected under Plant Breeders' Rights (PBRs) and Protection of Plant Varieties and Farmers' Rights (PPV&FR) Act, but no such effective policy has been since long for the protection and over-exploitation of traditional resources.
5. Very less incentives have been given to the indigenous communities to protect and sustain utilisation of these traditional resources.
6. With adoption of modern culture and inception of Western culture, the living, cultural values and traditions of the people have been changed. People from the rural areas rich in traditions start moving to cities to change their occupations. All these urban styles directly and indirectly lead to change in preferences of the people which significantly affected the conservation and sustainable use of PGR and TK.

Moreover, for the protection of traditional resources, no effective policy has been framed by the international organisations. Recently, some policies have been framed, but their effective implementation is not there. It is estimated that the erosion of PGR also leads to the erosion of TK and its related resources.

1.10 Key Common Factors Sustaining PGR and TK

Recently, keeping in view the importance of PGR and TK to secure food, health and traditional cultures, a number of demands are being going on all over the world. With the inception of CBD in 1992, scientists and policymakers have given due consideration for the protection of these valuable resources. To protect and sustain these resources, the following are the key identified factors which need to be implemented successfully:

1. With the inception of green revolution, modern crop varieties have been introduced and the local landrace varieties wiped out of cultivation. For protection of PGR and TK, there is a need to revive these landraces for cultivation so that the biological diversity can be maintained.
2. There is a need to revive the traditional festivals and cultural events related to preserve TK and associated knowledge in the rural areas.
3. Reform should be made to avoid fragmentation of landholding of indigenous communities so that the traditional PGR would not be disturbed.

4. Regular demonstrations and field days should be organised by national and international agencies related to protection, conservation and their sustainable utilisation.
5. Emphasis should be given for exploration, collection and in situ conservation of PGR and associated knowledge.
6. There is a need for the farming communities to be aware about the importance and values of PGR and traditional resources. They should also be aware about their rights for commercialisation of these resources.

1.11 National and International Policies on PGR and TK Management

With the rise of global trade, PGR and TK come under threat due to over-exploitation of these resources. The economic pressure and privatisation have a great demand for PGR. The developed countries are trying to use these genetic resources for commercial purposes. The major global issues impacting PGR and TK for sustainable utilisation include the international treaties, conventions and agreements; global climate change; use of biotechnology and other technological advances in agriculture; and biosecurity and biosafety. The rights and access to these resources are provided by the CBD (Byerlee and Fischer 2002). Member countries of CBD including India brought new acts establishing their sovereignty on PGR and TK occurring within their geographical boundaries. The USA is not a signatory to CBD.

The WTO deals in the formulation of rules and regulations related to commercialisation of goods among the different countries. TRIPS and Sanitary and Phytosanitary (SPS) agreements under WTO have direct impact on PGR activities. IP on PGR, microbiological processes and microorganisms are covered under TRIPS rules and regulations. Amendments to the existing patent act and the protection of PGR in WTO are directly controlled by TRIPS. The preservation and sustainable utilisation of Plant Genetic Resources for Food and Agriculture are included under GPA. These resources are maintained and conserved as in situ or on farm to preserve genetic diversity in genetic resources.

The main mandate of ITPGRFA is the conservation and sustainable use of PGRFA and the fair and equitable sharing of the benefits arising out of their use, in harmony with the CBD, for sustainable agriculture and food security. It has been emphasised and recognised that the indigenous communities and local farmers have significantly contributed and are contributing towards collection of economical important PGR, their preservation and sustainable utilisation. ITPGRFA has provided recognition to the farmers and sharing of benefits arising out of the utilisation of these resources under revised Standard Material Transfer Agreement (SMTA) that has relevance to the food security. For protection of TK and to discuss IPR-related issues of accessing the PGR, the WIPO has set up ICGTK in 2001 (Varaprasad and Sivaraj 2010). TK is legally protected in association with the biological resources.

To secure the food security in India, the PGR and TK values are recognised in harmony with several international treaties; the government has come up with the initiatives such as enactment and implementation of the Biological Diversity Act (BDA) in line with CBD and PPV&FR Act, providing unique rights to farmers on par with qualified breeders. The Patent Act was modified suitably ensuring approval for all the patents involving biological resources from the National Biodiversity Authority (NBA). Provisions to protect and conserve threatened plant species through notification are being implemented through BDA. The National Bureau of Plant Genetic Resources (NBPGR) issued a plant quarantine order under the Destructive Insects and Pests Act which facilitates the safe import of plant material authorising for PGR. Inter-ministerial committees such as Genetic Engineering Approval Committee (GEAC) are functioning to facilitate genetic transformation research and import of genetically modified plants (Varaprasad and Sivaraj 2010).

1.12 Conclusion

The improvement of crops largely hinges on immediate conservation of PGR and TK for their effective and sustainable utilisation. A vast amount of PGR is threatened and endangered, and some have even gone extinct mostly due to genetic erosion and environmental transformation by anthropogenic effect. In order to meet current global challenges, all countries and institutions must as a matter of primary obligation discover, collect and conserve valuable and potentially valuable PGR and sustainable utilisation. In order to reduce the potential losses of PGR and TK resources, the knowledge of distribution of biodiversity in species as well as in ecosystems is important, and this can be achieved through efficient survey, inventory, appropriate research, field studies and analysis. Sustainable agriculture should be promoted by using diverse crop varieties in farming system, bringing more underutilised crops for cultivation and their commercialisation. The management of PGR should be integrated with the best utilisation of indigenous knowledge. Modern technologies should be combined with traditional practices for the protection, conservation and improvement of the resources.

For PGR and TK more natural reserved areas should be created, and those existing should be properly managed and financially supported, and an effective enforcement of laws should guard them. In order to make PGR more accessible to farmers and the plant breeders, proper documentation of these resources is required. Multi-institutional support is needed for conservation and potential use of indigenous knowledgeable medicinal plants for health and related purposes. Moreover, sharing of benefits to access traditional practices and other genetic resources is imperative for uniform development and to secure the livelihood of the farmers. Such access to PGR and TK and equal sharing of benefits with the indigenous and local people should be provided under certain norms and conditions. PGR and TK should be recognised on priority basis by the world communities to secure the world food security. More awards and respectful recognition should be given to the indigenous people and farmers who are really involved in conservation of valuable genetic resources. There is a need to create more awareness among the farmers for proper conservation of these resources to save the biodiversity. Approaches for better management of PGR and TK need to look for more species and genes to provide bio-alternatives and use both traditional approaches and modern technologies.

References

Agrawal A, Gibson CC (1999) Enchantment and disenchantment: the role of community in natural resource conservation. World Dev 27(4):629–649

Alston JM, Chan-Kang MC, Marra PP, Wyatt TJ (2000) A meta-analysis of rates of return to agricultural R&D: ex pede herculem? Research report 113. Intl Food Policy Res Institute, Washington, DC, p 118

Ameenah GF (2011) Traditional roles and future prospects for medicinal plants in health care. Asian Biotechnol Dev Rev 13(3):77–83

Anuradha RV (2007) Biopiracy and traditional knowledge. http://www.hinduonnet.com/folio/fo105/01050380.htm

Baillie JE, Hilton-Taylor C, Stuart SN (2004) IUCN red list of threatened species. A global species assessment. IUCN- World Conservation Union, Gland, p 25

Bhullar NK, Zhang Z, Wicker T, Keller B (2012) Wheat gene bank accessions as a source of new alleles of the powdery mildew resistance gene *Pm3*: a large scale allele mining project. BMC Plant Biol 10:88

Bonham C, Dulloo E, Mathur P, Brahmi P, Tyagi V, Tyagi RK et al (2010) Plant genetic resources and germplasm use in India. Asian Biotechnol Dev Rev 12(3):17–34

Briggs KG (2000) Plant genetic resources. Manag Agric For Fish Enterp 1:1–8

Brown AHD (1989) The case for core sets. In: Brown AHD, Frankel OH, Marshall DR, Williams JT (eds) The use of plant genetic resources. Cambridge University Press, Cambridge, pp 136–155

Byerlee D, Fischer K (2002) Accessing modern science: policy and institutional options for agricultural biotechnology in developing countries. World Dev 30(6):931–948

CBD (2005) Invasive alien species. Convention on biological diversity. http://www.biodiv.org/programmes/cross-cutting/alien/

Chandra R (2010) Knowledge as property, issues in the moral grounding of intellectual property rights. Oxford University Press, New Delhi, p 292

Dowd D (2009) Inequality and the global crisis. Pluto Press, London, pp 144–145

Erstling J (2011) Using patent to protect traditional knowledge. Tex Wesleyan Law Rev. www.wmitchell.edu/cgi/viewcontent.cgi?article=1187&context=facsch

Evenson RE, Gollin D (2003) Assessing the impact of the green revolution, 1960 to 2000. Science 5620:758–762

FAO (1996) Report on the state of the world's plant genetic resources for food and agriculture. International technical conference on plant genetic resources, Leipzig

FAO (1998) The state of the world's plant genetic resources for food and agriculture. FAO, Rome, p 89

FAO (2001) International Treaty on Plant Genetic Resources for Food and Agriculture. http://www.fao.org/ag/cgrfa/it/ITPGRe.pdf

FAO (2012) Food and Agriculture Organization. The State of Food and Agriculture. ftp://ftp.fao.org/docrep/fao/009/a0800e/a0800e.pdf

Frankel PH (1984) Genetic perspective of germplasm conservation. In: Arber W, Llimensee K, Peacock WJ, Starlinger P (eds) Genetic manipulations: impact on man and society. Cambridge University Press, Cambridge, pp 161–170

Gangadharan GG (2005) Potential of traditional knowledge in the development of healthcare products. Indian J Tradit Knowl 4(2):139–142

Gepts P (2006) Plant genetic resources conservation and utilisation: the accomplishments and future of a societal insurance policy. Crop Sci 46:2278–2292

Greaves (1996) Tribal rights. In: Brush, Stabinsky (eds) Valuing local knowledge: indigenous peoples and intellectual property rights. Island Press, Covelo

Gupta VK (2006) Documentation of traditional knowledge for protection: preservation and wealth creation. In: Proceedings of the international conclave on traditional medicine, New Delhi

Gupta BB, Salgotra RK, Bali AS (2009) Status paper on rice in Jammu and Kashmir. Rice Knowledge Management Portal (RKMP). http://www.rkmp.co.in

Hammer K (2003) A paradigm shift in the discipline of plant genetic resources. Genet Resour Crop Evol 50:3–10

Hammer K, Teklu Y (2008) Plant genetic resources: selected issues from genetic erosion to genetic engineering. J Agric Rural Dev Trop Subtrop 109:15–50

Hirwade M, Hirwade A (2012) Traditional knowledge protection: an Indian prospective. J Libr Inf Technol 32(3):240–248

Hoisington D, Khairallah M, Reeves T, Ribaut JM, Skovmand B, Taba S, Warburton M (1999) Plant genetic resources: what can they contribute toward increased crop productivity? Proc Natl Acad Sci U S A 96:5937–5943

Hood L (2003) Principles, practice and future of systems biology. Drug Disco Today 8(10):436–438

Hvalkoff S (2001) Outrage in rubber and oil. In: Zerner C (ed) People, plants and justice. Columbia University Press, New York, p 56

Hyten DL, Smith JR, Frederick RD, Tucker ML, Song Q, Cregan PB (2009) Bulked segregant analysis using the Golden Gate assay to locate the *Rpp3* locus that confers resistance to soybean rust in soybean. Crop Sci 49:265–271

IUPGR (1983) International Undertaking on Plant Genetic Resources for Food and Agriculture. http://www.fao.org/ag/cgrfa/IU.htm

Koo B, Pardey PG, Wright BD (2004) Saving seeds: the economics of conserving crop genetic resources *ex situ* in the future harvest centers of the CGIAR. CABI Publishing, Wallingford, p 67

Leipzig Declaration (1996) International technical conference on plant genetic resources, Leipzig. http://www.fao.org/FOCUS/f/96/06/more/declar-f.htm

Makombe K (1993) Sharing the land: wildlife, people and development in Africa. IUCN, Regional Office of Southern Africa, Harare

Mandal PK (2011) "Challenges to India's Patent Regime-"Traditional Knowledge sharing". www.papers.ssrn.com/sol3/papers.cfm?abstract_id=1393363

McNeely JA, Mooney HA, Neville LE, Schei P, Waage JK (2001) Global strategy on invasive alien species. IUCN – the World Conservation Union, Gland, p 46

Ogwu MC, Osawaru ME, Ahana CM (2014) Challenges in conserving and utilizing plant genetic resources. Intl J Genet Mol Biol 6(2):16–22

Olubisi F, Oluduro OF (2012) Nigeria: in search of sustainable peace in the Niger Delta through the amnesty programme. J Sustain Dev 5(7):4–7

Patwardhan B, Chopra A, Vaidya ADB (2003) Herbal remedies and the bias against Ayurveda. Curr Sci 84(9):1165

Pisupati B, Warner E (2003) Biodiversity and the millennium development goals. IUCN/UNDP, Colombo, p 8

Prada D (2009) Molecular population genetics and agronomic alleles in seed banks: searching for a needle in a haystack? J Expt Bot 60:2541–2552

Rosendal GK (1995) Genbanker-bevaring av biologisk mangfold (Genebanks-conservation of biodiversity). In: Stenseth NC, Paulsen K, Karlsen R (eds) Afrikanatur, samfunn og bistand. Ad Notam Gyldendal, Oslo, pp 375–392

Salgotra RK, Gupta BB (2005) Variability and correlations among different quantitative traits of rajmash (*Phaseolus vulgaris* L.). Abstracts of fourth international food legumes research conference, New Delhi, India, p 502

Salgotra RK, Gupta BB, Bhat JA, Sharma S (2015) Genetic diversity and population structure of basmati rice (*Oryza sativa* L.) germplasm collected from North Western Himalayas using trait linked SSR markers. PLoS One 10(7):e0131858. doi:10.1371/journal.pone.0131858

Sanchez PA, Swaminathan MS (2005) Cutting world hunger in half. Science 307:357–359

Sinclair FL, Walker DH (1999) A utilitarian approach to the incorporation of local knowledge in agroforestry research and extension. In: Buck LE, Lassoie JP, Fernandes ECM (eds) Agroforestry in sustainable agricultural systems. CRC Press LLC, Boca Raton, pp 245–275

Singh S, Sidhu JS, Huang N, Vikal Y, Li Z, Brar DS, Dhaliwal HS, Khush G (2001) Pyramiding three bacterial blight resistance genes (*xa5, xa13* and *Xa21*) using marker-assisted selection into *indica* rice cultivar PR106. Theor Appl Genet 102:1011–1015

Smale M, Koo B (2003) Genetic resources policies: what is a gene bank worth? Research at a glance. Briefs 7–12. IFPRI, IPGRI, and the Systemwide Genetic Resources Program, p 34

Sodhi NS, Erlich PR (2010) Introduction. In: Sodhi NS Erhlich PR (ed) Conservation biology for all. Oxford University Press, Oxford, p 1

Sundaram RM, Sakthivel K, Hariprasad AS, Ramesha MS, Viraktamath BC, Neeraja CN, Balachandran SM (2010) Development and validation of a PCR-based functional marker system for the major wide-compatible gene locus *S5* in rice. Mol Breed 26:719–727

Tanksley SD, McCouch SR (1997) Seed banks and molecular maps: unlocking genetic potential from the wild. Science 277:1063–1066

Ulukan H (2011) Plant genetic resources and breeding: current scenario and future prospects. Intl J Agric Biol 13:447–454

van Hintum TJL (1999) The general methodology for creating a core collection. In: Johnsons RC, Hodgkin T (eds) Core set for today and tomorrow. International Plant Genetic Resources Institute (IPGRI), Rome, pp 10–17

Varaprasad KS, Sivaraj N (2010) Plant genetic resources conservation and use in light of recent policy developments. J Plant Breed 1(4):1276–1293

Varkey E (2011) Traditional knowledge-the changing scenario in India. www.law.ed.ac.uk/ahrc/files/67varkeytraditionalknowledgeinindia

Wani SH, Wani SA, Ishfaq Abidi, Dar ZA, Lone A, Gull Zaffar, Khan MN (2003) Plant genetic resources of Kashmir region – prospect for crop improvement. National Conference on Status and Conservation of Biodiversity in India, Kashmir University, Kashmir

Xu K, Xu X, Fukao T, Canlas P, Maghirang-Rodriguez R, Heuer S, Ismail AM, Bailey-Serres J, Ronald PC, Mackill DJ (2006) *Sub1A* is an ethylene response factor like gene that confers submergence tolerance to rice. Nature 442:705–708

Plant Genetic Resources, Traditional Knowledge and Their Use in Crop Improvement

2

Gopal Katna and Vinod Kumar Sood

Abstract

Plant genetic resources and traditional knowledge offer opportunities for economic growth and sustainable food security for developing world. The landraces, modern varieties, genetic stock and advanced breeding material are the wealth of plant genetic resources. All these have contributed towards achieving food security and protection of biodiversity for sustainable development. The plant genetic resources have been used for the development and improvement of cultivars. The crop improvement by using these genetic resources is considered as one of the most important methods to ensure food security on a sustainable basis. This will also help in conserving the genetic diversity for future increase in agricultural crop production. Documentation and sustainable utilisation of plant genetic resources for crop improvement and equal sharing of benefits are the main objectives of crop improvement. The plant genetic resources are the rich source of important resistance genes for biotic and abiotic stresses. The rich source of genes can be transferred to the cultivated varieties by using conventional as well as modern plant breeding methods. Both conventional and biotechnological research inputs such as pure line selection, fine mapping of quantitative trait loci, marker-assisted selection, advanced backcross quantitative trait loci analysis and introgression libraries, association studies, direct allele selection and gene transfer have contributed in solving some of the constraints limiting crop productivity for sustainable food security.

2.1 Introduction

Plant Genetic Resources (PGR) and Traditional Knowledge (TK) are important components of food security. To feed the growing world population is really a challengeable task to the scientific community. The world population will go more than eight million by 2050, and to feed this ever-increasing population is a daunting task

G. Katna, Ph.D. (✉) • V.K. Sood, Ph.D.
Department of Crop Improvement, CSK Himachal Pradesh Krishi Vishvavidyalaya,
Palampur 176062, Himachal Pradesh, India
e-mail: gkatna@rediffmail.com; vkspbg23@rediffmail.com

R.K. Salgotra, B.B. Gupta (eds.), *Plant Genetic Resources and Traditional Knowledge for Food Security*, DOI 10.1007/978-981-10-0060-7_2

(FAO 2013). We have to keep pace in food production with the growing population which requires an astonishing increase in productivity. Moreover, there is very little chance to bring more land under cultivation. The PGR and TK are being continuously eroded due to over-exploitation and urbanisations. Feeding this population will require an astonishing increase in food production. The increase in food production should be achieved keeping in view the factors such as soil erosion, genetic erosion, water scarcity and limited labour. The PGR and TK of developing and least developed countries are being over-exploited by the developed nations after the inception of globalisation of trade (Rome Declaration on World Food Security 1996). Moreover, there is no any financial support provided by the international organisations for the rejuvenation of these valuable resources in the rural areas. For the foreseeable future, traditional methods of farming, conservation and sustainable utilisation of PGR and TK for crop improvement are our primary objectives for future food security.

The importance of PGR and TK to humanity stems from their use as an important source of genetic material for developing crops and medicinal plants fundamental to the world population's nutrition and health. The diversity of PGR is threatened by a complex of factors, inter alia, poverty, economic development and market integration. Before the discovery of scientific methods of crop improvement in agriculture, indigenous people were depended on landraces and nomadic farming. Traditional varieties/landraces are genetically more diverse than modern varieties. But to meet the demand of the growing population, the high-yielding varieties (HYVs) or modern varieties are introduced. The traditional landrace varieties are being replaced by these HYVs which narrow down the genetic diversity among the species. With the inception of green revolution, the HYVs have been introduced coupled with the use of more chemicals. To enhance the genetic diversity in crop species, these traditional landrace varieties must be put in cultivation in association with HYVs. PGR and TK are the basic resources used for the development of crop varieties, and the future food security depends on the conservation and sustainable utilisation of these resources.

Agriculture is the main occupation and source of income for the livelihood of more than 50 % world population in developing and least developing countries (SoW2-PGRFA 2010). With the introduction of modern HYVs and use of chemicals for production of these varieties, the genetic diversity of plant species deteriorates drastically. The severe loss in genetic diversity has not only affected the livelihood of indigenous people but there is significantly reduction in traditional practices of agricultural production. Moreover, the traditional landrace varieties and their wild relatives have played a significant role in providing food, fuel and medicines to these people. The monoculture of modern cultivars will reduce the genetic diversity in plant species in near future. The indigenous people have lost their adaptability to the changing conditions. Today about 30 cultivated plant species are providing food to the humankind, and the Food and Agriculture Organization (FAO) is trying to provide and satisfy the basic needs of the human (SoW1-PGRFA 1996). Out of these 30 cultivated plants species, 12 plants provide about 60 % of energy to human mostly from wheat, rice, potatoes and maize crops. Most of the nations are interdependent for food and other PGR (Frison and Halewood 2005). Due to this most of countries are growing the same variety over years to feed the growing population which reduces the genetic diversity in plant species. This will also lead to genetic vulnerability of diseases and insect-pests. The genetic diversity has been drastically reduced with the process of domestication in agriculture (Gepts 2006). The scientific community became aware after the genetic erosion of PGR and TK and tried to evaluate the genetic resources in gene banks (Hoisington et al. 1999). The characterization of PGR by comparisons of plant morphology, such as yield, colour, texture, taste, etc., is the simplest and easiest approach (Gilbert et al. 1999; Hoisington et al. 1999). In addition to these qualitative/quantitative phenotypic traits, pedigree analysis and geographical distribution are also helpful for measuring genetic diversity (Hammer 2003).

PGR constitute a major global heritage, and their conservation and utilisation are an immediate concern. There have been tremendous efforts in ex situ conservation of PGR. Despite its overall advantage and promotion by the international community, in situ conservation is still inadequate. With the introduction of HYVs developed through advanced breeding methods, the genetic diversity among the genotypes is decreasing (Tanksley and McCouch 1997; Hyten et al. 2009). In order to prevent the genetic erosion of these crop species, there is need to collect, evaluate, conserve and sustainably utilise. The core sets are developed to avoid the repetition of genotypes and represent the maximum genetic diversity (Frankel 1984; Brown 1989; Van Hintum 1999). Characterisation and quantification of genetic diversity have long been a major goal in evolutionary biology. Information on the genetic diversity within and among closely related varieties is essential for a rational use of genetic resources. The analysis of genetic variation both within and among elite breeding materials is of fundamental interest to plant breeders. It contributes for monitoring germplasm and can also be used to predict potential genetic gains. Diversity based on phenological and morphological characters usually varies with environments, and evaluation of these traits requires growing the plants to full maturity prior to identification. Protein or isozyme marker studies are also influenced by environment and reveal low polymorphism. Further, these markers are not reliable because many characters of interest have low heritability and are genetically complex. Now, molecular markers have proven to be powerful tools in the assessment of genetic variation and in the elucidation of genetic relationships within and among species. Earlier, the characterisation was done on the basis of morphological traits only. With the advent of molecular marker, now characterisation of germplasm can be done (Tanksley and McCouch 1997). Molecular markers are unaffected by the environmental factors and can be used at anytime (Andersen and Lubberstedt 2003). Several molecular markers, viz. restriction fragment length polymorphisms (RFLP), Random Amplified Polymorphic DNA (RAPD), simple sequence repeats (SSRs), Amplified Fragment Length Polymorphism (AFLP) and single nucleotide polymorphisms (SNPs), are presently available to assess the variability and diversity at a molecular level. Among these, SSRs or microsatellite markers are highly polymorphic, more reproducible, codominant and distributed throughout the genome (Gupta et al. 2001). All these markers are applied in crop improvement for Marker-Assisted Selection (MAS), construction of maps, QTLs mapping, etc. (Moose and Mumm 2008).

2.2 Importance of PGR in Crop Improvement

PGR are the genetic materials which are used for the development or improvement of crop varieties (Becker 1993). PGR are also used in the improvement of crop varieties which are susceptible to diseases and insect-pests. In conventional crop improvement programmes, PGR may not be useful for plant breeders without selection (Hallauer and Miranda 1981). Gene pool of a crop includes all cultivars (obsolete and current), wild species and wild relatives containing all genes available for breeding use. PGR has been divided into different gene pools such as primary gene pool, secondary gene pool, tertiary gene pool and isolated genes (Harlan and de Wet 1971; Becker 1993). In primary gene pool, crop plants can be easily crossed among themselves and produce fertile seeds, whereas secondary gene pool consists of wild relative plant species which can be crossed with plant species of primary species but produces sterile hybrids. In tertiary gene pool, plants produce sterile hybrids, but special gene of interest can be transferred by using protoplast fusion or embryo rescue techniques. In the isolated genes, the genes can be isolated from unrelated plant species, microorganisms or animals and transferred to primary plant species. The importance of the different classes of genetic resources for crop improvement depends on the target crop species. For example, the maize genetic diversity in the primary gene pool is so large that the secondary or tertiary gene pools are rarely used. In rapeseed, on the other hand, genetic diversity in the primary gene pool is small, and breeders have to transfer important traits from *Brassica* species of the secondary and tertiary gene pool into the cultivated species (Hu et al. 2002).

Due to genetic erosion and narrow genetic base, PGR can be completely vanished by biotic and abiotic stresses. When there will be complete destroy of PGR, the plant breeders must try to identify new source of resistance genes for the diseases and insect-pests. In addition, the introduction of commercial varieties into traditional farming systems leads to decrease in genetic diversity. The genetic erosion is also caused by weeds, growing of single crop on large areas over years and deforestation.

India is the home of rice cultivars, landraces and many lesser known types that have been under cultivation since ages by indigenous farmers as well as local entrepreneurs (Vinita et al. 2013). Among the landrace diversity, Gujinina, Kalonunia, Red hira sail, Kanakchur, Radhuni tilak, Dangi basful, Kala mogha and Gujanonia are aromatic types collected from West Bengal. Different landraces such as Tulaipanji, Swarna, Purnima, Red hira sail, Kanakchur, Nagra dhan and Gogal sail are with awn. Red rice (Karad) is popular in the Himachal Pradesh. India has also contributed important in form of wheat crop germplasm in various countries (Zeven and Zeven-Hissink 1976). Diversity of maize is impressive in the North-Eastern Himalayan region, specifically in Sikkim, including Murli makai (Sikkim Primitive), Kali makai (with dark purplish-black kernel type), Rathi makai (with dark red kernels), Paheli makai (with yellow/orange flint kernel type), Seti makai (with white kernel type), Putali makai (with transposon-induced pericarp variegation), Chaptey makai (with white, dent-type kernels), Gadbade makai (predominantly white kernels with some purple flint kernels), Bancharey makai (a high-altitude maize with yellow, flint kernel type), Kukharey makai (with short-statured plants), Kuchungdari (with orange-coloured popcorn-type kernels) and Kuchungtakmar (with a mix of yellow, white, purple and red kernels) (Anonymous 2009).

To meet the demand of growing population and sustainable utilisation of PGR, plant genetic diversity is the prerequisite to secure livelihood and ensure food security (FAO 1996b). The concrete aims of Global Plan of Action (GPA) are (1) the development of crop varieties for marginal or stress environments; (2) production of HYVs through reduced utilisation of fertilisers, nutrients, water and agrochemicals for control of diseases and insect-pests; and (3) alternative farming for production of pharmaceutical crops. To achieve these aims, extensive in situ and ex situ preservation of PGR must be assured. Evaluation of conserved accessions and their use by plant breeders or farmers needs to be supported and facilitated. The aim should be to exploit intraspecific variation within a crop and to increase interspecific diversity in agriculture through genetic improvement and promotion of less popular, neglected or underutilised crop species (Padulosi et al. 2002). Many underutilised species can be used for small holdings for selecting biotic and abiotic stress-tolerant plants for sustainable production. The PGR need to be evaluated for their outcrossing rates, yield potential, response to inputs, agronomic value and the amount of genetic variation for specific traits, to allow more efficient genetic improvement and promotion.

2.3 Classical Methods of Using PGR in Crop Improvement

Before the development of modern varieties, farmers cultivated landrace varieties. The landrace cultivars have more genetic diversity compared to modern cultivars. Moreover, the landrace varieties are specific to particular environments and thus these are more diverse. In the release of modern cultivars, the emphasis has been given for specific traits and genes only. The breeders have crossed different parental material to get HYVs with desirable traits. Besides increasing the yield potentials of crop varieties, quality traits have also been improved. Traditional crop varieties have been developed by using the local germplasm and wild relatives of plant species. In conventional plant breeding methods, the emphasis was given for high yield and other quality traits by using the available PGR in region. With the inception of globalisation, the breeders can import the important genetic material from one country to another to be used in breeding programmes. The breeders are using the diverse genetic material for the development and improvement of crop varieties resistance to biotic and abiotic stresses. The breeders are utilising diverse plant material and genetic stock in crop improvement programmes for the development of varieties resistant to diseases and insect-pests and tolerant to salinity, heat and drought (Duvick 1986). Most of the new crop varieties take 8–11 years for development which lasts only for 5–6 years under cultivation (USDA 1990). After that, these varieties become susceptible to a number of diseases and insect-pests. The landraces and wild relatives of plant species are the main source of genes resistant to various biotic and abiotic stresses, and plant breeders are introducing these resistance genes in the cultivated ones. The newly improved varieties have significant effect to enhance the yield production (Wilkes 1991).

With the advancement of modern technologies such as genetic engineering and biotechnology, the PGR has been used to increase the efficiency of conventional crop improvement methods. These techniques have been used for introduction of gene of interest into cultivated varieties which are otherwise not possible to transfer through conventional breeding methods. The genetic engineering methods can be used for transferring the gene of interest from the unrelated and wild species to the cultivated ones. The biotechnological approaches are used to overcome the difficulties arise in the conventional breeding programmes. Therefore, the landraces and the wild relatives of plant species are the rich sources of important genes. The conservation and sustainable utilisation of these genetic resources are priority of each nation. PGR should be preserved to meet the challenges being posed on crop production from time to time, as these are the rich sources of resistance genes for various biotic and abiotic stresses besides various quality traits. Therefore, conservation of PGR (in situ and ex situ) is required for sustainable agricultural crop production (Fig. 2.1). Generally, there are three ways of using PGR in plant breeding (Simmonds 1993):

1. Transfer of gene of interest from PGR to popular landrace cultivated varieties or modern HYVs cultivars
2. Development of new genetic stocks as a source of resistance to numerous diseases, insect-pests and other environmental factors and quality traits
3. Pre-breeding materials developed for research purpose to enhance the efficiency of conventional plant breeding methods

2.3.1 Introgression and Pyramiding

These approaches are deployed to introgress few genes and quantitative trait loci (QTLs) of interest to the cultivated varieties. The backcross breeding methods are used for the transfer of resistance genes from landraces to cultivated varieties. These methods are also used to remove defects in the widely cultivated varieties. To introgress specific traits from PGR into high-yielding genotypes, breeders would principally consider PGR of the primary gene pool, followed by the secondary gene pool and eventually the tertiary gene pool (Becker 1993). Due to Darwin's observations on 'parallel variation' which correspond to Vavilov's law of 'homologous series', it is possible that a certain allele will be found in a cultivated form if genetic variation for it exists among wild relatives (Becker 1993).

2.3.2 Incorporation

The main aim of the genetic enhancement and broadening the genetic base in the breeding materials is to enhance the genetic diversity. The selection intensity is maximum in the initial stage due to genetic diversity and recombination, and subsequently decreases. In this regards, Cooper et al. (2001) identified three methods:

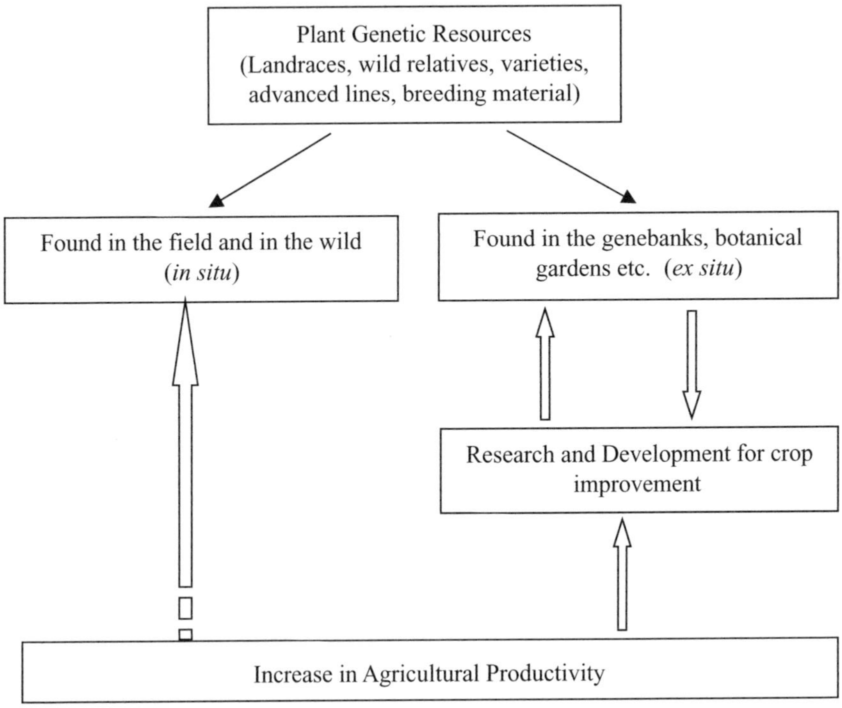

Fig. 2.1 The relationship between genetic resources and agriculture

1. Development of synthetic or composite-cross populations (long term)
2. Incorporation of PGR in a region's breeding materials to reduce the effects of historical bottlenecks during the evolutionary spread of the crop (medium term)
3. To increase the genetic variation in breeding population (short term)

2.3.3 Pre-breeding and Wide Crosses

In pre-breeding, the crosses are attempted by using landraces, wild relatives and exotic materials for the crop improvement. The crosses are made for the improvement of both qualitative and quantitative traits. These genetic materials provide the plant breeders with resourceful germplasm which can be utilised for the development of new cultivars, improvement in well-adapted varieties, pyramiding of resistance and quality traits and development of hybrids.

2.3.4 Breeding of Varieties for Specific Environments

Development of new variety shows variations in yield production over environments due to genotype x environment interactions. Due to genotype x environment interactions, the varieties perform differentially. In extreme stress conditions, the yield of the variety is reduced significantly. This genotype x environment is the main hindrance for the release of widely grown varieties (Annicchiarico 2002). Landrace varieties cultivated to different stress conditions such as salinity, drought and heat can serve as a good source of genetic materials for resistance breeding for these traits (Hawtin et al. 1997). These landraces can be used as donors of resistance genes for the improvement of crop varieties. The development of varieties for specific environments will include:

1. Newly developed stock of materials should be evaluated at local environment or under stress conditions at farmers' field.

2. The varieties should be crossed with elite breeding materials.
3. Breeding the genetic material for different environments and locations.
4. Early segregating materials should be grown at different locations and areas.
5. Farmers' participatory approach for selection of genetic materials so that it can be easily transferred to farmers' fields (Ceccarelli et al. 2001; Witcombe 2001).

Thus, the utilisation of these diverse PGR in crop improvement will help to ensure the future food security of the world.

2.4 Genomic Approaches for Using PGR in Crop Improvement

With the advances in techniques in modern biotechnology, different types of molecular markers were developed and used to increase the selection efficiency of desired trait in the segregating and advanced breeding generations. Today, to increase the efficiency of conventional plant breeding methods, molecular markers play an important role for the improvement of crop varieties. Molecular markers are used in numerous ways such as development of linkage maps, association mapping, MAS, marker-assisted backcross breeding (MABB), etc. All these techniques are directly and indirectly contributing in crop improvement. The molecular markers help in the identification of important genes of interests to be used in breeding population.

The quantitative trait loci (QTLs) are the traits controlled by number genes which are associated with a particular trait of interest. The identification of QTLs through conventional breeding methods on the basis of morphological and phenotypic evaluation is impossible. With the use of molecular markers, linkage maps have been developed for identification of genes of interest and QTLs by using QTL analysis (Mohan et al. 1997), and identification of genomic regions associated with this is termed as QTL mapping (McCouch and Doerge 1995; Mohan et al. 1997). The identified markers which are tightly associated to gene of interest can be used for MAS in crop improvement (Ribaut and Hoisington 1998). The use of molecular markers in MAS has increased the efficiency of conventional plant breeding methods (Rafalski and Tingey 1993).

The importance of molecular markers in crop improvement has been envisaged by linkage analysis, construction of linkage maps and identification of genomic regions linked with trait of interest (Collard et al. 2005). The construction of linkage maps, QTL analysis and their application in crop improvement through MAS has been studied by various researchers (Haley and Andersson 1997; Jones et al. 1997; Paterson et al. 1991; Paterson 1996a, b; Staub et al. 1996; Tanksley 1993; Young 1994).

2.4.1 Genetic Diversity Study

Genetic diversity study is very useful for development and improvement of crop species. The diversity analysis is the basic and foremost important for the development of crop varieties. In plant breeding, the crossing programmes are designed on the basis of genetic variation present in the germplasm. The genetic diversity among the genotypes, varieties and landraces is studied on the basis of morphological, biochemical and molecular traits. Diversity is assessed among varieties and landraces of rice, barley, buckwheat, winter wheat, beans, etc. by using morphological, isozyme, RAPD and SSR markers. Clustering based on the genetic distance is the common method of showing relationship which is very useful in further plant breeding methods (Salgotra et al. 2015).

2.4.2 Marker-Assisted Selection (MAS)

MAS method helps to increase the efficiency of conventional crop improvement programmes. It helps in the selection of gene of interest closely linked with a maker. MAS can be used for selection of suitable genotype in the early segregating generations. The pyramiding of genes can be done in a well-adapted variety through marker-assisted backcross (MAB). MAB has been widely used to improve the traits of interest in a variety and can save 2–3 generations. MAS is used (1) to identify suitable genotypes possessing molecular markers linked with the trait (foreground selection) and (2) to avoid the linkage drag of the donor parents (background selection). Foreground selection requires a tight linkage between the trait of interest and its flanking markers that are being selected for. Background selection necessitates genotyping with a larger number of markers representing each chromosome that cover the whole genome (Haussmann et al. 2004).

MAS has been proved to be an important tools for crop improvement in rice (Mackill et al. 1999; McCouch and Doerge 1995), wheat (Eagles et al. 2001; Koebner and Summers 2003; Van Sanford et al. 2001), maize (Stuber et al. 1999; Tuberosa et al. 2003), barley (Thomas 2003; Williams 2003), tuber crops (Barone 2004; Gebhardt and Valkonen 2001), pulses (Kelly et al. 2003; Muehlbauer et al. 1994), oilseeds (Snowdon and Friedt 2004), horticultural crop species (Baird et al. 1996, 1997; Mehlenbacher 1995) and pasture species (Jahufer et al. 2002). The molecular markers have been identified as the important tools to enhance the efficiency of traditional plant breeding programmes (Kasha 1999). No doubt that the initial outcomes of DNA marker technology may not be so effective, but number of researchers are using the molecular marker technology effectively in MAS (Eagles et al. 2001; Kelly and Miklas 1998; Lee 1995) (Table 2.1).

2.4.3 QTL Mapping and Gene Pyramiding

The yield and yield components, and other quality traits are controlled by number of genes. In MAS or MAB, the identified QTLs can be used in crop improvement programmes for resistance to various biotic and abiotic stresses and quality traits (Table 2.1).

2.5 Traditional Knowledge

TK is the local knowledge that is unique to a given culture or society (Warren 1991). TK has played a significant role in indigenous peoples' livelihood and food security. But till date no scientific record and documentation has been done. There is need for the development of database of TK for its potential uses. Moreover, this information helps in numerous decision-making processes and other communications (Flavier 1995).

The researchers have different views of TK, but theses researchers have covered only some aspects of it. Ellen and Harris (1996) have given following characteristics of TK:

1. TK has an origin of place and set of experiences which is generated by local and indigenous people living in those places.
2. TK is transmitted through imitation and demonstration, or orally.
3. TK is the result of experiences of everyday life and is constantly reinforced by experience and trial and error.
4. It is experimental rather than the knowledge developed through theory.
5. Repetition is the characteristic of tradition even when new knowledge is added (Hobsbawm and Ranger 1983).
6. TK is a continuous process and changing constantly, being discovered as well as lost and produced as well as reproduced.
7. TK is characteristically shared to a much greater degree than other forms of knowledge, including global science. This is why it is sometimes called 'people's science', an appellation which also arises from its generation in contexts of everyday production (Hobart 1993).
8. TK has focused on individuals and may attain a degree of consistency in rituals and other symbolic constructs but its distribution is always fragmentary.
9. Although the claims for the existence of culture-wide abstract classifications of knowledge are based on non-functional criteria (Berlin 1992; Atran 1990), TK is at its densest and directly applicable to its organisation.

Table 2.1 DNA marker technology for development/improvement of various crops

Crop	Breeding strategies	Objective	Type of marker
Maize	AB-QTL	Improvement of performance of hybrids for yield, grain moisture and plant height	SSR
	Introgression, MAS	Transfer of transgene for insecticidal protein genes into elite inbred line	RFLP
	MAS	Development of high-quality protein maize (QPM)	SSR
	MAS	Enhancement of drought tolerance	RFLP, PCR-based markers
Barley	Introgression, MAS	Enhancement of boron toxicity in barley	SSR
	Introgression, MAS	Resistance to barley yellow mosaic virus (BYMV)	RFLP
Rice	Introgression, AB-QTL	Identification of yield-improving QTLs from *O. rufipogon*	RFLP, SSR
	Introgression, MAS	Bacterial leaf blight (BLB) resistance (*X. oryzae* pv. *oryzae*) from elite restorer line of hybrid rice 6078	AFLP
	Pyramiding	Bacterial leaf blight (BLB) resistance (*X. oryzae* pv. *oryzae*) into elite rice variety PR 106	SSR
	Pyramiding	Bacterial leaf blight (BLB) resistance (*X. oryzae* pv. *oryzae*) into elite rice variety Samba Mahsuri	SSR
	MAS	QTLs for drought tolerance	RFLP, SSR
	MAS	Introgression of 'FR13A' into 'Swarna'	SSR
Potato	Introgression, MABC	Potato late blight (*Phytophthora infestans*) resistance derived from *Solanum bulbocastanum*	PCR marker
	Pyramiding	Resistances to potato virus Y, root cyst nematode (*Globodera rostochiensis*), potato virus X and potato wart (*Synchytrium endobioticum*)	CAPS SCAR-based PCR markers
Tomato	Introgression, MABC	Acylsugar-mediated pest resistance from *Lycopersicon pennellii*	RFLP, RAPD
	MAS	Resistance to tomato spotted wilt virus (TSWV)	CAPS

(continued)

Table 2.1 (continued)

Crop	Breeding strategies	Objective	Type of marker
Wheat	Introgression	Stacking of QTL for *Fusarium* head blight (FHB) resistance from non-adapted sources in an elite spring wheat background	SSR
	MAS	Leaf resistance, stripe rust resistance and stem rust resistance gene complex	SSR
	MAS	Introduction of six *Fusarium* head blight QTLs, orange blossom wheat midge resistance (*Sm1*) and leaf rust resistance (*Lr21*)	SSR
Vegetables	MAS	Resistance to *Uromyces appendiculatus* and *Colletotrichum lindemuthianum* in *Phaseolus vulgaris*	RAPD
	MAS	Common bacterial blight (*X. axonopodis* pv. *phaseoli*) resistance in dark red kidney bean (*P. vulgaris* L.)	SCAR
	Pyramiding	Resistance to *U. appendiculatus*, *C. lindemuthianum* and *Phaeoisariopsis griseola* into cultivar 'Perola' (*P. vulgaris* L.)	RAPD, SSR, SCAR
Apple	Pyramiding	Resistance to apple scab (*Venturia inaequalis*) and powdery mildew (*Podosphaera leucotricha*)	SSR

Source: Salgotra et al. (2015) Journal of Plant Science and Research

10. TK is situated within broader cultural traditions; separating the rational from the non-rational and the technical from the non-technical is problematic (Scoones and Thompson 1994).

2.6 Genetic Erosion of PGR and TK Systems

The loss of diversity is termed as genetic erosion. The genetic erosion is mainly caused by the introduction of HYVs in place of traditional varieties (FAO 1996a). The modern varieties are developed by taking only few genes and traits of interests. These modern varieties are grown on a large scale replacing the local landraces. On the other hand, landrace varieties possess more number of genes, and these varieties have region-specific adaptation. With the inception of green revolution, HYVs are introduced with more inputs of agrochemicals such as chemical fertilisers, pesticides, fungicides and weedicides. More area has been covered with one or two crops which led to the reduction of landrace varieties. Other causes of genetic erosion are urbanisation and environmental degradations.

More genetic erosion has been observed in main staple crops of the world such as wheat, rice, maize and potato. With the introduction of modern cultivars, there has been continuous removal of landraces and their wild relatives. Landraces become out of cultivation with the introduction of HYVs. Moreover, farmers prefer to grow more HYVs than the indigenous landrace cultivars because of higher productivity. The areas under landraces and wild-related species are reduced day by day with the urbanisation. Moreover, the governments did not show any interest in the preservation of these valuable resources.

In order to counter the loss of genetic erosion caused by a number of factors such as introduction of modern varieties, use of agrochemicals and urbanisations, the traditional methods of agricultural production can concentrate on the preservation of these PGR. The landraces and wild-related plant species should be protected and conserved in an area or a region. This will help to enhance the genetic diversity of plant species. It has observed that both the ex situ and in situ conservation of plant species will help to enhance the genetic diversity. Farmers should be encouraged to cultivate newly released varieties with the landrace varieties. This will not only enhance the genetic diversity but also protect the modern cultivars from the infestations of diseases and insect-pests. This will help the farmers to grow diverse cultivars and avoid the practices of monoculture.

With the advancement in modern techniques, the traditional practices are eroded. TK of indigenous people has also been eroded with the introduction of modern techniques. Indigenous knowledge of local people for conserving and cultivating of PGR and TK is lost with the inception of modern technologies. Indigenous people are creating more interests in the adoption of modern and easy-going techniques compared to traditional ones. Moreover, TK is lost or modified over time naturally, but there is need to conserve the traditional practices to ensure food security (World Bank 1997). The international and national organisations should frame the policies for conservation and sustainable utilisation of PGR and TK. Some incentives should be given to the farmers for preserving the genetic diversity of these resources. The programmes should be framed for farm conservation of landraces and wild-related species. This conserved biodiversity will provide resistance sources of genes to plant breeders for the development and improvement of cultivars.

2.7 Threats Posed by Intellectual Property Rights (IPRs) to PGR and TK

With the introduction of HYVs, the landrace varieties and the wild relatives are continuously diminishing from their natural habitats. Intellectual Property Rights (IPRs) give the rights of ownerships to the developers of varieties. But the developers are not providing any incentives to the local communities who are conserving these germplasm. Moreover, IPRs are helping to spread these modern varieties to wider areas. The private companies have taken away the ownership of PGR from the indigenous communities and get the IPRs. The landraces are maintained by the indigenous farmers; they collect and conserve them by using their experiences and knowledge. The researchers get these materials from the gene banks and use them in research programmes to develop new varieties and make money after patenting. But no incentives are given to the farmers for further improvement and maintenance of these PGR and TK. In this system, farmers are not preserving and saving the seeds for the next season and totally depending upon the private companies for procuring the seed materials which is a big problem for the farmers.

To meet the criteria to obtain the IPRs, the genotype must fulfil the criteria of Distinctness, Uniformity and Stability (DUS). Evaluation of genotypes for DUS tests of landraces is very difficult as landraces possess a wide range of genes. The plant breeders breed more uniform varieties with narrow genetic base and varieties have few numbers of genes.

2.8 Conservation of PGR and TK for Food Security

Today, PGR and TK are at high risk due to climate change, change in economy and cultures. Traditional cultures and knowledge are vanishing with the adoption of new technologies. With the introduction of HYVs and modern crop production practices, the landraces as well as the knowledge of traditional production are reducing drastically. PGR are preserved as ex situ and in situ gene banks. Both ex situ and in situ conservation methods are important for conservation and sustainable utilisation of PGR and TK.

The CBD acknowledged the contribution of TK which help in protecting the important plant species. Indigenous people are using the traditional culture, traditional songs, traditional knowledge of the use of medicinal plants and traditional practices since long. The traditional cultures of the local communities are involving the preservation of PGR and TK. They are daily using diverse plant species in food, medicines and other purposes. They are involved in the conservation and sustainable utilisation of PGR and TK for their livelihood and to ensure food security.

The growing population, climate change and genetic erosion have shifted the attentions of international communities for conservation and sustainable utilisation of PGR and TK. The plant breeders and the local farmers are involved in Participatory Plant-Breeding (PPB) in the selection, conservation and sustainable utilisation of PGR. The farmers are being rewarded for conservation of landraces and sharing of benefits. The local landrace varieties are rich in diverse genetic materials and are the good source of important genes such as resistance to biotic and abiotic stresses and quality traits. For protection of these important germplasm and traditional varieties rich in genetic diversity, the Indian Parliament, adopting a 'sui generis' system, enacted the 'Protection of Plant Varieties and Farmers' Rights (PPV&FR) Act, 2001', and thereafter the rules were notified in 2003. The PPV&FR Authority is also giving awards to the farmers who are conserving and depositing the diverse genetic materials for conservation.

To encourage the farming community, the PPV&FR Authority in consultation with the Government of India had confirmed the modalities and criteria for 'Plant Genome Saviour Community Award'. The award consists of rupees ten lakhs in cash, a citation and a memento. The award is given to all farming communities, indigenous people, rural people and tribes who are actively engaged in preservation and improvement of PGR. The applicant is required to submit information in support of their claims and a brief proposal for the utilisation of award money towards community welfare measures/development schemes including establishment of local seed/grain bank, water conservation facilities, facilities for grain/seed threshing/postharvest processing, farm schools or other such activities. With the approval of the Competent Authority for the institution of 'Plant Genome Saviour Community Award', the two societies were awarded—Kopatgiri Nandiveerimath Seva (KNS) Foundation, Karnataka, and Panchabati Gramya Unayana Samiti, Koraput, Odisha.

2.9 Other Uses of PGR and TK in Food Security

Traditional practices (or *purane nuskhe*) have home remedies for every ailment possible. There are so many traditional methods in history to solve the problems with the TK which are moving from generation to generation. If there is a problem of coughing or sniffling, they use a concoction with grinding of some leaves or seeds for immediate relief. By applying a little mustard oil on the child's belly button after bath every day, they will never have chapped lips. If there is problem of burning, then mash a ripe banana and apply on burns and bandage with betel leaves for cure. To relieve children from worm infestation, give them a drink of a little buttermilk mixed with a pinch of hing (*Ferula assafoetida*) fried in ghee. To relieve teething pains in babies, mix clove oil with sunflower oil and massage gently on the gums. To treat skin rashes and itching, combine a small amount of turmeric powder with aloe vera or water to form a paste and apply on the affected area. If the child has excessive dry and flaky skin, then use a paste made of chickpea flour (besan), camphor (a pinch) and milk/rose water before bath. People try to use neem (*Azadirachta indica*) and mint/pudina (*Mentha* L.) leaves to protect their food grains and seed from storage pests. Even in field, farmers used ash to maintain soil fertility and to protect the crops from frost.

Since time immemorial, the red rice landraces are gifted on various occasions such as marriages, religious ceremonies, child birth and community lunches and to a sick person. The excess water of thick consistency decanted after cooking of red rice is considered very useful for pregnant ladies and children.

1. Bael (*Aegle marmelos*) is useful for intestinal problems such as dysentery and diarrhoea.
2. Garlic (*Allium sativum*) is used to control excess conservation of lipids and cholesterol.
3. Indian aloe (*Aloe vera*) is useful in ophthalmic disorders, increasing semen quality, spleen and liver disorders and anomalies of skin and blood.
4. Asparagus (*Asparagus racemosus*) used to nourish the female reproductive system and promote lactation.
5. Cucumber (*Cucumis sativus*) is used as a food to renal stone patient.
6. Cumin (*Cuminum cyminum*) seeds are used to increase taste, as an appetiser and for the treatment of abdominal colic, flatulence, dysentery, diarrhoea, indigestion, nausea, vomiting, etc.
7. Carrot (*Daucus carota*) roots are used to treat physical weakness, problem before delivery, cardiac problems, piles, colic burning sensation, asthma, etc. and as mouth freshener.
8. Citron (*Citrus medica*) is useful in heart disorders, abdominal colic, vomiting, nausea, indigestion, etc.
9. Jamun or jambolan (*Syzygium cummini*) is very useful for diabetic patients.
10. Dates are traditional crops in Arab and other Middle East countries. It is highly nutritious and full of simple sugars, and in periods of lengthy fasting, it provides energy throughout the day. It prevents different diseases like constipation, anaemia, heart diseases and strokes. They are good for overall body health because of their countless dietary nutrients. High fibre content in it prevents high cholesterol levels. Vitamin C, potassium and calcium are essential for bone and teeth development especially in children. They can be given to children as sweet dish after main meal instead of toffees and chocolates, which are of low or no nutritious value.
11. Mango (*Mangifera indica*) is beneficial for people suffering from acidity. The bioactive elements such as aldehydes, esters, terpenes, etc. help in easy digestion. As it is a rich source of iron, it is useful for anaemic people also.
12. Boil mixture of amla (Indian gooseberry) pieces in coconut oil is used to cure premature greying of hair naturally at home.
13. Malabar nut (*Adhatoda vasica*) commonly known as vasaka is used for comfort from bronchitis, cough and bronchial asthma. It also has immune system-boosting properties.
14. Neem (*Azadirachta indica*) is used for skin disorders. It is antifungal and antibacterial and has blood-purifying and immune-boosting properties.

2.10 Conclusion

To meet the increasing demand of food, agriculture growth must grow at faster rate to feed the world population. PGR and TK are the basic contributing factors for securing the food security. Most of the developing and least developing countries are gifted with diversified agroecosystems in which crop diversity plays an integral role. There is a growing challenge to ensure that these resources are sustainably used and conserved for future generations. To achieve future aims, extensive in situ and on farm/ex situ conservation of PGR must be assured. The conservation of PGR, TK and traditional practices can provide more resilience to numerous factors of changing environments. We should explore their potentials and conserve them for sustainable utilisation in future breeding programmes for the improvement of varieties to ensure food security. The international and national communities must encourage the indigenous communities for the preservation of these landraces and provide some benefits to them. There is a need to frame and implement more programmes linked with the conservation of landraces and their wild relatives. The research institutions must be involved in basic and strategic research for potential conservation of PGR and TK. The policy makers and planners should frame such laws which can provide some incentives to the indigenous communities who are involved in the preservation of these valuable resources. Incentives and facilities must be provided by the governments for on farm/in situ conservation of landraces and wild-related species of crops. The capacity-building programmes should be launched for indigenous people for the potential preservation and sustainable utilisation of these resources for food security.

References

Andersen JR, Lubberstedt T (2003) Functional markers in plants. Trends Plant Sci 8:554–560

Annicchiarico P (2002) Defining adaptation strategies and yield stability targets in breeding programmes. In: Kang MS (ed) Quantitative genetics, genomics and plant breeding. CAB International, Wallingford, pp 365–383

Anonymous (2009) www.icar.org.in/files/newsletters/icar.../ICAR-News-Jan-March-09.pdf

Atran S (1990) Cognitive foundations of natural history: towards anthropology of science. Cambridge University Press, Cambridge

Baird WV, Ballard RE, Rajapakse S, Abbott AG (1996) Progress in Prunus mapping and application of molecular markers to germplasm improvement. HortSci 31:1099–1106

Baird V, Abbott A, Ballard A, Sosinski B, Rajapakse S (1997) DNA diagnostics in horticulture. In: Gresshoff P (ed) Current topics in plant molecular biology: technology transfer of plant biotechnology. CRC Press, Boca Raton, pp 111–130

Barone A (2004) Molecular marker-assisted selection for potato breeding. Am J Potato Res 81:111–117

Becker H (1993) Pflanzenzu chtung. Verlag Eugen Ulmer, Stuttgart

Berlin B (1992) Ethnobiological classification: principles of categorization of plants and animals in traditional societies. Princeton University Press, Princeton

Brown AHD (1989) The case for core sets. In: Brown AHD, Frankel OH, Marshall DR, Williams JT (eds) The use of plant genetic resources. Cambridge University Press, Cambridge, pp 136–155

Ceccarelli S, Grando S, Amri A, Asaad FA, Benbelkacem A, Harrabi M, Maatougui M, Mekni MS, Mimoun H, El-Einen RA, El-Felah M, El-Sayed AF, Shreidi AS, Yahyaoui A (2001) Decentralized and participatory plant breeding for marginal environments. In: Cooper HD, Spillane C, Hodgkin T (eds) Broadening the genetic base of crop production. CABI Publishing in co-operation with FAO and IPGRI, CAB International, Wallingford, pp 115–135

Collard BCY, Jahufer MZZ, Brouwer JB, Pang ECK (2005) An introduction to markers, quantitative trait loci (QTL) mapping and marker-assisted selection for crop improvement: the basic concepts. Euphytica 142:169–196

Cooper HD, Spillane C, Hodgkin T (2001) Broadening the genetic base of crops: an overview. In: Cooper HD, Spillane C, Hodgkin T (eds) Broadening the genetic base of crop production. CABI Publishing in cooperation with FAO and IPGRI, CAB International, Wallingford, pp 1–23, Division, Office of Public Affairs

Duvick DN (1986) Plant breeding: past achievements and expectations for the future. Econ Bot 40(3):289–297

Eagles H, Bariana H, Ogbonnaya F, Rebetzke G, Hollamby G, Henry R et al (2001) Implementation of markers in Australian wheat breeding. Aust J Agric Res 52:1349–1356

Ellen R, Harris H (1996) Concepts of indigenous environmental knowledge in scientific and development studies literature – a critical assessment; draft paper East-West Environmental Linkages Network Workshop 3, Canterbury

FAO (1996a) Report on the State of the World's Plant Genetic Resources for Food and Agriculture, prepared for the International Technical Conference on Plant Genetic Resources, Leipzig, Germany. Food and Agricultural Organization of the United Nations, Rome

FAO (1996b) Global plan of action for the conservation and sustainable utilization of plant genetic resources and the Leipzig Declaration, adopted by the International Technical Conference on Plant Genetic Resources, Leipzig, Germany. Food and Agricultural Organization of the United Nations. FAO and IPGRI, CAB International, Rome, pp 385–398

FAO (2013) Food and Agriculture Organization. The State of Food and Agriculture. ftp://ftp.fao.org/docrep/fao/009/a0800e/a0800e.pdf

Flavier JM (1995) The regional program for the promotion of indigenous knowledge in Asia. In: Warren DM, Slikkerveer LJ, Brokensha D (eds) The cultural dimension of development: indigenous knowledge systems. Intermediate Technology Publications, London, pp 479–487

Frankel OH (1984) Genetic perspectives of germplasm conservation. In: Arber W, Llimensee K, Peacock WJ, Starlinger P (eds) Genetic manipulation: impact on Man and Society. Part III, Paper No. 15. Cambridge University Press, Cambridge

Frison C, Halewood M (2005) Annotated bibliography addressing the international pedigrees and flows of plant genetic resources for food and agriculture SGRP, Information document submitted by the System-wide Genetic Resources Programme of the CGIAR to the 8th conference of the parties to the Convention on Biological Diversity (COP 8) and the Ad Hoc Open-ended Working Group on Access and Benefit-sharing, International Plant Genetic Resources Institute (IPGRI), Rome, Italy

Gebhardt C, Valkonen JPT (2001) Organization of genes controlling disease resistance in the potato genome. Annu Rev Phytopathol 39:79–102

Gepts P (2006) Plant genetic resources conservation and utilization: the accomplishments and future of a societal insurance policy. Crop Sci 46:2278–2292

Gilbert JE, Lewis RV, Wilkinson MJ, Caligari PDS (1999) Developing an appropriate strategy to assess genetic variability in plant germplasm collections. Theor Appl Genet 98:1125–1131

Gupta PK, Roy JK, Prasad M (2001) Single nucleotide polymorphisms: a new paradigm for molecular marker technology and DNA polymorphism detection with emphasis on their use in plants. Curr Sci 80:524–535

Haley C, Andersson L (1997) Linkage mapping of quantitative trait loci in plants and animals. In: Dear P (ed) Genome mapping – a practical approach. Oxford University Press, New York, pp 49–71

Hallauer AR, Miranda JB (1981) Quantitative genetics in maize breeding. Iowa State University Press, Ames

Hammer K (2003) A paradigm shift in the discipline of plant genetic resources. Genet Resour Crop Evol 50:3–10

Harlan JR, de Wet JMJ (1971) Toward a rational classification of cultivated plants. Taxon 20:509–517

Haussmann BIG, Parzies HK, Prester T, Susic Z, Miedaner T (2004) Plant genetic resources in crop improvement. Plant Genet Resour 2(1):3–21

Hawtin G, Iwanaga M, Hodgin T (1997) Genetic resources in breeding for adaptation. In: Tigerstedt PMA (ed) 18 B. I. G. Haussmann et al. Adaptation in plant breeding. Amsterdam: Kluwer Academic Publishers, pp 277–288

Hobart M (1993) Introduction: the growth of ignorance? An anthropological critique of development. Routledge, London, pp 1–30

Hobsbawm E, Ranger T (1983) The invention of tradition. Cambridge University Press, Cambridge

Hoisington D, Khairallah M, Reeves T, Ribaut JM, Skovmand B, Taba S, Warburton M (1999) Plant genetic resources: what can they contribute toward increased crop productivity? Proc Natl Acad Sci U S A 96:5937–5943

Hu Q, Hansen LN, Laursen J, Dixelius C, Andersen SB (2002) Intergeneric hybrids between Brassica napus and Orychophragmus violaceus containing traits of agronomic importance for oilseed rape breeding. Theor Appl Genet 105:834–840

Hyten DL, Smith JR, Frederick RD, Tucker ML, Song Q, Cregan PB (2009) Bulked segregant analysis using the GoldenGate assay to locate the Rpp3 locus that confers resistance to soybean rust in soybean. Crop Sci 49:265–327

Jahufer M, Cooper M, Ayres J, Bray R (2002) Identification of research to improve the efficiency of breeding strategies for white clover in Australia: a review. Aust J Agric Res 53:239–257

Jones N, Ougham H, Thomas H (1997) Markers and mapping: we are all geneticists now. New Phytol 137:165–177

Kasha KJ (1999) Biotechnology and world food supply. Genome 42:642–645

Kelly JD, Miklas PN (1998) The role of RAPD markers in breeding for disease resistance in common bean. Mol Breed 4:1–11

Kelly JD, Gepts P, Miklas PN, Coyne DP (2003) Tagging and mapping of genes and QTL and molecular marker-assisted selection for traits of economic importance in bean and cowpea. Field Crops Res 82:135–154

Koebner RMD, Summers RW (2003) 21st century wheat breeding: plot selection or plate detection? Trends Biotech 21:59–63

Lee M (1995) DNA markers and plant breeding programs. Adv Agron 55:265–344

Mackill DJ, Nguyen HT, Zhan J (1999) Use of molecular markers in plant improvement programs for rainfed lowland rice. Field Crops Res 64:177–185

McCouch SR, Doerge RW (1995) QTL mapping in rice. Trends Genet 11:482–487

Mehlenbacher SA (1995) Classical and molecular approaches to breeding fruit and nut crops for disease resistance. HortSci 30:466–477

Mohan M, Nair S, Bhagwat A, Krishna TG, Yano M, Bhatia CR, Sasaki T (1997) Genome mapping, molecular markers and marker-assisted selection in crop plants. Mol Breed 3:87–103

Moose SP, Mumm RH (2008) Molecular plant breeding as the foundation for 21st century crop improvement. Plant Physiol 147:969–977

Muehlbauer F, Kaiser W, Simon C (1994) Potential for wild species in cool season food legume breeding. Euphytica 73:109–114

Padulosi S, Hodgkin T, Williams JT, Haq N (2002) Underutilized crops: trends, challenges and opportunities in the 21st century. In: Engels JMM, Rao VR, Brown AHD, Jackson MT (eds) Managing plant genetic diversity. IPGRI/CABI Publishing, Wallingford, pp 323–338

Paterson AH (1996a) Making genetic maps. In: Paterson AH (ed) Genome mapping in plants. R. G. Landes Company/Academic Press, Austin/San Diego, pp 23–39

Paterson AH (1996b) Mapping genes responsible for differences in phenotype. In: Paterson AH (ed) Genome mapping in plants. R. G. Landes Company/Academic Press, Austin/San Diego, pp 41–54

Paterson A, Tanksley S, Sorrels ME (1991) DNA markers in plant improvement. Adv Agron 44:39–90

Rafalski J, Tingey S (1993) Genetic diagnostics in plant breed: RAPDs, microsatellites and machines. Trends Genet 9:275–280

Ribaut JM, Hoisington D (1998) Marker-assisted selection: new tools and strategies. Trends Plant Sci 3:236–239

Rome Declaration on World Food Security and Plan of Action (1996) World Food Summit, 13–17 November 1996, Rome

Salgotra RK, Gupta BB, Sood M (2015) Biotechnological interventions and their role in sustainable hill agriculture. J Plant Sci Res 2(1):1–8

Scoones I, Thompson J (eds) (1994) Beyond farmer first. Intermediate Technology Publications, London

Simmonds NW (1993) Introgression and incorporation. Strategies for the use of crop genetic resources. Biol Rev 68:539–562

Snowdon R, Friedt W (2004) Molecular markers in *Brassica* oilseeds breeding: current status and future possibilities. Plant Breed 123:1–8

SoW1-PGRFA (1996) The First Report on the State of the World's Plant Genetic Resources for Food and Agriculture, FAO Commission on Genetic Resources for Food and Agriculture, Rome

SoW2-PGRFA (2010) The Second Report on the State of the World's Plant Genetic Resources for Food and Agriculture, FAO Commission on Genetic Resources for Food and Agriculture, Rome

Staub JE, Serquen F, Gupta M (1996) Genetic markers, map construction and their application in plant breed. HortSci 31:729–741

Stuber CW, Polacco M, Senior ML (1999) Synergy of empirical breeding, marker-assisted selection, and genomics to increase crop yield potential. Crop Sci 39:1571–1583

Tanksley SD (1993) Mapping polygenes. Annu Rev Genet 27:205–233

Tanksley SD, McCouch SR (1997) Seed banks and molecular maps: unlocking genetic potential from the wild. Science 277:1063–1066

Thomas W (2003) Prospects for molecular breeding of barley. Ann Appl Biol 142:1–12

Tuberosa R, Salvi S, Sanguineti MC, Maccaferri M, Giuliani S, Landi P (2003) Searching for quantitative trait loci controlling root traits in maize: a critical appraisal. Plant Soil 255:35–54

U. S. Department of Agriculture (USDA) (1990) New germplasm helps assure food for the future. USDA, Washington, DC

Van Hintum TJL (1999) The general methodology for creating a core collection. In: Johnsons RC, Hodgkin T (eds) Core set for today and tomorrow. International Plant Genetic Resources Institute (IPGRI), Rome, pp 10–17

Van Sanford D, Anderson J, Campbell K, Costa J, Cregan P, Griffey C, Hayes P, Ward R (2001) Discovery and deployment of molecular markers linked to fusarium head blight resistance: an integrated system for wheat and barley. Crop Sci 41:638–644

Vinita P, Taneja N, Vikram P, Singh NK, Singh S (2013) Molecular and morphological characterization of Indian farmers rice varieties (*Oryza sativa* L.). Aust J Crop Sci 7(7):923–932

Warren DM (1991) Using indigenous knowledge in agricultural development, World Bank Discussion Paper No.127. The World Bank, Washington, DC

Wilkes G (1991) In situ conservation of agricultural systems. In: Alcorn JB, Oldfield M (eds) Biodiversity: culture, conservation, and eco-development. Westview Press, Boulder

Williams KJ (2003) The molecular genetics of disease resistance in barley. Aust J Agric Res 54:1065–1079

Witcombe JR (2001) The impact of decentralized and participatory plant breeding on the genetic base of crops. In: Cooper HD, Hodgkin T (eds) Broadening the genetic base of crop production. CABI Publishing in co-operation with FAO and IPGRI, CAB International, Wallingford, pp 407–417

World Bank (1997) Knowledge and skills for the information age, the first meeting of the Mediterranean Development Forum; Mediterranean Development Forum, URL:http://www.worldbank.org/html/fpd/technet/mdf/objectiv.htm, pp 39–57

Young ND (1994) Constructing a plant genetic linkage map with DNA markers. In: Ronald IKV, Phillips L (eds) DNA-based markers in plants. Kluwer, Dordrecht

Zeven AC, Zeven-Hissink, Ch.N (1976) Genealogies of 14,000 wheat varieties. CIMMYT, Mexico, D.F., and the Netherlands Cereal Center, Wageningen

3 Global Strategies for Sustainable Use of Agricultural Genetic and Indigenous Traditional Knowledge

Zoran Jovovic and Suzana Kratovalieva

Abstract

Mankind is becoming more aware that all life on Earth is significantly threatened and has begun to accept this threat. Thus, we seek to restore damaged natural resources and preserve those still existing. The survival of humans on our planet is directly related to genetic resources, and thus the rational exploitation of genetic resources is necessary to continually raise public awareness on their manyfold importance. From the aspect of feeding the growing human population, plant genetic resources for food and agriculture are invaluable, making them increasingly important for world food security. Therefore, plant genetics are significant in the implementation of many strategies designed for recovery of degraded ecosystems and natural habitats and for the conservation and protection of endangered plant and animal species. Sustainable use of agricultural genetic resources is not limited only to counteracting the loss of germplasm but also to maintaining the traditional knowledge related to agricultural plants. In searching for the concept of sustainable development, humans realized long ago that the thousands of years of knowledge and experience of indigenous people contributed significantly to solving the many challenges created by natural phenomena and human activities. More active participation of local communities in biodiversity conservation programs would contribute to more efficient and cost-effective conservation and biodiversity management. There is no doubt that indigenous knowledge is more important today than ever before, and in this sense, its preservation may be considered as necessary to modern society.

Z. Jovovic, Ph.D. (✉)
Biotechnical Faculty Podgorica, University of Montenegro, Mihaila Lalica 1, Podgorica 81000, Montenegro
e-mail: zoran.jovovic.btf@gmail.com

S. Kratovalieva, Ph.D.
Ministry of Agriculture Forestry and Water Economy, State Phytosanitary Laboratory,
Aleksandar Makedonski blvd, Skopje 1000, Former Yugoslav Republic of Macedonia
e-mail: suzanakratovalieva@gmail.com

R.K. Salgotra, B.B. Gupta (eds.), *Plant Genetic Resources and Traditional Knowledge for Food Security*, DOI 10.1007/978-981-10-0060-7_3

3.1 Introduction

There are about 7 billion people on our planet today, and it is estimated that this number could surpass 9 billion by 2050. The fastest population growth is happening on the African continent, where, by the middle of this century, the population could be doubled. For these reasons, provision of sufficient quantities of food for so many people in the world is undoubtedly one of the greatest global challenges.

World agriculture has progressed so rapidly that traditional methods of production have been very quickly and aggressively replaced with modern technological approaches and solutions. Intensive agricultural development through the irrational use of natural resources has led to significant changes in the environment, such as lack of water, soil erosion, air pollution, pollution of rivers, seas, and oceans, disappearance of plant and animal species, forest degradation, and desertification, and has seriously disrupted the balance of our planet. Man is becoming more aware of this imbalance and now accepts that life on Earth is significantly threatened. This awareness orients us towards restoration of the damaged environment and preservation of the remaining natural resources. The survival of humans on the Earth is directly related to genetic resources, and in addition to rational exploitation of these resources it is necessary to constantly raise social awareness of their significance. It is clear that resources must be used but also that they cannot be endlessly exploited. Therefore, raising the question "In what way and to what extent?" is justifiable. The limits of many human activities are determined by the physical endurance of the environment. If man really wants to pass on this planet to future generations in a state in which it will be able to continue to support human life, then we must live within these limits and significantly reduce our consumption of natural resources.

The level of social concern about the status of disappearing plant species, as well as other existentially endangered species, is closely connected with local culture, ethnic characteristics, folklore, traditional pharmacopoeia, the level of environmental knowledge, and so on (Kratovalieva et al. 2010; Stefanova et al. 2012). While searching for the concept of sustainable development, mankind realized long ago that the thousands of years of knowledge and experience of indigenous peoples can contribute significantly to solving these problems. Indigenous traditional knowledge (ITK), especially concerning the environment, natural resources, and climate, allows many local communities around the world to provide stable living conditions in concord with conservation and the sustainable use of biological diversity. Traditional knowledge and habits gained in the long history of agricultural development have been of great help for farmers for conservation and the proper selection of genetic resources from different natural habitats such as fields, meadows, pastures, and forests in their immediate environment (Zachariev et al. 2012).

There are many indications that biodiversity has become significantly compromised under the influence of climate change and the uncontrolled exploitation of natural resources, and plant and animal species today are disappearing at a much faster rate than in the past. In this time of increased impacts of both climate change and growing human needs, significant social efforts must be undertaken to ensure a long-term and sustainable food supply. Thus, we must strive to establish a flexible production system that will enable the production of healthy and safe food for all people in the world while preserving biodiversity and natural resources. Conservation and sustainable use of Plant Genetic Resources (PGR) and Traditional Knowledge (TK) associated with genetic resources is a long-term need that surpasses national interests. With the aim of stopping further erosion, 20 years after the Rio Earth Summit, the United Nations (UN) declared the period 2011–2020 as the "UN Decade of Biodiversity" (UN Resolution 65/161 of 22 December 2010), in the spirit of the Convention on Biological Diversity (CBD).

3.2 Overview of the Sustainable Use of Plant Genetic Resources for Food and Agriculture (PGRFA)

The concept of biodiversity includes all the different ecosystems, species, genes, and their relative abundance. In other words, it represents the overall diversity of life on Earth and all its processes. Over many millennia of human history, the emergence of new species was a significantly more intense process compared to their disappearance, which led to the great wealth of biodiversity in nature. However, under the impact of climate change and the uncontrolled exploitation of natural resources, genetic diversity is rapidly decreasing, and thus the ability of the environment to respond to any noticeable distortion of biodiversity is also decreasing. The extinction of species now takes place very rapidly, and each day 74 species disappear; that is, 27,000 species are lost annually. If the extinction of species continues with this intensity during the next 30 years, 20 % of the species that are alive today on the planet could disappear (Hammond 1995). In recent decades, significant focus has been placed on the active use of local genetic resources and associated TK. The correlation between biology and agriculture, and with anthropology, is of great importance and gives certain advantages in research on the general interest for conservation and use of genetic resources. Many such studies have shown that, despite intense technological development, many rural regions have preserved their traditional and local values to a significant extent, whereas in urban areas these values have completely disappeared (Giovannini and Heinrich 2009).

Agrobiodiversity (AgBD) is certainly an important component of overall biodiversity. According to CBD, it implies the variability of all animals, plants, and microorganisms that are directly or indirectly involved in agriculture and the production of food. In a broader context, AgBD can be divided into two main categories. The first consists of plant genetic resources for food and agriculture (PGRFA), Wild Relatives (CWR), domestic animals (including fish and other managed aquatic animals), and the genetic resources of fungi and microorganisms. The second category of AgBD includes all other species that in some way contribute to the production of food, such as, for example, microorganisms in the soil, pollinators (bees, butterflies, etc.), and predators. This category also includes uncultivated species from the environment that have an impact on the agroecosystem and its productivity. Indigenous knowledge and culture are integral aspects of agrobiodiversity.

Sustainability is used as a concept in a variety of contexts (development, agriculture, economics, technology, environment, etc.) and for these reasons is defined in different ways. In a broader sense, sustainability is the ability to maintain the balance of certain processes or conditions. Seen from the ecological point of view, it is a way of long-term preservation of diversity and productivity based on respect for nature. One of the most important characteristics of the economics of natural resources and the environment is sustainability or sustainable development (Kratovalieva et al. 2012a). Sustainability, or sustainable development, represents not only an essential prerequisite but also the ultimate goal of many human activities on the Earth, the key to survival and duration of the human species. Sustainability should not be necessarily a current need but rather a clearly defined goal for the future. There should be a fair, balanced compromise between the environmental, social, and economic dimensions of sustainability, providing a balanced meeting of the needs of both present and future generations. Unfortunately, today, under the pressure of global capital, "sustainability" is an overemphasized issue of the economy in relationship to other dimensions of sustainability, and thus it becomes a major cause of damage to the natural balance, leading to withdrawal from the postulated goals of equality (Adams 2006). From the aspect of the human population, sustainability represents an instrument for long-term prosperity based on rational and responsible use of natural resources. In contrast to other components of biodiversity, one of the basic characteristics of agricultural biodiversity is that farmers actively manage it and that many of its components could not survive without constant human impact. In traditional systems of crop production, to ensure their own existence, farmers are much more dependent on the availability of natural resources (quantity and quality of soil, water resources, agricultural biodiversity) than on external inputs for agricultural production, that is, fertilizers and plant protection products (PPP). This system involves the cultivation of a wide range of crops for human food, animal feed, medicines, building materials, etc. In such systems, more attention is paid to avoiding various production risks than to maximizing production.

The disappearance of traditional systems of production led to the disappearance or endangering of a large number of cultivated species, cultivars, varieties, and locally adapted populations from the farm (Scarascia-Mugnozza and Perrino 2002). The beginning of the erosion of PGFRA is considered to be the moment when modern selection began significantly suppressing traditional cultivars. Replacing traditional varieties with new, high-yield, and genetically uniform selections is the main reason for genetic erosion: this concept was implemented with the "Green Revolution." Although these measures were fully justified from the economic point of view, they caused the disappearance of a huge number of local plant varieties and populations (Pencic et al. 1997). The "Green Revolution" led to a significant increase in food production, even in countries where starvation and food production deficiency was very pronounced, but this was a result of using newly created secondary varieties with high production potential but requiring great use of agrochemicals (Ejeta 2009). At the same time, this 'revolution' significantly decimated resources of the traditional cultivars that were vital for small-scale farmers and the future of plant breeding.

Intensive development of agriculture in Europe and North America, and in some less developed countries in other parts of the world as well, has led to significant depletion of genetic resources (Brush 2000). According to estimates by the FAO (1999), replacement of indigenous PGRFA with modern varieties and hybrids in the past century caused about 75 % of PGR to be permanently lost in the world. Viewed from this perspective, the so-called green revolution can be considered a synonym for the loss of agrobiodiversity. In the period of the green revolution disappearance of many primary sources of biodiversity, numerous potential offspring were lost as well. Such fragmentation of populations and genetic drift caused increasing homogeneity of populations and narrowing of genetic variation and heterozygosis, respectively. Reduced genetic diversity decreased the ability of populations and species to respond effectively to any noticeable changes in the external environment. Thus, it further reduced their chances of survival.

Genetic resources for food and agriculture (GRFA) are the most important link between man and nature. They occupy an important place in human life, especially in the efforts to survive and build a better life for humans and generations to come (Plotkin 2000). GRFA are resources of great importance that are used in food, textiles, the construction industry, and the energy industry, as well as in the production of feed. Global food production is based on these resources and represents an excellent tool in the fight against hunger and poverty. In addition, these resources are significant in the implementation of many strategies designed for renewal of disrupted ecosystems and natural habitats, as well as the conservation and protection of endangered plant and animal species. One of the biggest challenges faced by countries rich in biodiversity today is the preservation, collection, characterization, evaluation, documentation, and sustainable use of these resources. Sustainable use of GRFA represents exploitation of biological diversity in a way that does not cause disruption of biodiversity and allows maintaining that level of diversity and productivity which meets the needs of present and future generations (Kratovalieva et al. 2012b). It must be based on sustainable agricultural practices that reduce genetic erosion and conserve genetic diversity. Agricultural land covers approximately 38 % of the total land area on Earth, and thus the responsibility of farmers around the world to protect biodiversity is huge (Strajeru et al. 2009).

Traditional agriculture, and many other areas of nature, are under increasingly strong pressure from the dynamics of economic development

(Teklu and Hammer 2006). To increase total production, farmers are stimulated through various forms of incentives to change their old, locally adapted varieties and populations to improved, high-yielding selections. Although the permanent loss of agrobiodiversity on farms remains a serious concern, recent efforts in this sector demonstrate that there is a firm determination to build clear and effective mechanisms of access and sustainable use of GRFA in the world. Preservation of agricultural diversity should provide enough stable production of high-quality food and plant products and mitigate the risks associated with intensive cropping systems. Increasing diversity of cultivated crops ensures the greater safety of food production. Bear in mind that the increasing attention devoted to genetic resources in recent years seems to have finally matured the conviction that the preservation of life on Earth depends on the preservation of biodiversity. Thus, activities to raise awareness of the importance of the conservation of biological diversity and the impact it has on the well-being of the human population in the future should be significantly intensified (Jovovic et al. 2013).

The general framework for the conservation and sustainable use of biological diversity and fair and equitable distribution of profits arising from the use of genetic resources is represented through the Convention on Biological Diversity (CBD). The Nagoya Protocol further regulates the conditions for access to genetic resources and the fair and equitable division of monetary and nonmonetary benefits raised from their use. This protocol regulates the rights and obligations of entities that use genetic resources for research, development, and commercialization, or which use traditional knowledge associated with genetic resources. To achieve full implementation of the protocol, it will be very important to eliminate all ambiguity concerning access and benefit-sharing rules so that all interested parties are able to make decisions under conditions of full legal safety (Frese et al. 2015). The beliefs related to genetic resources and knowledge arising from their use have matured and these resources must be used in an environmentally sustainable and socially equitable manner, which implies a fair distribution of the subsequent profits.

Support for the conservation and sustainable use of GRFA in many countries of the world is an integral part of the support for the development of rural areas. To ensure sustainable agricultural production, it is necessary to promote activities of traditional cultivation and underused plant species on farms. Therefore, it is necessary to find sustainable models of farm cultivation that would both stimulate producers to grow traditional crops and lead to an increase of farm income and sustainability of production. Particular attention should be given to species that are of particular importance for the local community: such an approach would lead to increased efficiency of the existing conservation on farms. For these reasons, we must work intensively to encourage farm cultivation with the aim of increasing the number of households that actively use genetic resources. Economic sustainability is one of the key factors on influencing choices made by farmers in engaging traditional breeding populations; thus, it is necessary to create conditions in which this work will be economically viable. For this reason, the European Union (EU) is introducing new mechanisms of support for agricultural producers for reimbursement of expenses and loss of profits resulting from growing underused and endangered plant species. As signatory to the International Treaty on Plant Genetic Resources for Food and Agriculture (ITPGRFA), the EU has committed to take all necessary efforts to stop the decline of agricultural biodiversity, not only in Europe but also internationally. Bearing in mind that the on-farm conservation of indigenous PGRFA is one of the most important priorities in plant genetic resource management (Dhillon et al. 2004), in recent years incentives for sustainable use of biodiversity on farm become an integral part of EU support for regional and rural development (Melozzi 2009). There is a clear intention to allocate even greater financial resources for this purpose in the future.

The market is an important factor in the disappearance of PGRFA. With the arrival of new cultivars, indigenous PGRFA began to disappear from production and from the local market. Yields of traditional cultivars were significantly lower compared to modern varieties, and their production has become less competitive and affordable. Because of inability to compete in local markets, many farmers completely replaced their traditional varieties with new varieties some time ago. Therefore, finding new markets is one of the priorities in the path of preserving PGRFA (Frank et al. 2002). To facilitate access of genetic resources to the market of commercial varieties, in EU has been announced a comprehensive audit of a large amount of legislation that treats the problem of seed and planting material: this would open possibilities for the production and sale of seed from less uniform varieties.

Approximately 7.4 million accessions are conserved in more than 1,750 gene banks that exist today in the world (The Second Report on the State of the World's Plant Genetic Resources for Food and Agriculture, FAO 2010), and more than 2,500 botanical gardens around the world represent reliable guardians of a large number of extremely valuable ex situ collections. All this is merely a symbolic sample of what actually exists in nature. This material is often well adapted to local conditions and is rich in genes that can facilitate the adaptation of crop varieties to environmental changes, including climate change, new diseases, pests, and many other challenges (Frese et al. 2015; Hammer and Teklu 2008). Plant Gene Banks are important as a source of genetic material for the restoration and enhancement of biodiversity, the survival of future generations, and the biosphere. However, because of incomplete or missing documentation, or unfinished characterization and evaluation, as well as difficulties in accessing stored material and data, despite numerous international strategies that address these issues a significant fraction of the stored material is not used in the extent to which it really should be because of its real or potential value (Thorn 2011). In addition, as a result of poor management, poor maintenance, and lack of funds, a significant number of accessions stored in many gene banks around the world are permanently lost (FAO 1998).

PGRFA have economic significance only if properly used. To increase the scope of their use, it is essential that this material is well preserved, documented, evaluated, with as much information as possible about the use value. Information on the genetic, phenotypic, and comparative features contributes to better ex situ and in situ conservation, and to sustainable use of genetic resources. Also, the foregoing information is of particular importance for the institutions and individuals who are engaged in or have an interest in PGRFA. The research community is actively working to improve our knowledge of PGR and TK. This intensified research, on a national level, will give a clear assessment and provide sufficient knowledge about the extent and distribution of agricultural genetic diversity in each specific crop gene pool. The existence of an active system of communication between research institutions, production practices, and decision makers contributes to the rapid and continuous flow of knowledge and other relevant information. In this way, new knowledge would become more accessible to end users, and the possibility of overlapping and dissemination of unreliable and inaccurate information would be largely avoided.

New ways of transferring knowledge from the research sector to production practices should be adapted to the requirements of end users and the level of TK that they possess. Such a system of professional assistance would lead to technological and technical improvement of agricultural holdings, and thus to greater competitiveness. To achieve economic competitiveness for traditional and underutilized plant species, optimization of existing production technologies and introduction of new ones should be our goal. In addition to increased productivity, such technologies would lead to stronger expression and other agronomic, organoleptic, and nutritional properties, and in the end to fulfilment of consumer requests.

As PGRFA often show a high degree of genetic variability, they are an important building material for creating new varieties (Simeonovska et al. 2013). Bearing in mind that the PGRFA mainly appeared in environments with low availability of nutrients, they can be a very valuable source of variability to create varieties adapted to production systems with low fertilizer input. Through the establishment of a strong public–private partnership (PPP), new possibilities for the valuation of these resources appear. Such partnership would create an environment in which the leading breeding companies, on the basis of fair and clearly defined criteria, establish a priority list of "long-term crop-specific pre-breeding programs" that would be able to respond to future needs and challenges in the field of plant breeding.

Agriculture and tourism are intertwining industry branches that complement each other in many areas. Development of alternative forms of tourism (agrotourism, hunting and fishing, eco-tourism, etc.) and everything else that accompanies these activities is one model to improve the quality of life of the local population in less developed areas. The production of unique traditional products of high quality appears as a real chance for the development of rural areas, along with a larger selection of agricultural products tourist offer will improve. One of those possibilities is growing PGRFA, especially in areas where the cultivation of most other crops would not be profitable. Lower yields from this production would be compensated through tourism service and significantly higher selling prices. For these reasons, we must create a suitable environment where parallel with the development of specific agricultural production, specific tourist offers would be developed, and vice versa.

Although this is in great contrast to current practices, restoring biodiversity on agricultural land is one of the most important strategies of sustainable agriculture. Organic production as part of a sustainable agricultural system is based on increasing biodiversity, and growing PGRFA in organic production is an additional way of evaluation. Using traditional cultivars in this way will enable the farmers to obtain additional financial resources, bearing in mind the higher price of organic products (Jovovic 2014). Organic agriculture is based on broad and intensive crop rotation, which requires the cultivation of a large number of plant species throughout the year. For development of organic production, traditional varieties and ecotypes adapted to growing conditions in a given climatic region have special importance, and part of the preserved plant material can be directly used in this production. For the successful cultivation of indigenous PGRFA, it is necessary to create optimal cultural practices based on TK on breeding those plant species. Growing indigenous genetic materials to produce food of high nutritional value, and authentic and natural taste, and also leads to the increase of agrobiodiversity. To grow these indigenous populations in organic production, it is necessary they possess resistance or lower susceptibility to biotic and abiotic factors. Growing resistant varieties is significant from both ecological and economic aspects.

Promotion of most expressed genotypes, as products with protected geographic origin, also has a positive impact on increasing the sustainable use of genetic resources in agriculture (Kratovalieva et al. 2012c). Establishing a sustainable system of government incentives that would stimulate farmers to grow traditional varieties, as well as the various support mechanisms for public consumption, would be an important step in that direction.

The inclusion of different types of processing capacity with relatively small scale of production, specialized sales channels, and restaurants can actively contribute to the revival of agricultural holdings and strengthening sustainable use of PGRFA. All this would lead to economic strengthening of local economies, and thus the greater valorisation of genetic resources. Promotion of new products derived from PGRFA or the use of TK associated with genetic resources is certainly important. Commercialization of these products and the expansion of their markets can significantly contribute to increasing the area under traditional cultivars and species at risk of extinction. However, such activities require the joint action of a large number of social agents (farmers, processors, retailers, and consumers), and the level of these synergies will influence the final results.

More than 90 % of plants live in symbiosis with microorganisms and fungi. The annual contribution of soil bacteria in the process of conversion of nitrogen into nitrates and nitrites in the

US is about 33 billion dollars, and that of insects as pollinators of cultivated plants about 30 billion dollars (Alonso et al. 2001). The importance of microorganisms, pollinators, and other organisms in food production is enormous. Although their importance was neglected for years, lately it is increasingly becoming the subject of global attention. Therefore, in the next period more attention should be paid to the conservation and sustainable use of these resources, especially if taken into account that they are still largely underexplored.

3.3 Value of Indigenous PGRFA

It is estimated that there are about 350,000 different species within the plant kingdom, and of these about 70,000 are considered edible. Of these, only 7,000 species are used in some manner in the human diet. It is an amazing fact that 95 % of the food in the world is produced from just 30 plant species and that more than half of the total amount refers to three cultivars: wheat, corn, and rice (Morico et al. 1999; Wilson 1992). Although world food production is based on a few plant species, for conservation of agricultural biodiversity diverse production at the regional, national, or local level is crucial. In the development of civilization, thinking primarily of their own needs and present well-being, humans have largely disrupted the balance that existed in nature for centuries. Intensive infrastructural projects undertaken by man have led to significant damage of the ozone layer, general environmental pollution, and global climate change. All this has a negative impact on biodiversity in all its elements (genes, varieties, populations, species, ecosystems) and PGRFA are by far the most exposed to these changes. Because of habitat destruction and the introduction of elite germplasm of wild relatives, local populations and traditional varieties of major crops are the most vulnerable and most threatened aspect of biodiversity that is at the highest risk of disappearing (Dimitrijevic et al. 2011). It is estimated that the cultivated species today contain only a small fraction of the genetic variability of their wild ancestors (Glogovac and Takac 2010).

Biodiversity encompasses the entire genetic material of all organic species existing on Earth. Created in the early period of the establishment of life on Earth, species survived and changed over a long period of evolution in the space and time continuum. The use of biological diversity is as old as the history of agriculture. PGRFA are the basis of world food production and thus their significance for the survival of the human population is priceless. In many stages of human development, they had very important functions: PGRFA rescued many nations from starvation, and led to population growth and economic development (Jovovic et al. 2013). The conservation of ecosystems and their services is the basis of sustainable economy and social development in the world. For these reasons, awareness of the need and importance of preserving genetic diversity as a main prerequisite for biodiversity is continually and rapidly growing worldwide. In addition to their large role in increasing production and food safety, PGRFA have significant influence on reducing pressure on other ecosystems, including vulnerable and endangered species. Although PGRFA have great significance for the development of rural areas, their valorisation, because of the still insufficiently active role of the rural population in their use and economic evaluation in agrobiodiversity, is not at the level at which it indeed should be. However, many rural regions of the world have particularities that are tied to different events wherein the skills and knowledge of agrobiodiversity, as part of their own tradition, are passed on to younger members of the community. Today, modern society is increasingly accepting traditional values and knowledge. Numerous discussions and workshops are held in which TK related to AgBD is transferred to the younger generation (Thomas et al. 2009).

Through intensive development of technology, new plant species are being constantly discovered that can be used in agriculture, while at the same time, under pressure from cheaper synthetic products, the exploitation of many species is being abandoned (Jovovic et al. 2012). The introduction of new, uniform, resistant, and high-yielding varieties and hybrids has led to a significant increase in food production in the world, transforming some countries from importers of food into significant suppliers of the world market. From the aspect of nutrition of a constantly growing population, PGRFA are invaluable, and their importance as a basis for world food security must be strongly emphasized. The growing risk of the irreversible loss of agrobiodiversity requires much greater effort in the preservation of local ecotypes and traditional varieties. Such an approach will not only preserve diversity but also increase the economic potential of local communities (Pieroni et al. 2004).

Because PGRFA are invaluable in human nutrition, more attention is paid to the preservation of the variability of cultivated species than that of natural communities today. The more pronounced impact of climate change, significant loss of diversity of agricultural crops, and rising problems caused by plant diseases, pests, and weeds have led to a more serious approach to the indigenous PGRFA and the need to preserve the most expressive genotypes that nature has created. However, it is not an easy job. Therefore, it is very important to collect and preserve all that is left, and to be committed to work to revitalize lost genetic variability.

An effective system of PGRFA conservation requires the possession of wide knowledge of important historical facts, cultural development, economic conditions, the development of agriculture, the level of development of society, local and religious customs, and climate change. Awareness of the high degree of erosion of genetic resources in recent decades has led to the establishment of a large number of seed and field gene banks in developed countries (poor in genetic resources) and in poor and developing countries (rich in genetic resources). In addition to ex situ conservation, where preserved variability remains unchanged (Dhillon et al. 2004), a very important way of controlling genetic diversity is in situ protection and conservation of agrobiodiversity in cultivated areas or natural habitats. This type of dynamic conservation, through the process of adapting crops to environmental conditions, allows continuation of the evolution processes without removing genetic diversity (Maxted and Guarino 1997).

In addition to the major crops, that is, those to which is attributed the greatest importance for food security in many regions of the world (wheat, rice, corn, potatoes, sweet potatoes, beans, coconut, tropical fruit), significant attention recently has been paid to the preservation of poorly used or unused PGRFA, as in certain areas these significantly contribute to the improvement of food safety, nutritional value (vegetables, fruit), as well as diversity of diet and the health of humans and animals (herbs). This enormous wealth of nature, whether genetic material of actual or potential value, can and must be permanently preserved. To increase agrobiodiversity in the forthcoming time period, attention should be directed to the preservation of valuable indigenous species (local populations or ecotypes) that have been grown for centuries in certain areas and which are adjusted to the climate and growing conditions.

The importance of PGRFA is huge because they serve as raw materials for food production, the production of construction materials, energy, fiber, clothing, and medical treatment, satisfying the culinary, aesthetic, and health needs of the people. As part of overall biodiversity, PGRFA contributes to greater security of food production, sustainable rural development, and poverty reduction in the world. From the earlier discussion, it is evident that indigenous PGRFA have enormous importance for humanity, for its economic and social development. In addition, PGRFA have other nonmonetary values for humans. TK and the cultural heritage of rural areas are a very important part of agrobiodiversity. There is a strong link between the agricultural practice of some traditional communities and their cultures and religions. As a consequence of the great diversity of land, climate, and plants in the world, indigenous PGRFA have contributed to the creation of diversity of food cultures. The variety of nutrition is a consequence of the diversity of PGRFA. Thus, the countries rich in biodiversity are distinguished by a much diversified diet.

In addition to their actual importance, special attention in the coming period should be focused on the potential value of PGRFA. There is no doubt that the potential value of these resources is enormous, but the extent to which it will be implemented will significantly depend on the extent to which PGRFA will be used in agriculture. To approximate their level of use to the true potential, it will be necessary to intensify research activities. If it is taken into account that the PGRFA potential is a source of new resistance, and of new commercially valuable ingredients, etc., then continuing their study becomes important for their recognition and intensive use (Djuric 2010).

3.4 Significance of Indigenous PGRFA for Sustainable Development

There is no doubt that the first conservators of genetic variability were farmers (Zeven 1998). During the long period of human civilization, in long selection processes, nature and people have left, to coming generations, thousands of locally adapted populations typical for the areas from which they originate (Myers 1994). Generations of farmers and breeders around the world, by using the genetic diversity of wild and cultivated plants, promoted their own crops and thus preserved and expanded genetic variability (Hammer 2004). Manipulations with genetic diversity constantly led to increased biodiversity. In addition, biodiversity was also increased by natural mutations and crossings, migration, and conscious and unconscious selections. The wealth of genes preserved in primitive varieties in farmers' fields around the world is a precious reservoir of genetic diversity and an important building material on which basis new varieties were created. This spread of genetic diversity lasted for millennia, until modern science and technology started to dominate agriculture.

With the intensification of agricultural production, which was reflected in introduction of new and homogeneous varieties, farming became more uniform and the role of farmers in conservation PGFR was smaller (Jovovic et al. 2013). Modern agriculture, which is based on the use of elite germplasm and free use of inputs, resulted in a significant loss of genetic diversity. Indigenous populations are increasingly being replaced with modern varieties that are less resistant to plant diseases, pests, and abiotic stresses. In this way, valuable sources of germplasm are rapidly lost and the basis for sustainable development of agriculture is endangered (Newton et al. 2010). Even in areas where indigenous plant populations are grown on a larger scale, fear that their potential is not fully realized is quite justified. In addition to the rapid disappearance of traditional varieties and locally adapted populations, modernization of agriculture led to a significant increase in the use of PPP. Thus, the "chemical revolution" quickly mastered modern agriculture, which today is almost impossible to imagine or sustain without the heavy use of PPP.

The disappearance of PGFR during the modernization of agriculture is intensive, and therefore ex situ conservation emerges as the only sure way to protect them (Zeven 1996). For these reasons, cultivation of PGRFA on farms should receive special attention because growing indigenous populations in the traditional way, in the climatic and soil conditions to which these cultivars are most suitable, significantly promotes their conservation and sustainable use. It is therefore necessary, as soon as possible, to find sustainable models of on-farm conservation that, on one hand, stimulate farmers to grow PGRFA, and on the other, lead to improvement of life on the farm and the sustainability of production. Today indigenous PGRFA are mainly grown on small plots in remote rural areas. They have enormous significance for traditional small-scale farming, as 1.4 billion people worldwide provide for their own existence in this way (Andersen 2003). To prevent decline of its yield and quality, traditional small-scale farmers exchange their own seed and planting material with farmers both nearby and at a distance. In addition, they are deeply aware that genetic diversity between and within cultures significantly increases the resistance of their crops to pests, diseases, and climatic conditions, for example, drought. Today 75 % of the poorest people in the world live in rural areas and survive thanks to traditional small-scale farming (IFAD 2001), and therefore appropriate access to genetic resources is vital for their survival. In recent years, cultivation of CWR is increasingly promoted, and that of other useful species in agricultural, horticultural, and silvicultural systems, which thus encouraged their conservation and preservation even for small areas—on farms, in gardens, or just with several plants in situ. Many countries in the world provide subsidies for these types of preservation (Maxted and Kell 2009). With further development of science and research methods, a need for various aspects of genetic resources will grow, and this will decrease the intense process of genetic erosion (Hammer and Teklu 2008).

The main reason for growing PGRFA is very good organoleptic and culinary qualities, as well as possession of certain specific characteristics (disease resistance, resistance to high and low temperatures, drought, the possibility of growing in poor production conditions and at high altitudes, etc.). In many cases, there is also sentimentality, but also inability to procure seed material for reasons of poor financial situation and distance from urban areas. Indigenous PGRFA are especially suitable for growing in areas where the cultivation of most other crops would not be economically feasible. Traditional cultivars are suitable for cultivation in areas with lack of arable land and excessive physical labour, which is common in most of the underdeveloped and developing countries. These resources are usually grown in traditional production systems, mainly for domestic consumption. Smaller quantities of products are intended for sale in local markets. Distance from large markets and the relatively favourable selling prices allow small farmers to be competitive in local markets, despite the considerable volume of work invested. The aforementioned points clearly show the importance of PGRFA to the lives of many families in rural areas around the world (Jovovic et al. 2013).

We live in a time of global climate disturbances caused by the constant increase in the numbers of the human population on the planet and its natural urge to survive. The rapid population growth in the world has caused a global increase in food demand, which has resulted in increasing interventions in nature, and thus greater degradation and impoverishment of ecosystems. PGRFA, as undoubtedly the most important resource for life and survival of human civilization, are under great pressure of various threat factors. Numerous and very different reasons for the loss of biological diversity include, most importantly, the conversion of natural habitats, intense urban development, infrastructure construction, the impact of various industrial plants, the use of dirty technologies, conversion of grasslands and shrubs in deserts, overexploitation and/or complete abandonment of farming activities, and climate change. Thus, biodiversity is now seriously threatened, particularly in the areas of farmland, mountain regions, and coastal areas. The most reliable indicator of the vulnerability of biodiversity is the disappearance or the endangerment of plant and animal species. Numerous data indicate that each year a large number of plant species disappears from the planet, both old and new varieties, thereby significantly destroying a huge genetic potential for plant life and people. Therefore, growing concern for the future survival of humanity seems quite justified. The international community is facing a number of challenges:

1. To ensure the survival of growing populations on the planet Earth;
2. To provide enough food for so many people and at the same time providing returns to biodiversity for its renewal;
3. To organize dynamic and sustainable development of society and at the same time to preserve biodiversity and so on.

A significant number of authors believe that the problem of hunger in the world is conditioned more by the unequal distribution of food than by its amount, and that in the world today enough food is produced to feed the current world's population (Douthwaite 2001; Moore-Lappé et al. 1998). However, even assuming that the current distribution problem can be solved, a need remains to produce much larger quantities of food for the growing population of tomorrow. Providing enough food for people and animals in the future will be achievable in two ways: one is increasing arable land, and the other is intensifying production on existing arable land. The fight against poverty is, beyond doubt, one of the top priorities of the modern world; however, either alternative may be a new cause of significant loss of biodiversity. Fortunately, increased awareness of the seriousness of this problem and the need for urgent action for the protection of biodiversity caused the situation to begin to change significantly in recent years. It is the impression that biodiversity conservation is slowly becoming a new philosophy of life of modern man, and that he accepts basic ecological principles and is becoming aware of the importance that the conservation of biological diversity has for the survival of life on Earth. The problem of lack of food in the world and climate change is emphasized in the second Global Plan of Action (GPA) for PGRFA, which was approved by the FAO Council in 2011. This plan strengthens the commitment of the international community to PGR that have vital importance for human prosperity.

Most of today's studies show that climate change is becoming an increasingly serious source of threat to biodiversity, and also a growing challenge for agriculture. It is quite certain that the adjustment of these challenges will to a large extent depend on the genetic diversity of crops and domestic animals, as well as services provided by other components of agricultural biodiversity (McCouch 2004). Success in these activities will largely depend on the level of international cooperation and the free exchange of the existing gene pool, because no country is independent at this point but depends on the diversity that comes from other parts of the world. A significant contribution in this direction represents ITPGRFR, which obliges member states to undertake work to promote the positive effects of growing PGRFA and reducing negative impacts of agricultural practices on biodiversity in agroecosystems. Bearing in mind the importance of PGRFA for sustainable agricultural development for current and future generations, strengthening the capacities of developing countries (particularly those in centres of origin of biological diversity), and countries with economies in transition, is urgently needed so that these countries can devote themselves to preservation, improvement, and management of their gene pools. Aware of the need to conserve the world's diversity of PGRFA, members are committed to promote an integrated approach to the exploration, conservation, and sustainable use of PGRFA. This approach requires intensive work on inventorying and studying PGRFA, investigating and recording plant genetic resources for food and agriculture, taking into account the status and degree of variation of existing populations and populations that have potential use value (Diamond 2002). Monitoring the degree of endangerment of these species must be permanent. Significant support will be directed towards farmers and local communities to establish a more efficient system of managing these resources and ensure their sustainable use on farms.

With the aim to ensure the sustainable use of PGR and the establishment of sustainable systems of crop production, communities today can use a significant number of different mechanisms:

1. Management of an agricultural policy that promotes the development and maintenance of various agricultural practices which enhance the sustainable use of PGRFA and other natural resources.
2. Support to farmers growing local and locally adapted crops and underutilized species.
3. Support for farmers who create new and use their own varieties and apply ecological principles during production for control of disease, weeds, and pests, and maintain soil fertility.
4. Support for pre-breeding programs that use PGRFA to reduce crop vulnerability and genetic erosion.
5. Support for programs to create new varieties that are adapted to the given environmental conditions, and those based on the use of traditional varieties, landraces, and wild relatives of cultivated plants.
6. Promote and support for the development and transfer of technologies that will lead to improved and sustainable use of PGRFA.

7. Support for the wider use of diversity of varieties and species in agricultural production and activities aimed at the expansion of the genetic base of cultivated crops.
8. Promotion and support for in situ conservation of wild crop relatives and wild plant species, including protected areas.
9. Promotion and support of an effective and sustainable system of ex situ conservation.
10. Tracking the vitality, the degree of genetic variation, and the integrity of the existing collection.
11. Intensification of research aimed at the preservation and increase of biological diversity that will be at the service of agricultural producers.
12. Support activities that will lead to a global increase in food production compatible with sustainable development.
13. Adjustment of legislation relating to the registration of varieties and seed distribution.

The use and sustainable use of PGRFA are not only limited by the loss of germplasm but also by the TK related to them. This knowledge arises from the use of these resources and is related to plant species, varieties, cultivation, production technology, harvesting technology, storage, use, processing, etc. Therefore, future activities should be concentrated on their study and a detailed analysis of their real value and potential for improved deployment and exploitation. Intense involvement of PGRFA in existing breeding programs (whether by means of farmer's selection, classical plant breeding, or modern biotechnologies) will create conditions for more sustainable agricultural production. It will also lead to increased awareness about the importance of PGRFA in adapting to the increasingly prominent changes of the environment, establishing sustainable agriculture, and achieving sustainable food security. It is clear that we will face a period of scarcity of food in the world, rapid climate changes, and natural disasters (floods, strong winds, erosion of land). The solution to these challenges is the creation of tertiary genotypes that will be able to adapt to new changes. For that, it will be necessary to define new selection programs and strategies and provide adequate capacities and approaches for crop biotechnology and breeding (Tester and Langridge 2010).

3.5 Indigenous Knowledge in the Light of Ethnobotanical Context

Traditional rural communities for centuries guarded and improved their indigenous knowledge, especially as related to natural resource management and agriculture. Among many cultural traditions, ITK has a special place in people's lives. There are several terms for indigenous knowledge often used as synonyms: traditional knowledge, local knowledge, indigenous technical knowledge, traditional environmental knowledge, indigenous knowledge system, folk knowledge, indigenous science, people's science, etc. Regardless of the term used, indigenous traditional knowledge (ITK) represents a very wide range of skills that cover all aspects of life: food, agriculture, hunting, medicine, arts, and technologies that many traditional societies use worldwide.

There is no generally acceptable definition that could cover all existing forms of ITK. According to the definition proposed by the World Intellectual Property Organization (WIPO), traditional knowledge refers to the knowhow, skills, innovations, practices and procedures, learning, and perceptions that arise as a result of intellectual activity and that are developed within the traditional context. Traditional Knowledge (TK) is the common name for knowledge, the skills (know-how) of traditional populations coming from and related to the environment, consisting of a codified, oral, or other form of knowledge system (Kiene 2006). TK is an important economic resource and an important component of the identity of a community. It is a way of life and it reflects its values. There are clear distinctions in the level of TK between different social groups. With the development of human society, this knowledge was actively exchanged between different communities. During this period, the traditional communities were accepting knowledge from other systems of knowledge and thus perfected their knowledge, and as well other communities adopted components of traditional knowledge. In addition to the importance that TK has for the local community, it plays a special role in life of every individual of the society, because it represents the culmination of innovative achievements of ancestors and valuable gifts from past generations. The type of TK that members of the community have depends on a number of factors, particularly age, sex, social status, intellectual ability, and interest. TK represents knowledge created under certain cultural, environmental, and social circumstances and is a characteristic of a particular community. It is closely related to the existence of traditional communities that to survive were forced to develop specialized knowledge about natural phenomena (Lasisi et al. 2011). In this sense, TK entails a comprehensive system of knowledge necessary for adjustment to changes in the environment, as well as for dissemination of new knowledge among family members, and within and between communities (Ragupathy et al. 2009). TK must not be understood as static and unchanging but rather as a highly specialized and flexible system. TK and skills were created for centuries in various cultural practices and under the umbrella of a large number of different ancient languages. Therefore, the interpretation and understanding of TK is one of the most important scientific priorities in the pursuit of responses concerning the sustainability of human survival and its ability to adapt to a number of negative consequences caused by changes in the environment.

Ever since it can be remembered, man has used the products found in the wild or in the community where he lived. To provide food and meet other needs necessary for life, traditional communities have developed specific knowledge and skills and in this way survived. At the same time, their lives were easier and more enjoyable. Local communities exchanged these know-how practices with other communities and passed them on to younger generations, who in many ways improved them. Numerous literary data indicate that peoples' knowledge and skills, as part of the traditional habits, were first gained in regard to plant species in their immediate environment (meadows, pastures, forests). Following their growth and development they have gained new knowledge about nature (which part of the plant to use in the diet, when to collect, how to keep it, etc.) that had enormous significance for the life of the local population (Karg et al. 1999). Based on this knowledge, a new scientific discipline was created: ethnobotany. Ethnobotany studies the complex relationship between human culture and plant life (Ford 1994). It particularly focuses on the study of ethnobotanical knowledge and the ways in which people of different cultural and environmental conditions use PGR (food, pharmaceuticals, cosmetics, production of natural dyes, textiles, building materials, rituals, etc.) (Deepak and Anshu 2008). Also, ethnobotany records the ways in which TK is classified, generated, and transmitted in the communities (Newmaster et al. 2007). Ethnobotany requires specialized knowledge in many different areas: botany (to identify and correctly collect plant species), anthropology (understanding the culture of the people working in the field of research), linguistics (local terms for morphological features of plants), and others. Knowledge of the local language is crucial for determining the density and distribution of cultural diversity and the preservation of TK. Ethnobotanists represent a separate group of scientists, who have a completely different view of plants: according to them, the plant world is considered an integral part of human culture. The first ethnobotanist was Carl Linnaeus, and Richard Evans Schultes is considered as a founder of ethnobotany. In the book *Plants of the Gods: Their Sacred, Healing and Hallucinogenic Powers*, which Schultes wrote with Albert Hofmann and Christian Rätsch, he detailed the numerous ways in which the indigenous people of Mexico and the Amazon used plants in everyday life. For an ethnobotanist to meet the real potential of certain plant species, he has to be familiar with morphology, systematics, and chemical properties for its exploitation, as well as the relationship that these plant species have to a particular social group. To preserve the efforts that traditional communities for centuries invested in the creation, preservation, popularization, and transfer of experiences and skills, the close cooperation of local ethnobotanists and local connoisseurs of plants is required (Koushalya 2012). For these reasons, the ethnobotanical approach was and remains the best way of studying traditional knowledge. Today, in the world, there are a number of centres for ethnobotanical research and education, which gives hope that the preservation of TK in the future will be much more intensive and efficient.

With the development of human civilization, ways of thinking were developed and thus TK as well. TK about plants and methods for their use is the result of millennial experience of many generations. The ethnobotanical aspects of TK had enormous significance in the daily lives of people and certain social and religious groups. TK considers plants used, and which are still in use, in nutrition and preservation of the health of people and animals (McClatchey et al. 1999), then those plant species that are used to obtain the fibres, produce dyes for textiles, tea preparation, control of harmful insects in the household (moths, flies, wasps), as well as those that are used in carrying out various religious rituals and so on (Djuric et al. 2007). People across the planet, since prehistory, have used indigenous plants for medicinal purposes. Traditional ways of treatment are encountered in many parts of the world, and they are an almost universal phenomenon in unindustrialized societies. The World Health Organization (WHO) listed more than 20,000 plant species that are globally used for the treatment of various diseases in humans and animals. The main task of modern ethnobotany is to provide reliable knowledge about the traditional use of plants. However, providing such information is not always an easy task because this knowledge is often kept as a big secret and spreads very slowly beyond the boundaries of individual local communities or ethnic groups.

The disappearance of a large part of the TK on plants is a consequence of the abandonment of the traditional system of crop production, but also, and more prominently, socioeconomic changes. In the study of TK, a large problem is the disappearance of many indigenous communities around the world. Many reasons have led to these losses. Traditionally, these communities are increasingly exposed to the pressures of economic globalization, which promotes intensive agriculture and industrial development. Also, an important factor in the disappearance of TK is considered to be a system of education in contemporary society, which favours the modern approach and devalues the importance of indigenous knowledge. The curricula of many countries in the world prioritize urban concepts of life and work, where TK and skills, as well as ancient languages, are often perceived as a step backwards. With further development of biotechnology and the increasing trend of depopulation of rural areas, the loss of traditional knowledge will be even more pronounced (Coelho 2012; Lasisi et al. 2011). Increases in the living standards of the population in rural areas are causing the disappearance of TK (Benz et al. 2000). Therefore, it is important that this knowledge is preserved before permanently disappearing along with the current generation of old people who are committed to its preservation. TK is not a static system that is easily documented and as such preserved for future generations; on the contrary, it is a very dynamic system that is characterized by a steady inflow of new knowledge important for people who use it in everyday life. By establishing control over their own education system and resources, traditional communities will be able to define the position and the importance that TK should hold in the future development of the community (Briggs 2005).

In different regions of the world, TK is preserved to a different extent. The development of local communities in developing countries is based 65–80 % on TK, especially concerning the prevention and protection of health of humans and domestic animals. Most of the small households, whose production is mainly intended for their own consumption, rely on the local population, traditional varieties, and knowledge within the same family for centuries, passed on from generation to generation (Mesfin et al. 2013). In the production of medicinal plants, CWR are commonly exploited resources (Emiru et al. 2011). Besides its importance in the treatment and prevention of diseases, traditional medicine offers many other benefits. In places where a developed market for medicinal products exists, traditional communities can generate significant financial income. The use of herbs for these purposes is associated with sustainable forest management, because traditional communities must preserve the ecosystems from which the sources of their existence come (Martin 1995). If we know that the survival of people in these areas is mainly based on their own work and the available natural resources, then initiatives for the collection and conservation of local genotypes are justified and function in the survival of these communities (Uprety et al. 2010). In this way, in addition to conservation of biodiversity, TK is also preserved.

3.6 Role of Local Communities

Indigenous people live in all regions of the world, inhabit smaller or larger spaces, and sometimes can be encountered in very remote and inaccessible areas, often with minimal living conditions. They live in territories that cover about 22 % of the world's land, land which includes 80 % of the planet's biodiversity (Sobrevila 2008). In addition, they are the owners of 11 % of the world's forestland (White et al. 2004). Indigenous people represent 95 % of the world's cultural diversity. There are no precise data on the number of indigenous societies that exist today in the world, but it is assumed that the number is between 6,000 and 7,000. More than 6,000 languages are used in the world today, and 4,000 to 5,000 of these languages belong to indigenous nations. Around 300 million indigenous people live in 90 countries around the world, accounting for about 5 % of the world's population and about 15 % of the world's poor people.

Today we talk about a continuing learning process even though we know that this process is actually taking place continuously all the time as humankind evolves. During this long period of development, people have survived thanks to natural resources, which are largely used in a sustainable manner. Thanks to them, the benefits that nature has to offer is this interactive approach in a complex ecosystem was built for centuries and constantly upgraded (Berkes et al. 2003). Using such knowledge and everyday practices, local people have built and maintained a certain level of social order. Through the link of environment and society, knowledge of local communities becomes part of everyday life, and these communities are societies which possessed a certain quantum of knowledge and skills. TK used in many sectors (agriculture, forestry, fisheries) was developed and tested over a long period of time. In this way, slowly but surely, the local communities began to differ and to have a different social impact (Berkes and Folke 2002). Although the models of storage, upgrading, and transmission of TK were different, all local communities have significantly contributed to enhancing understanding of traditional values and the adoption of sustainable and proven practices (Armitage 2003).

Their knowledge, traditional practices, and experience means that local communities are vital in current sustainable ecological development (Ryser 2011). Therefore, the importance of the application of these skills in the modern context is increased day by day. However, it was not always easy to integrate local communities into a modern society. Although in this effort were many similarities and common-built platforms, there were some disagreements as well and even conflicts. As much as science was able to find a common practice of management and penetrate local communities, in these efforts there was always idleness (Nabhan 2000). Simply speaking, it took time for scientists, backed by local people, who introduced them to the mentality of the local communities, to achieve a satisfactory level of communication. By ensuring full use of the indigenous languages, efficient communication between different cultures could be achieved. It is very difficult to translate words from indigenous languages to a modern language, just as it is difficult to translate modern scientific and technical terms into the language of traditional people. Language is synonymous with identity and a means of communication with the surrounding world. It would be good if scientists could even understand the oral language of the traditional communities they are studying and ideal if they know its written language. Finding bilingual people in local communities is invaluable in eliminating language gaps. Otherwise, to overcome language barriers, recruitment of a experienced (certified) interpreter remains the only solution (Hart 1995).

In working with local communities, a collaborative approach must be expressed and full respect for their language and culture must be demonstrated. Only in such ways can TK researchers provide full support and involvement in the local community. Only on these principles is it possible to provide interest, the active participation of the community, and mutual trust. Thus, it is very important to communicate on a more personal level, which implies participation in various community events. This kind of social interaction can significantly improve the relationship with the local community and the results of joint missions. Therefore, the establishment of an effective communication system is essential for obtaining the support of the local community and its participation in TK-related initiatives. In this regard, it is necessary to build mechanisms that will allow the TK of local communities to be treated in the right way and ensure that every aspect of its use is in line with the establishment, not only of protocols, but also with the wishes of people who possess these skills. Many traditional societies in the world are trying to keep an inalienable right to their own knowledge, experience, cultural practices, and traditions. Often they fear a potential loss of control over the information and they do not want to share this with others. Therefore, to establish mutual trust it is necessary to prevent any form of expropriation, misuse, or misinterpretation by individuals or groups outside the local community. In this sense, it is necessary to provide such access to local TK holders to provide some level of permanent control over their knowledge.

Before starting the mission of collecting and using this knowledge, a TK researcher must have consent in advance from both the local authorities and individuals who possess this knowledge. This consent is usually done in writing, although it can be in the form of an oral agreement. It is very important that these initiatives are acceptable to the indigenous people and culturally appropriate. If certain elements of TK are considered as confidential information, it is necessary to ensure the full protection of TK and intellectual property. All this must be agreed in advance so that the knowledge of traditional communities entrusted can be treated in accordance with established protocols on the control of TK information. After that, contacts are intensified, transfer of information is in both directions, and work on several tracks is stimulated by common interest in TK (Becker and Ghimire 2003). To make this complex knowledge more readable, it is often necessary to separate it from the traditional context. However, in removing TK from its cultural environment, the true meaning is lost; hence, the involvement of TK holders in its interpretation and application is of utmost importance.

The role of local communities in the preservation of TK is reflected in the fact that they are cooperating among themselves in establishing a platform on which they built their policy. The high esteem and relationship that they had with TK, as well as the correct way in which they used local resources, has enabled local communities to maintain and use biodiversity in a sustainable manner. Although much has changed since then, civilization seems to be the strongest link between the local communities remaining in the area of TK (Cunningham 2001). The development of modern society brought a need for to find more effective ways of communication between urban society and local communities. The importance and relevance of TK in solving more pronounced problems caused by natural phenomena and human activities are increasingly recognized. Such a common approach to indigenous knowledge becomes a chain linking "molecules" of sustainable development, conservation of resources, environmental protection, and various forms of sustainable agriculture, ethnopharmacology, and anthropology. In fact, new technologies are built on primitive but functional solutions that are owned by local communities (Ford 2001).

TK means more than the preservation of knowledge: it is the transfer of skills and close cooperation between local communities. Its long existence is possible because it is so deeply entered into the sphere of many sciences (biology, ecology, agriculture, anthropology, philosophy) and makes a strong connection between the physical and biological world. The achievements of complementary ethnoscience seem to be the strongest bond today. The credit belongs to the wise men – "fathers" of local communities, who made smart decisions in significant moments of history and led their communities forward. Such TK consistency and compliance with the various scientific disciplines, in different periods of time, could be held steady only by continuous introduction of new information and knowledge, which supports the assertion that it is not only the academic community who always recognized the true values of different times. The creativity with which endangered local communities managed to preserve TK, as well as heritage and above all ethnocultural treasures, is now particularly appreciated in developed countries. In fact, it represents national identity and recognition of certain countries in the world even though it may never be noticed from an economic and commercial point of view (Quinn 2001). Thus, TK and missions of the local communities have an invaluable ethnocultural value that continues even today. Efforts made to promote TK have led to its increasing use over time and upgrades through merging and finding common values of local communities in different geographic areas (Hiwasaki et al. 2014).

Indigenous knowledge is a collective heritage and the essence of identity of indigenous people. TK is mainly of a practical nature, particularly related to agriculture, fisheries, health, horticulture, and forestry. Knowledge of traditional communities about biological relationships, which include rare and endangered plant and animal species, may help in matters of identification, protection, and recovery of the habitats of these species. In addition, indigenous people have an important role in mitigating the effects of climate change. The territories where indigenous people and communities live are better preserved than those in their neighbourhood. Indigenous people are neglected partners in the preservation of biodiversity, and without their full engagement, many initiatives related to biodiversity and climate change will be significantly affected. More active participation of local communities in biodiversity conservation programs would contribute to the more efficient and cost-effective conservation and management of biodiversity.

TK has enormous significance for modern man and his efforts to contribute to a better life in rural areas. Thus, every day interest in the integration of local knowledge into the scientific management of ecosystems is growing. Reflected in the increasing number of new TK-related initiatives and projects, it sends an encouraging message to people in remote and underdeveloped areas to preserve their traditions even more carefully and try to find more information about local knowledge, culture, and values. Bearing in mind that the preservation of TK is related to the local community, special attention should be focused on their conservation, as they are the least developed and thus most vulnerable. Therefore, it is important to support the capacity building of local communities through training, equipping, and participatory workshops with community members. An important aspect of support is the employment of their members. Real possibilities for that exist and are most notable in the areas of language and culture. As experts in the protection and management of biodiversity and natural resources, they can be of great benefit to many TK-related initiatives as translators, interpreters, consultants, pollsters, researchers, project managers, and so on. These contributions need to be adequately rewarded. Given that TK belongs to a community, all these activities should be carried out through local authorities.

The main problems that the society faces today, and related to the promotion of the sustainable management of resources based on TK, are the lack of institutional engagement, rapid disappearance, marginalization, and still insufficiently recognized value and significance of TK. To overcome this situation it is necessary to initiate a broad set of measures aimed as follows:

1. Build and strengthen national institutions and the inclusion of TK in relevant national policies.
2. Provide the widest forms of support for TK communities.

3. Protect the cultural heritage of TK holders.
4. Intensify the approach to the collection and promotion of TK.
5. Globally support scientific research of TK and its use in the prevention of certain diseases, risks, and disasters, especially in areas vulnerable to climate change.
6. Support local communities for biodiversity conservation activities in the territories of their ancestors.
7. More intensive research in the regions where almost isolated communities live.
8. Include TK in education curricula.
9. Establish a uniform system for the interpretation and scientific explanation of the data and information that have been found.
10. Establish a system of TK transfer to younger generations and other local communities in geographically remote regions.
11. Promote the market value of TK-based goods and services, etc.

Although these activities are already performed to some extent, it must be acknowledged that this is not an easy job. Approaching local communities requires persistence, openness, and dedicated work, and this is the only way to reach the "LIVE Scientific Knowledge" (Berkes et al. 2007). Today NGOs and other forms of associations with the primary objective to protect the environment and prevent the loss of biodiversity declare some areas as protected areas, and these are similar to the local communities in which the people built a system of education (Folke 2004). Nowadays this is done in a more modern way and with the use of very diverse tools of modern technology.

3.7 Comprehensive View of the Use of TK and Its Involvement in Recent Technological Developments

TK has accompanied the lives of indigenous people from time immemorial. It was important in all stages of human history and its value is still high even today. This comprehensive system of knowledge is integrated in many fields important for human survival and quality of life. It offers numerous opportunities for a better understanding of social and natural phenomena and represents the potential for sustainable development of society. For these reasons, interest in integration of TK in the scientific management of ecosystems is increasing daily.

It is evident that many strategies for reducing poverty in the world have proven to be extremely unsuccessful, and for this reason, the attention given to indigenous knowledge and practices is increasing. The inclusion of TK in the modern concepts is associated with many difficulties and thus many good ideas remained unrealized. Although the tendency to involve local knowledge in modern concepts is ever present, it can be concluded that the impact of modern science is still predominant. Many scientists around the world continuously insist on differences between science and TKs. They believe that scientific knowledge is universal, whereas TK is related to certain local people and their understanding of the world. TK represents a holistic view of the world, and as such does not fit the criteria of modern science. Implementation of TK was often hindered by poor communication on both sides, which was the dominant characteristic of development models in the past. Also, an additional difficulty is that sometimes developers use IK in a way that can be offensive (Novellino 2003). Sometimes the mere mention of traditional knowledge creates confusion among decision makers and developers of TK. They continue to believe that science is the right path of development, and that TK and culture are a great impediment to their rational development, which causes a reduced sense of value in indigenous people for their own TK (Skinner 2003). Often problems raised by poor communication between cultures represent an obstacle to the efficient use of TK. Even in areas where this knowledge is used it is still treated in a manner appropriate to modern science. In general, the wider use of this knowledge is still hampered by various stereotypes and contradictions and thus indigenous ways of life and knowledge systems remain quite unknown.

Despite the significant achievements of science and progressive development of technology, studying numerous events in an ecosystem is not an easy task. To realize and understand the dynamic effects of climate change and its impact on the functioning of an ecosystem, knowledge of numerous traditional communities around the world will be of great benefit. TK is used primarily in the community in which it was created and is the primary means of understanding the immediate environment with all its risks. In the management of natural resources, TK has a vital role for the survival of many generations. Therefore, TK is correctly considered a basis for sustainable life of the community, and it is very important to examine the extent to which it is used today, and what can be done to use it more efficiently. TKs related to environmental management, food safety, agricultural production, and biodiversity represents hidden value that science has yet to explore. When we talk about TK usually we think of ecology, agriculture, health, and sustainable development, although it is deeply connected with other areas of human activity such as cultural anthropology, sociology, ecology, education, mathematics, geography, and informatics.

Indigenous people were forced to develop a number of agricultural techniques to survive: shifting cultivation, crop rotation, mixed crops and intercropping, anti-erosion techniques, clearing and burning of forests. Optimal use of local resources, small dependence on external inputs, and use of available local knowledge and experience were the basic characteristics of agriculture. In an ongoing effort to survive, farmers have used various traditional systems of land preparation, sowing of seeds, plant propagation (cutting off the mother plant to a larger number of parts), seed production, preparation and cultivation of seedlings, and crop protection from pests and diseases. They possessed a wide knowledge on storage and seed processing (drying, threshing, cleaning, grading), as well as processing, packaging, and sale of agricultural products (Nakashima 1990). They were able to classify soil on the basis of texture and possessed knowledge about soil horizons. Based on established indicators, they knew how to differentiate land based on its fertility. Also, they know that external factors significantly affect the properties of the soil, as the environmental characteristics depend on the quality of the soil. Indigenous farmers have implemented many practices for the conservation and improvement of soil fertility and applied different irrigation and water conservation techniques. Their fields were fertilised using animal dung, ash, and organic household waste. Different crop residues were used for increasing soil fertility and as mulch and thus increased the resistance of crops to drought. They knew that crop residues on the production plots significantly increased soil fertility, organic carbon content and infiltration and reduced water runoff. As an indicator of lack of water in the soil they used individual plants found in their fields. They owned other techniques that were focused on the preservation of water and increasing soil fertility. They had established various systems of indicators and were able to determine the optimal time to cultivate land, sow, and harvest. Their production systems were mainly based on agricultural diversity, which reduced risks associated with weather fluctuations. Even today, many indigenous farmers around the world provide their own development and survival with the use of systems of knowledge established a long time ago. Many of these tools, in some form, are encountered in conventional agriculture. However, these innovative approaches created by traditional communities are still difficult to fit into certain modern frames of agriculture and other spheres of life. But as the pressure of climate change and world population grows, it is certain that this innovation will increasingly gain importance.

Climate change, including drought, significantly affects crop production and food safety in many parts of the world. Biodiversity conservation and cultivation of traditional varieties is of great importance for the adaptation and survival of indigenous farmers. In addition to resistance to numerous climatic stresses and new pests and diseases, traditional varieties have the advantage of being already available and inexpensive (farmers' own saved seeds). Numerous studies have shown that many indigenous farmers around the world are aware that their survival in climate change conditions relies on planting a wide variety of traditional varieties, as this variety of cultivated plants increases the availability of resources in the field, reduces risks, and increases adaptive capacities (Trakansuphakon 2010). All data mentioned indicates the need to preserve a strong local landraces and local seed production. Preserving the diversity of crops in addition to providing food for the traditional community often represents the satisfaction of certain social and cultural needs.

With the constant increase of input costs, modern conventional agriculture is becoming increasingly expensive, particularly for small-scale farmers. To overcome these problems, they are forced to rely much more on other practices, and on those practices that are more environmentally friendly. One of these possibilities is organic agriculture, that in addition to locally available resources is based on the use of local knowledge (choice of crops, tillage system, system of planting, crop protection, storage methods, etc.). This choice implies the cultivation of local traditional crops and varieties, which are usually better adapted to local conditions and provide high and stable yields. There is justified concern that a significant part of traditional crops and varieties and associated knowledge is permanently lost by replacement with modern selections. Numerous initiatives implemented in indigenous communities related to organic production indicate the rising awareness of farmers on the importance of traditional crops for their sustainable development.

The use of medicinal plants is as old as human civilization. Primitive man in the distant past used medicinal plants for treating many diseases of humans and domestic animals, and cosmetic use of these plants was known almost in all ancient cultures. According to the WHO about 4 billion persons, or 80 % of the world population, particularly in underdeveloped and developing countries, use traditional healing methods based on the use of medicinal plants (Singh 2009), and about 74 % of plant origin products are used in contemporary medicine in a manner very similar to that practiced by ancient civilizations (Igdir et al. 2013). Today, unfortunately, almost 30 % of the world population has no regular access to modern medicine, and in some parts of Africa, Asia, and Latin America about half the population is deprived of regular medical care. The continuous presence of infectious and chronic diseases caused by their lifestyle has serious consequences for the people in these regions. Lately, rapid change in ways of thinking has caused treatments sometimes described as primitive to become increasingly important for modern man, with the result that today nearly 50 % of the population in developed countries use different forms of complementary and alternative medicine (Busia 2005).

The number of medicinal species used in various parts of the world is impressive. In India, the country with most likely the oldest and the richest traditions in the use of medicinal plants, 7,500 species are used in folk medicine, representing about half of the total number of its indigenous plant species (Shankar and Majumdar 1997). In China, the figure is around 6,000 plants (He and Sheng 1997) and in Africa more than 5,000 species (Iwu 1993). In Europe, where there is a long tradition of using medicinal plants, about 2,000 species are commercially exploited (Lange 1998). The enormous importance of the use of medicinal plants in the treatment of people was also recognized by the WHO. This organization has issued a number of publications on commonly used medicinal plants with the aim of popularizing traditional medicine, and new publications are being prepared.

Medicinal aromatic plants represent an important part of international trade (Handa et al. 2008). In international trade nearly 3,000 species are present (Schippmann 1997), whereas their number is even higher in local, national, and regional trade. TK represents an important basis for sustainable use of resources and balanced development. By defining the conditions for access and use of natural resources, including a fair share of the benefits coming from their use, conditions for the preservation of cultural and biological diversity in the world would be met. For example, of the estimated 43 billion dollars that is the annual profit on global market of medicines derived from plants, local communities receive only 0.001 % (Melchias 2001). Bearing in mind the importance that traditional plants have for the development of local communities, in many developing countries pharmaceutical companies are established today that commercialize benefits from PGR and TK. In addition, traditional clinics are established where in addition to indigenous people, populations from developed countries are treated as well. Traditional medical practices apply some methods that are used in modern medicine (hydrotherapy, heat therapy, different spinal treatments, treatments for cancer, obesity, diabetes, etc.), and for these reasons, many foreign pharmaceutical companies send their agents to grasp the secrets of TK doctors. In addition, billions of dollars are invested in numerous studies of plant species that could be commercialized.

Indigenous people possess exceptional knowledge on plants that have multiple applications and great economic value in the world market. These resources also represent an important source of income for many undeveloped countries around the world. These plants are used in the treatment and diet of humans and domestic animals and are an important source of proteins, essential amino acids, and minerals. They are used as condiments but also for the treatment of many diseases. However, their overexploitation and commercialization leads to their rapid disappearance from their natural habitats (Shiembo 1999), which can have multiple harmful consequences for local communities whose territories are exploited.

Traditional Medical Practitioners (TMP) over a long period of human development have had a central role in providing primary health care for the population. They inherited their knowledge from previous generations of practitioners and, before they were able to begin to work independently, they had to spend some time in training. Each TMP for certain diseases or disorders had their own lists of medicinal plants, the method and dosage of their application. Over a long period of time, with a number of successful and unsuccessful attempts, they developed their medicinal formulations and methods of treatment (Nasir and Nur 2014). In many countries of the world traditional medicine is still an integral part of the national medical system, and in many of them the number of TMPs is still significantly higher than the number of modern practitioners. TK has always served in the fight against many diseases and has helped many people around the world to be healed.

Medicinal plants and TK today are important in the fight against major diseases that strongly affect the population of many underdeveloped countries, especially those who live in the African countries south of Sahara, but also in countries of Asia and South America. In the treatment of infectious diseases such as malaria, HIV, and schistosomiasis, traditional medicine has made significant contributions. A large number of modern drug products are actually traditional medical knowledge. Malaria represents one of the biggest global problems on the planet (Wilcox and Bodeker 2007). Every year more than 800,000 people die, primarily children in Africa. Malaria control is performed with quinine and artemisinin today. These drugs are based on TK of local communities of Peru and China and were isolated from the plant *Artemisia annua*. A large number of other traditional plants are used to control malaria. Beside malaria, schistosomiasis represents a huge health problem for many African countries. This disease, transmitted by freshwater snails, causes enormous problems to citizens in 76 countries of tropical Africa. More than 200 million people are infected with this vicious disease and about 200,000 die each year. The discovery of a low-cost and biodegradable molluscicide made a big step in its treatment. This molluscicide is derived from the traditional African plant endod (*Phitolacca dodecandra*), cultivated for centuries in many countries of Africa. This widely known herb has the ability to control the disease vector. This discovery represents a much cheaper means for treating schistosomiasis compared to expensive molluscicides of chemical origin. Experience of the application of these drugs increased confidence in the traditional way of treatment (Patwardhan 2005). Many reported data indicate that traditional medicine and traditional plants are gaining more importance and can be a good basis for further research and the discovery of new and more effective treatments.

The rapid population growth on the planet and the growing consumption of fossil fuels are important factors in the increase of concentration of greenhouse gases and warming on a global scale. As a result of these activities, the world is facing a number of challenges: global increase of temperature, altered rainfall patterns, global warming, global sea level rise, and so on. Further, floods, droughts, heat waves, and strong winds are becoming more common. All this disturbance decreases agricultural production and causes water shortages and more intense occurrence of infectious diseases with reduction of the quality of life of local residents. Although their impact on climate change is low, small indigenous communities are the most exposed to its negative impact because of their small population size, isolation, and unrecognized rights over their territory and resources (Adger et al. 2004).

TK and its use have particular importance in the field of climate change. Local people, thanks to their daily activities, spend much time in the natural world, covering a vast area of territory. In this way they perceive and monitor daily changes that are rapidly occurring in the environment. Based on their own observations of weather and climate, numerous indigenous communities from around the world inform us every day that the weather and climate are changing intensively. The consequences of rapid global warming are well documented in the Arctic in particular. Indigenous people in the Arctic realized a long time ago that the weather is not what it used to be and that they can no longer predict the weather as they once could do. Such unpredictability of weather (sudden appearance of storms and blizzards) increases the risks of local people, especially hunters (Gearheard et al. 2010). These communities are an important in situ source of confidential information on the impact of climate change. These data are very important for science, because they complement scientific data. However, the data obtained from the local communities often conflict with those presented by science. The reason for such differences is probably the different approaches to these observations. While indigenous people make their own conclusions on the basis of the maximum and minimum values of a number of environmental and social parameters, scientists work on the basis of the average value of individual environmental indicators (Weatherhead et al. 2010). Despite these differences, a common and constructive approach to the representatives of both groups of knowledge represents an important step in addressing the challenge of climate change.

Living for centuries in close contact with nature, indigenous peoples have gained a deep understanding of natural phenomena and also the interplay between the various elements of a habitat. The knowledge of traditional people on ecology and weather-induced changes can provide much other information about the complex relationships within the ecosystem about which contemporary science has not yet offered answers. Local people have gained new knowledge over time about the numerous interactions between the various components of the ecosystem to which scientists objectively cannot come during scientific expeditions. The knowledge of indigenous people about long-term changes in the environment is of vital importance for understanding the climate change and building adaptive mechanisms that are in line with their priorities. Many local communities in different parts of the world rely even today on TK related to conservation of the environment and natural disasters (especially droughts and floods). This knowledge represents an important basis for the creation of successful and inexpensive adaptation measures. For these reasons, governments, environmental, and other institutions should strongly support the efforts of local communities in activities related to climate change assessment and adaptation. Particular attention should be paid to the knowledge of the people living in particularly vulnerable areas, such as the Arctic, isolated islands, high-altitude regions, desert margins, and coastal areas (especially the lower coastal regions). That is why modern society needs to realize, as soon as possible, that indigenous communities, because of their practical knowledge, are of prime importance in planning activities related to the adaptation and reduction of risks of climate change. Integration of TK with science will allow more intensive research and thus finding the most appropriate solution. In this way local communities would be stimulated to use their knowledge, enriched with scientific knowledge, in creating new strategies for disaster risk reduction and adaptation to climate change. This integrated approach would lead to increase resilience to these challenges. The lack of local people who have modern research experience can be a significant problem in the implementation of these efforts.

Indigenous communities have always possessed well-developed systems of local knowledge for environmental management that allowed them to easily adapt to various changes. In many of them are people able to read the signs and predict natural events. Based on the behaviour of individual animals they were able to predict certain natural phenomena (overflow of water in rivers, poor harvest, good or bad rainfall season, epidemics, etc.), which was of invaluable importance for designing an appropriate protection system. Based on the position of the sun and the cry of a specific bird, they were able to predict the beginning of the rainfall season, which was of great importance for agriculture. By the colour of the clouds, they knew to predict the appearance of hail and they knew that after a long dry period a storm is coming. TK on the movements of gales and rain helped them to build adequate shelter, windshields, walls, fences, and the like in advance for their protection. Use of TK in weather prediction is different in different regions of the world. Although a significant portion of the rural population in the developing world has the awareness of the importance of scientific weather forecasts, only a small number of farmers use such information in planning activities at the farm (Orindi and Murray 2005). Numerous studies conducted in recent years indicate that the number of farmers who use TK-based weather forecasting significantly decreased, because of unreliability. Yet many farmers in the world and predicting the seasonal climate use TK based on a variety of local indicators (plants, animals, insects, wind, astronomy, etc.). The disappearance of some local attributes has caused this system serious problems. Climate change and deforestation have caused the disappearance of many birds and plants that were reliable indicators for predicting climate events (Kijazi et al. 2012). All the aforementioned methods for weather forecasting were of great benefit in preventing and adapting to emergencies. This knowledge still has great value and a great degree of acceptance in the communities from which it originates as a basis for their sustainable living. Nevertheless, it is still not integrated into the scientific systems of sustainable development and management of natural disaster.

TK also provides much information that does not belong to the domain of research. For example, this knowledge was of great benefit to wildlife management in identifying priorities and the adoption of appropriate legislation. Many regulations in this area (e.g., reporting of catches after the hunt) created jointly by local people and scientists have contributed significantly to the conservation of populations of many wildlife species, especially endangered ones. In addition, local knowledge is important for identifying potential impacts of development activities. Local people in their territories registered many changes in animal behaviour caused by human activities, such as migratory routes changes. Many TK related to ecology, climate change, agriculture and alike, were also confirmed by science. Such acknowledgments to local knowledge encourage, and can be a stimulus for, more active involvement of traditional communities in research and environmental management. However, special attention should be paid to this area as excessive involvement of local people can lead to decline of their interest, and appropriate awards for the provided expertise should stimulate their participation. The sustainable use of TK is essential for the sustainable future of many indigenous people and therefore should involve intensive work to strengthen regional and global awareness of the importance of this knowledge and traditional ways of life of indigenous people (Brunk 2013).

3.8 Involvement of Preserved Local Practice in Everyday Life and Compilation of Strategies That Exceed the Time

TK includes accumulated and transferred knowledge and skills of many communities that lived in a particular area in a historic period of time. This knowledge is created in local conditions and was mostly related to local resources: land, water, wildlife, and culture. These resources represent a dynamic system that is continuously changed according to the changing needs of humans. TK has always been in a well-balanced relationship with nature, the surrounding environment, and natural resources characteristic of certain areas. Safety of food and medical care for a large number of people is significantly affected and still depends on them. They are the foundation to long-term processes, fatefully important for the survival of indigenous communities. Culture and tradition are an integral part of local knowledge and determine the behaviour codex of local communities; only full optimization of these processes ensures their balance. TK belongs to everyone, to no one more or less. TK cannot be attributed only to one nation and only to one period. Thus, it is not known when these processes started, but it is known that they will never stop.

TKs and practices passed down for generations leave a strong mark in the lives of people who, while respecting the principles of TK, lived according to unwritten laws and rules. The process of knowledge transfer was carried out steadily and with varying intensity among individuals, families, tribes, and nations. Many local communities have passed their knowledge, skills, and culture testimony "from mouth to mouth," which has always been a safe way of preserving them. For some other communities engraved records are linked or objects with picturesque views of uses, and some communities left to human kind written, undisputed evidence (Shemdoe and Mhando 2012). TK faithfully represents the characteristics of local communities and indigenous people and their lifestyles. By setting social and community standards, TK became heavily protected and respected as a cultural value.

Either way, it is likely that the transition from one period of time to another took place through the TK and local practices, which on the way toward the future had a bypass function. The traditional people created traditional concepts in the whole period of development of human society and in time they became important social standards as appropriate and expedient solutions (Wekundah and Joseph 2012). That is why today preservation of local practices, as sources of knowledge and cultural diversity, is seen as a positive accomplishment of modern times.

TK helps in optimization of available natural resources and the creation of sustainable systems. Its full involvement in the everyday life of ordinary people held it alive during many periods of history. Maybe we need TKs today more than ever; therefore, preserving the knowledge and wisdom of local people can be considered as necessity of modern society (Ellen and Harris 1996). Their noncommercial value is essentially the reason for their long existence and sustainability. TK as collective treasure belonging to all is skilfully kept from any form of commercialization (Kratovalieva et al. 2013).

Even though the so-called broad knowledge of indigenous people and communities is very little present in formal education, more attention is paid to find modern systems and tools for its efficient transfer and presentation. But despite all that, thanks to many initiatives taken in the various branches of ethnoscience, much of this knowledge has been successfully preserved in the daily life of many indigenous people (Kim et al. 2014). The growing trend of urbanization and faster development of technologies encourages thousands of young people in developing countries to give up the knowledge of their ancestors, with the conviction that with new specialized knowledge they will be able to find better possibilities in the industrialized world. In search of new life, many of these young people end up without TK, but without modern skills as well, necessary for life in an urban environment. These trends lead to increased poverty and contribute to the rapid disappearance of TK and practices.

However, along with these processes, there are still many communities in the world that heritage and preserve this treasure as the basis of their sustainable living (Msuya 2007). Young generations in these communities relate with due respect to the knowledge which they inherited from their ancestors, trying to prolong its duration because it is a symbol of culture and ethics of their ancestors who have found ways to leave their knowledge and skills to their followers.

Thanks to the efforts of many previous generations, TK is currently available in all areas of human life and work. Numerous restaurants with the prefix "ethno" or "traditional" are opening around the world in recent years, providing more sustainable use of PGRFA (Huntington 2000). These restaurants are being built in the ethnic style and remind us of houses from past centuries. They are only preparing dishes from ancient menus and serve them in wooden plates. Also, in industrialized countries, young women are increasingly resorting to the preparation of meals in ways their grandmothers and great-grandmothers did. In this way the tradition is slowly returning in many homes, with the increasing focus placed on the family and more frequent gathering of family members, which is essentially a legacy of traditional life (Kratovalieva 2011).

The impression is that the National Pharmacopoeia reached a high level of social significance and that in this way it has no longer any obstacles is getting even more significance. Confirmation of this statement are the many universities, hospitals, and clinics around the world that study only traditional healthcare and treatments carried out in ancient methods and with use of traditional techniques (Thornton and Maciejewski 2012). The number of pharmacies preparing medicines and providing advice based entirely on herbal preparations is increasing as well. Numerous scientific publications, papers, and research programmes that are opposed to the modern lifestyle and the way various diseases are controlled indicate the need for common use of modern and traditional medicine. Traditional and scientific knowledge have much to offer to each other despite the fact that no one denies that they differ significantly. In fact, nobody today denies the power of modern technology, but only seeks to highlight the importance of TK that might fill many gaps as they arise (Nakashima et al. 2000).

In many areas of the economy today use of raw materials of natural origin is increasingly in vogue. It is interesting to mention that the construction industry is increasingly using stone and other natural materials, and design of plants is done on the model of the buildings from past centuries (De Guchteneire et al. 1999). Natural materials and colours are once again in focus of modern interests, and that is why indigenous plants are increasingly used in the production of natural dyes; their use is growing in the production of textiles and organic clothing. Natural colours are becoming increasingly important in art academies. The offer of furniture and other household items protected from deterioration with various types of wax, resin, and other essential oils is increasing day by day (Maulik et al. 2014), as well as requirements for this raw material in timber industry (Baumflek et al. 2010).

Increased interest for TK and its integration in modern streams is reflected through the increased number of new TK-related initiatives and projects, sending an encouraging message to people in remote and underdeveloped areas to better preserve their traditions and work on finding information about local knowledge, culture, and values. Bearing in mind that the preservation of TK is in relationship to the local community, special attention should be focused on preserving these communities because they are the least developed and thus most vulnerable. For such a task, it is necessary to set up an appropriate legal framework and to establish closer cooperation with the owners of this knowledge.

The implementation of the foregoing activities shall be carried out through a number of promising ways of cooperation, such as summer schools, camps, training programs, and projects. It is very important that all the strategies relating to the inclusion of TK are based on respect and strengthening of TK and culture, which is only possible with mutual respect and sincere cooperation of modern society and local communities. It is vital that participants in TK-related initiatives spend some time together with the individuals who own and operate the transfer of knowledge to select the most effective means of sustainable transfer of this knowledge. Also, it is necessary to demonstrate a sufficient level of respect for the role TK had played in the development of society, as well as the ways in which knowledge and skills are transmitted and stored from oblivion. Further efforts must be made in finding new ways for the implementation of TKs and keeping in mind that modern methods of education can sometimes have an opposite effect. Therefore, persistent work should identify opportunities for greater integration of TK in the modern training modules, as well as on the design of transitional modules and adapted curriculum.

TK and practices are present today in the daily life of many local communities, somewhere more and somewhere less. It is undisputed that TK and practices from community to community are quite different, but also they have much in common.

1. A practical approach that comes from experience and skills of previous generations and which is widely accepted in the community;
2. Easy-to-understand approach for all people who respect nature, work to protect the environment and its preservation, and who appreciate and sustainably use natural resources;
3. An integrated approach that cannot be separated from people who enforce them because of their philosophical perspectives on health and its preservation, which implies a balanced approach to all that is materialistic;
4. A true approach that involves tangible elements of a dynamic system and which carries the accumulated value of past times that merge tradition, science, and metaphysics in one single unit;
5. Applied strictly oriented approach to the concept which is based on existing and familiar systems or compatible and easily comparable.

In recent years TK have been considered more seriously in the world, so it is normal that interest in its implementation is a top priority of developed societies (Bridel 2003). Today in the world, there are a large number of TK-related initiatives, based mainly on the following:

1. Exploration of new plant components with the effect on specific diseases;
2. Standardization of herbal formulation and promotion of their already known therapeutic effects, as well as the commercialization of traditional medicines for specific diseases;
3. Cooperation with pharmaceutical companies on the development of a production line of herbal preparations;
4. Implementation of ethnobotanical studies and preparation of studies for plants that are widely used by the "folk doctors";
5. Intensifying activities for in situ conservation of medical plants and promoting their ex situ cultivation for use in traditional medicine;
6. Registration, mutual coordination, and cooperation of "folk doctors";
7. Publishing activities with an aim to enable easier access to medicinal plants, and other plants that can be used for something else;
8. Establishing networking and other mechanisms that will enable easier transfer of "folk doctors" knowledge to interested users (cooperation with research centres); and
9. Opening of hospitals with traditional medical practice and a holistic approach.

3.9 Tradition Today: From Now to Further

It is clear nowadays that tradition as a cultural heritage encompasses the highest values of a society and truly represents its collective spirit and identity. Indigenous peoples throughout the entire time of their existence used a number of divergent practices with the aim of conservation of biological populations and especially their productivity. Some of these have found application in modern times (Lodhi and Mikulecky 2010). TK has invaluable importance for the life of local communities, with significant contributions to food safety, preservation of health, improvement of education systems, and natural resource management. Unfortunately, many traditional practices are severely endangered today, and activities directed toward their preservation are more necessary than ever.

TK, as well as the knowledge that comes from tradition, is the basis of many innovations in the past, and certainly it will be in the future (Andanda 2012). There are currently a number of initiatives aimed for conservation and sustainable use of knowledge, related to several important areas: policy issues, joining of local organization, *sui generis* protection of local resources, and collaboration on projects.

3.9.1 Policy Issues

The focus of future policy for both international and many national organizations should be as follows:

1. Determination and putting into operation easy operational management structures and instruments for the identification, validation, and further transfer of TK from all spheres;
2. Greater involvement of TK in educational programs and various forms of mass media to keep the public informed on characteristics of local communities and the importance they have in contemporary society;
3. Creating an additional system of modifications and adaptations of TKs to make them more easily accepted, understood, and further developed;
4. Forcing the *sui generis* protection of IPRs of TK and securing preservation of the essence of their nature;
5. Providing national support for producers that will use the forgotten techniques of production, conservation, and drying of food;
6. The introduction of interregional curriculum for TK in existing universities with focus on developing a model for the preservation of their own TK using internationally proven methods and practices.

3.9.2 Joining of Local Organisations

Unfortunately, many people in poor countries have completely lost touch with the community from which they originated. Therefore, it is of great importance that this knowledge be written, as in this way it will be of benefit to many future generations. No one but the local people can feel the spirit of what is declared as TK or be able to transfer it in original manner or to control its further preservation. In this respect, national and local organizations involved in the protection and preservation of TK need to be more focused on connecting with local communities to ensure that the transfer of these skills is under the supervision and active participation of local people. As local know-how transfer skills are a solid guarantee of originality and authenticity, in the future it should be insisted on joint participation of all stakeholders. This approach would lead to the creation of a sustainable research platform leading to the discovery of other forgotten skills and knowledge. Realization of these initiatives through socialization with local indigenous people ensures that the use of TK takes place in partnership and fair relationship between the local society and other stakeholders (Nakashima 2000b).

3.9.3 *Sui Generis* Protection of Local Resources

There are a number of legal modalities for protection of IK. Some involve giving absolute rights to the community within which this knowledge is created, while others offer a variety of mechanisms of appropriate compensation and control over the means of application of traditional knowledge. In addition to these, there are other legal mechanisms for the protection of TK from abuse. However, in theory, we can find two main approaches of protection: the positive and defensive. Positive protection of TK provides to certain communities or individuals within those communities owning exclusive rights to the TK. This prevents misuse of TK, but also allows appropriate payment for its use by a person outside the community. In this form of legal protection, problems can arise when identifying the holder of that knowledge. Unlike the first approach, defensive protection is based on the principle of exclusivity, which prevents third parties to achieve legitimacy of certain TK (Andanda 2012), that might use this knowledge in inappropriate way, or to misuse it.

Considering that TK includes accumulated knowledge from all possible spheres, the protection mechanisms of some forms of TK depend on the types of knowledge and forms of its application. There are several possibilities for the legal protection of TK: intellectual property rights, *sui generis* legal protection of indigenous knowledge, as well as other protection mechanisms. Problems related to the protection of TK could be partly solved through a set of well defined property rights *sui generis*. Active participation of indigenous people in these processes would enable more efficient protection and preservation of the origin of TK. *Sui generis* system of legal protection implies creating a legal framework that would adequately define the issues of legal protection of some TK forms. However, few countries in the world today use a *sui generis* system of protection, while there are significantly more of those who still have doubts about ways to protect their rights. It is evident that indigenous people do not yet have a clear picture of the importance of this type of protection, and for these reasons, necessary steps should be taken to demonstrate the advantages and benefits stemming from the "*sui generis* rights." Because the TK has its own undeniable identity, the mode of inheritance and emphasized collective spirit of *sui generis* system of legal protection would be very satisfying. Nevertheless, there are countries in the world that have such systems (India, Philippines, Indonesia, etc.). However, there is a justified fear that such a system of protection can be the cause of a large number of legal problems, but also lead to significant incompatibilities in protection of traditional knowledge systems in the world. Therefore, in addition to its supporters, this system of protection has many opponents. The possibility of creating a *sui generis* system of legal protection of TK is announced by some international bodies. WIPO has prepared a whole different set of regulations that could be applied in such *sui generis* law. In this way, WIPO actively contributes to the development of the legislation that supports the *sui generis* protection. Today there are more and more initiatives that favour the recognition of intellectual property rights in the use of IK and products that originate from it.

A business-oriented approach is not a good approach to the preservation and protection of TK, as it represents a potential possibility of its transformation and loss of integrity. Therefore, it should not be allowed to place business-oriented initiatives that delve deeply into the core of IK and change its essence under the idea of environmental protection and the conservation of endangered species (Rahman 2000). From all these points mentioned, there is a strong impression that *sui generis* IPRs are a very complex area and that requires a serious scientific analysis.

3.9.4 Collaboration in Projects

In recent years, knowledge of local communities is getting greater attention and thus the number of programs aimed for their collection, preservation, and dissemination is growing as well. At the same time, the number of initiatives with the combined application of scientific information and TK is increasing. This approach is a good way to integrate local knowledge into modern development processes, because the knowledge of local people can significantly help with a better perspective on a number of social and natural phenomena, and especially more effective responses to many extreme events. New ways to access local knowledge are increasing their focus on establishing partnerships between local communities and their members, from one side and the scientific community on the other (Le Grange 2007). Through this kind of cooperation access to TK sources is facilitated, which leads to more efficient promotion and better understanding of its importance to both indigenous and nonindigenous people (Olaide 2012).

It is obvious that the awareness for the need of preservation of TK is higher today than ever before, and the conditions for its preservation are better. It is therefore important that TK is as soon as possible involved in national politics, and issues regarding its use, preservation, and distribution are clearly defined. To have the results achieved in line with the assumed objectives, it is essential that all TK-related initiatives be co-financed by the governments of the participating countries. In addition to the preservation and promotion of TK, there is also recognition of the country, its history, and cultural and historical wealth created for centuries. These processes should enable local communities to identify the knowledge that can be integrated with doctrine and which can later be distributed for further use, but also to preserve and use the one that can not be explained scientifically (Hiwasaki et al. 2014). TK research results obtained, approved for public dissemination by the local community, should be made available to all those who are interested in its use. Creating websites with links to specific areas would be an ideal way of their presentation. It is important that these initiatives include the observations of holders of indigenous knowledge, because in this way the confidence in the results achieved will increase, but also it will create a basis for better cooperation in the future.

This is just the beginning and there is still a lot of work ahead of us: that is why the society together with holders of local knowledge must design new methods to put this knowledge into the function of the common good.

3.10 Conclusion

Conservation and sustainable use of PGRFA require a holistic and comprehensive approach that will create conditions to put the abundance of available indigenous materials and TK into functional development through the application of acceptable development models. To be successful, it is necessary to actively involve all institutions concerned with genetic resources: universities, research institutions, plant breeders, seedlings producers, farmers, agrofood industry and other producers, consumers, NGOs, and others. There are significant opportunities for active involvement of agro-NGOs because of their role in conservation and the sustainable use of GRFA, and this role can certainly be significantly greater. In today's world, more attention is paid to establishment of a system that focuses on food safety, production according to the principles of good agricultural practices, conservation of natural resources, and biodiversity. An increasing number of NGOs is advocating more intensive use of GRFA as a source of food, including protection of the geographic origin of food produced from these resources, and even certain ingredients that have a characteristic and distinctive smell and taste. For sustainable development and implementation, the support provided by international competent authorities and other relevant stakeholders in these activities will have primary importance. More intensive use of genetic resources would be a very effective response to the growing global demand for food and other agricultural products.

References

Adams WM (2006) The future of sustainability: re-thinking environment and development in the twenty-first century. The World Conservation Union, p 19 http://cmsdata.iucn.org/downloads/iucn_future_of_sustanability.pdf

Adger WN, Brook N, Bentham G, Agnew M, Eriksen S (2004) New indicators of vulnerability and adaptive capacity. Tyndall Centre for Climate Change Research, Technical Report 7. School of Environmental Sciences, University of East Anglia, Norwich

Alonso A, Dallmeier F, Granek E, Raven P (2001) Biodiversity: connecting with the tapestry of life. Smithsonian Institution/Monitoring and Assessment of Biodiversity Program and President's Committee of Advisors on Science and Technology, Washington, DC

Andanda P (2012) Striking a balance between intellectual property protection and traditional knowledge, cultural preservation, and access to knowledge. J Intellect Prop Rights 17:547–558

Andersen R (2003) FAO and the management of plant genetic resources. In: Stokke OS, Thommessen OB (eds) Yearbook of international co-operation on environment and development 2003/2004. Earthscan Publications, London, pp 43–53

Armitage DR (2003) Traditional agroecological knowledge, adaptive management and the socio-politics of conservation in Central Sulawesi, Indonesia. Environ Conserv 30:79–90

Baumflek MJ, Emery MR, Ginger C (2010) Culturally and economically important nontimber forest products of northern Maine. Gen. Tech. Rep. NRS-68. US Department of Agriculture, Forest Service, Northern Research Station, Newtown Square

Becker CD, Ghimire K (2003) Synergy between traditional ecological knowledge and conservation science supports forest preservation in Ecuador. Conserv Ecol 8(1):1

Benz BF, Cevallos J, Santana F, Rosales J, Graff M (2000) Losing knowledge about plant use in the Sierra de Manantian Biosphere Reserve, Mexico. Econ Bot 54:183–191

Berkes F, Folke C (2002) Back to the future: ecosystem dynamics and local knowledge. In: Gunderson LH, Holling CS (eds) Panarchy: understanding transformations in human and natural systems. Island Press, Washington, DC, pp 121–146

Berkes F, Colding J, Folke C (2003) Navigating social–ecological systems: building resilience for complexity and change. Cambridge University Press, Cambridge, UK

Brekes F, Armitrage D, Doubleday N (2007) Synthesis: adapting, innovating, evolving. In: Armitrage D, Brekes F, Doubleday N (eds) Adaptive comanagement: collaboration, learning and multi-level governance. University of British Columbia Press, Vancouver

Bridel J (2003) Study of indigenous plants and non-timber products as related to traditional medicine in the Nuba Mountains and Southern Blue Nile Region of South Sudan http://cafnr.missouri.edu/iap/sudan/doc/study-indig.pdf

Briggs J (2005) The use of indigenous knowledge in development: problems and challenges. Prog Dev Stud 5:99–114

Brunk A (2013) Traditional knowledge. http://www.arcticpeople.org/focus-archive/item/598-traditional-knowledge%5B

Brush SB (2000) The issue of *in situ* conservation of crop genetic resources. In: Brush SB (ed) Genes in the field: on-farm conservation of crop diversity. IPGRI/IDRC/Lewis Publishers, USA

Busia K (2005) Medical provision in Africa: past and present. Phytother Res 19:919–923

Coelho N (2012) Our land our life. Taleemnet, Goa

Cunningham A (2001) Applied ethnobotany. Earthscan, London

De Guchteneire P, Krukkert I, von Liebenstein G (1999) Best practices on indigenous knowledge. Joint publication of UNESCO's Management of Social Transformations Programme (MOST) and the Centre for International Research and Advisory Networks (CIRAN). http://www.unesco.org/most/bpikpub.htm#ikbestpractices

Deepak A, Anshu S (2008) Indigenous herbal medicine: tribal formulation and traditional herbal practices. Aavishkar, Jainpur

Dhillon BS, Dua RP, Brahmi P, Bisht IS (2004) On farm conservation of plant genetic resources for food and agriculture. Curr Sci 87:557–559

Diamond J (2002) Evolution, consequences and future of plant and animal domestication. Nature (Lond) 418:700–707

Dimitrijevic M, Petrovic S, Cimpeanu C, Bucur D, Belic M (2011) Cereals and *Aegilops* genus biodiversity survey in the west Balkans: erosion and preservation. J Food Agric Environ 9(3-4):219–225

Djuric G (2010) The importance of plant genetic resources for food and agriculture. University book, Faculty of Agriculture, Banja Luka, Bosnia and Herzegovina

Djuric B, Gataric D, Radanovic D (2007) Wild medicinal herbs. Manual. Faculty of Agriculture, Banja Luka, Bosnia and Herzegovina, Glas srpske – Graphics AD, pp 1–130

Douthwaite B (2001) Enabling innovation. A practical guide to understand and fostering technological change. Zed Books Ltd, London

Ejeta G (2009) Revitalizing agricultural research for global food security. Science 327:831–832

Ellen R, Harris H (1996) Concepts of indigenous environmental knowledge in scientific and development studies literature – a critical assessment. Draft paper. East-West Environmental Linkages Network Workshop 3. International Development Research Centre, Canterbury

Emiru B, Ermias A, Wolde M, Degitu E (2011) Management, use and ecology of medicinal plants in the degraded dry lands of Tigray, Northern Ethiopia. J Hortic For 3(2):32–41

FAO (1998) State of the world's plant genetic resources. FAO, Rome

FAO (1999) Women: users, preservers and managers of agrobiodiversity. www.fao.org/FOCUS/E/Women/Biodive.htm

FAO (2010) The second report on the State of the World's plant genetic resources for food and agriculture. http://www.fao.org/agriculture/crops/core-themes/theme/seeds-pgr/sow

Folke C (2004) Traditional knowledge in social–ecological systems. Ecol Soc 9(3):7

Ford RI (1994) Ethnobotany: historical diversity and synthesis. In: Ford RI (ed) The nature and status of ethnobotany, vol 67, 2nd edn, Anthropological papers. Museum of Anthropology, University of Michigan, Ann Arbor, pp 33–49

Ford J (2001) The relevance of indigenous knowledge to contemporary sustainability. Northwest Sci. 7:185–190

Frank U, Tapia ME, Maass BL (2002) Challenges to organic farming and sustainable land use in the tropics and subtropics. *In situ* conservation of native potatoes in the Peruvian highlands. Deutscher Tropentagen, Witzenhausen (book of abstracts, p 73)

Frese L, Palmé A, Bülow L, Neuhaus G, Kik C (2015) On the sustainable use and conservation of plant genetic resources in Europe. Crop Wild Relatives 10:34–38

Gearheard S, Pocernich M, Stewart R, Sanguya J, Huntington HP (2010) Linking Inuit knowledge and meteorological station observations to understand changing wind patterns at Clyde River, Nunavut. Clim Chang 100:267–294

Giovannini P, Heinrich M (2009) Xki yoma' (our medicine) Xki tienda (patent medicine): interface between traditional and modern medicine among the Mazatecs of Oaxaca, Mexico. J Ethnopharmacol 121:383–399

Glogovac S, Takac A (2010) Use of old varieties and landraces of tomatoes as a source of genetic variability in breeding. Field Veg Crop 47:493–498

Hammer K (2004) Resolving the challenge posed by agrobiodiversity and plant genetic resources – an attempt. J Agric Rural Dev Trop Subtrop, Beiheft Nr. 76. Kassel University Press, Germany

Hammer K, Teklu Y (2008) Plant genetic resources: selected issues from genetic erosion to genetic engineering. J Agric Rural Dev Trop Subtrop 109(1):15–50

Hammond P (1995) The current magnitude of biodiversity. In: Heywood VH, Watson RT (eds) Global biodiversity assessment. Cambridge University Press, Cambridge, UK, pp 113–138

Handa SS, Khanuja SPS, Longo G, Rakesh DD (2008) Extraction technologies for medicinal and aromatic plants. International Centre for Science and High Technology, Trieste, pp 21–25

Hart E (1995) Getting started in oral traditions research. Occasional Papers of the Prince of Wales Northern Heritage Centre, no. 4, Government of the Northwest Territories, Yellowknife http://www.pwnhc.ca/research/downloads/OralTraditionsManual.pdf

He SA, Sheng N (1997) Utilization and conservation of medicinal plants in China with special reference to *Atractylodes lancea*. In: Bodeker G, Bhat KKS, Burley J, Vantomme P (eds) Medicinal plants for forest conservation and health care: non-wood forest products. FAO, Rome, pp 109–115

Hiwasaki L, Luna E, Rajib SS (2014) Process for integrating local and indigenous knowledge with science for hydro-meteorological disaster risk reduction and climate change adaptation in coastal and small island communities. Int J Disaster Risk Reduction 10:15–27, http://www.foodfirst.org/pubs/backgrdrs/1998/s98v5n3.html

Huntington HP (2000) Using traditional ecological knowledge in science: methods and applications. Ecol Appl 10(5):1270–1274

IFAD (2001) Rural poverty report 2000: the challenge of ending rural poverty. IFAD, Rome

Igdir HB, Rezaei RM, Hoseini M (2013) The packaging state of Iranian medicinal-plants in view of their production and exports. Adv Crop Sci 3(2):181–190

Iwu MM (1993) Handbook of African medicinal plants. CRC Press, Boca Raton

Jovovic Z (2014) Plant genetic resources in Montenegro and possibilities of their sustainable use. In: Agriculture and climate change. Centre for Development of Agriculture, Bijelo Polje, pp 55–65

Jovovic Z, Cizmovic M, Lazovic B, Maras V, Bozovic D, Popovic T et al (2012) The state of agricultural plant genetic resources of Montenegro. Agric For 57(1):33–50

Jovovic Z, Stesevic D, Meglic V, Dolnicar P (2013) Old potato varieties in Montenegro. Monograph, University of Montenegro, Biotechnical faculty Podgorica, pp 11–79

Karg AL, Tomasic S, Walker M, Bridson GDR, Burdet HM, Chautems M, Moruzzi-Bayo T (1999) Compilers of catalogue of portraits of naturalists, mostly botanists, in the collections of the Hunt Institute, the Linnean Society of London and the Conservatoire et Jardin Botaniques de la Ville de Genève. Part 3. Portraits of individuals. E-H. Hunt Institute for Botanical Documentation, Pittsburgh

Kiene T (2006) Traditional knowledge in the European Context. Iddri – Idées pour le débat, no. 02

Kijazi AL, Chang'a LB, Liwenga ET, Kanemba A, Nindi SJ (2012) The use of indigenous knowledge in weather and climate prediction in Mahenge and Ismani Wards, Tanzania. In: Proceedings of the first climate change impacts, mitigation and adaptation programme scientific conference, p 48

Kim DI, Dhungana SK, Zi PM, Lee JW, Moon HK, Hwang HS, Shin DH (2014) Addition of grape extracts can enhance quality of natural seasonings by changing their physicochemical properties. Afr J Biotechnol 13(15):1638–1644

Koushalya NS (2012) Traditional knowledge on ethnobotanical uses of plant biodiversity: a detailed study from the Indian western Himalaya. Biodivers Res Conserv 28:63–77

Kratovalieva S (2011) Ratkova skala – integrated review of the identified high nature values. IUCN & DEM, pp 1–51

Kratovalieva S, Popsimonova G, Tomich Z (2010) Legumes as pre-crop in organic production of potato in Republic of Macedonia. Biotechnol Anim Husbandry 26:649–655

Kratovalieva S, Mihovsky T, Pacinoski N (2012a) Nutritive value of grasslands in East Macedonia. J Mt Agric Balkans 14(3):576–588

Kratovalieva S, Pacinoski N, Mihovsky T (2012b) Grazing capacity of grasslands in west part of Macedonia. J Mt Agric Balkans 14(3):546–562

Kratovalieva S, Popsimonova G, Ivanovska S, Jankuloski L, Meglic V (2012c) Macedonian gene bank: seed protein content of wild red clover (*Trifolium pratense* L.) accessions. Agric Conspec Sci 77(4):199–202

Kratovalieva S, Selamovska A, Petrovic S, Dimitrijevic M (2013) Survey of seed germination at pear landraces. Bulg J Agric Sci 20(3):656–660

Lange D (1998) Europe's medicinal and aromatic plants: their use, trade and conservation. TRAFFIC International, Cambridge

Lasisi R, Ekpwynyong AS, Wabah GN (2011) Urbanization and loss of traditional ecological knowledge (TEK): lessons from Rumudomaya Community in Rivers State. Glob J Afr Stud 1:54–65

Le Grange L (2007) Integrating western and indigenous knowledge systems: the basis for effective science education in South Africa? Int Rev Educ 53(5-6):577–591

Lodhi S, Mikulecky P (2010) Management of indigenous knowledge for developing countries. In: Communication and management in technological innovation and academic globalization, pp 94–98
Martin GJ (1995) Ethnobotany. Chapman & Hall, London
Maulik SR, Bhowmik L, Agarwal K (2014) Batik on handloom cotton fabric with natural dye. Indian J Tradit Knowl 13(4):788–794
Maxted N, Guarino L (1997) Ecogeographic surveys. In: Maxted N, Ford-Lloyd BV, Hawkes JG (eds) Plant genetic conservation. The *in situ* approach. Chapman & Hall, London, pp 69–87
Maxted N, Kell SP (2009) Establishment of a global network for the *in situ* conservation of crop wild relatives: status and needs. FAO Commission on Genetic Resources for Food and Agriculture, Rome, p 266
McClatchey W, Paul A, Flaster T, McClatchey V (1999) An evaluation of educational trends in economic botany. Ethnobotany Educational Publication Series 1:1–21. Centre for International Ethnomedicinal Education and Research
McCouch S (2004) Diversifying selection in plant breeding. PLoS Biol 2:1507–1512
Melchias G (2001) Biodiversity and conservation. Science Publishers, Enfield
Melozzi L (2009) Incentives for agrobiodiversity within the European Union: the role of rural development plans. J Agric Environ Int Dev 103(1/2):11–30
Mesfin K, Tekle G, Tesfay T (2013) Ethnobotanical study of traditional medicinal plants used by indigenous people of Gemad District, Northern Ethiopia. J Med Plants Stud 1(4):32–37
Moore-Lappé FJ, Collins P, Rosset L (1998) 12 myths about hunger based on world hunger: 12 myths, 2nd edn. Grove/Atlantic and Food First Books, New York
Morico G, Grassi F, Fideghelli C (1999) Horticultural genetic diversity: conservation and sustainable utilization and related international agreements. Acta Hortic (ISHS) 495:233–244
Msuya J (2007) Challenges and opportunities in the protection and preservation of indigenous knowledge in Africa. Int Rev Inf Ethics 7(3)
Myers N (1994) Protected areas – protected from a greater what? Biodivers Conserv 3:411–418
Nabhan GP (2000) Interspecific relationships affecting endangered species recognized by O'odham and Comcaac cultures: the ethnoecology of interaction diversity. Ecol Appl 10:1288–1295
Nakashima DJ (1990) Application of native knowledge in EIA: Inuit, eiders and Hudson Bay oil. Canadian Environmental Assessment Research Council, Hull, PQ, Canada, p 27
Nakashima DJ (2000) What relationship between scientific and traditional systems of knowledge? In: Cetto AM (ed) Science for the twenty-first century: a new commitment. UNESCO Sources, Paris, pp 432–444
Nakashima DJ, Prott L, Bridgewater P (2000) Tapping into the world's wisdom. UNESCO Sources, p 12
Nasir A, Nur KA (2014) Traditional Knowledge and Formulations of Medicinal Plants Used by the Traditional Medical Practitioners of Bangladesh to Treat Schizophrenia Like Psychosis. Schizophrenia Research and Treatment, vol. 2014, article ID 679810, http://dx.doi.org/10.1155/2014/679810
Newmaster SG, Subramanyam R, Balasubramaniyam NC, Ivanoff RF (2007) The multimechanistic taxonomy of the Irulas in Tamil Nadu, South India. J Ethnobiol 27:233–255
Newton AC, Akar T, Baresel JP, Bebeli PJ, Bettencourt E, Bladenopoulos KV et al (2010) Cereal landraces for sustainable agriculture. A review. Agron Sustain Dev 30:237–269
Novellino D (2003) From seduction to miscommunication: the confession and presentation of local knowledge in participatory development. In: Pottier J, Bicker A, Sillitoe P (eds) Negotiating local knowledge: power and identity in development. Pluto Press, London
Olaide IA (2012) Management of indigenous knowledge as a catalyst towards improved information accessibility to local communities: a literature review. Chin Librarianship 35:87–98, www.iclc.us/cliej/cl35IO.pdf
Orindi VA, Murray LA (2005) Adaptation to climate change in East Africa: a strategic approach, international institute for environment development. Gatekeeper Series 117 http://pubs.iied.org/pdfs/9544IIED.pdf
Patwardhan B (2005) Traditional medicine: a novel approach for available, accessible and affordable health care. Paper submitted for Regional consultation on Development of Traditional Medicine in the South-East Asia Region, Korea, World Health Organization
Pencic M, Dumanovic J, Radovic G, Jelovac D (1997) The importance of plant gene bank for selection. Selection and seed production. Novi Sad IV(1-2):7–18
Pieroni A, Quave C, Villanelli M, Mangino P, Sabbatini G, Santini L et al (2004) Ethnopharmacognostic survey on the natural ingredients used in folk cosmetics, cosmeceuticals and remedies for healing skin diseases in the inland Marches, Central–Eastern Italy. J Ethnopharmacol 91:331–344
Plotkin MJ (2000) Medicine quest: in search of nature's healing secrets. Viking, New York, pp 48–57, issue 101
Quinn ML (2001) Protection for indigenous knowledge: an international law analysis. St Thomas Law Rev 14:287–313
Ragupathy S, Newmaster SG, Murugesan M, Balasubramaniyam V (2009) DNA barcoding discriminates a new cryptic grass species revealed in an ethnobotany study by the hill tribes of the Western Ghats in southern India. Mol Ecol Resour 9:164–171
Rahman A (2000) Development of an integrated traditional and scientific knowledge base. United Nations, Geneve

Ryser R (2011) Indigenous people and traditional knowledge. In: Berkshire encyclopedia of sustainability 5/10, ecosystem management and sustainability. Berkshire Publishing Group

Scarascia-Mugnozza GT, Perrino P (2002) The history of *ex situ* conservation and use of plant genetic resources. In: Engels JMM, Ramanatha RV, Brown AHD, Jackson MT (eds) Managing plant genetic diversity. CABI Publishing, Wallingford, pp 1–22

Schippmann U (1997) Plant uses and species risk. From horticultural to medicinal plant trade. In: Newton J (ed) Planta Europaea. Proceedings of the first European conference on the conservation of wild plants, Hyères, France, 2–8 Sept 1995. Plantlife, London, p 161–165

Shankar D, Majumdar B (1997) Beyond the biodiversity convention: the challenge facing the biocultural heritage of India's medicinal plants. In: Bodeker G, Bhat KKS, Burley J, Vantomme P (eds) Medicinal plants for forest conservation and health care, vol 11, Non-wood forest products. FAO, Rome, pp 87–99

Shemdoe SG, Mhando L (2012) National policies and legal frameworks governing traditional knowledge and effective intellectual property systems in Southern and Eastern Africa: the case of traditional healers in Tanzania. African Technology Policy Studies Network, Kenya, pp 1–42

Shiembo PN (1999) The sustainability of eru (*Gnetum africanum* and *Gnetum buchholzianum*): over-exploited non-wood forest product from the forests of Central Africa. In: Non-wood forest products of Central Africa: current research issues and prospects for conservation and development. FAO Forestry Department, Rome

Simeonovska E, Gadzo D, Jovovic Z, Murariu D, Kondic D, Mandic D et al (2013) Collecting local landraces of maize and cereals in South Eastern Europe during 2009 and 2010. Rom Agric Res 30:1–7

Singh KM (2009) Scope of medicinal and aromatic plants farming in Eastern India. http://ssrn.com/abstract=2019789

Skinner J (2003) Anti-social 'social development'? Governmentality, indigenousness and the Dfid approach on Montserrat

Sobrevila C (2008) The role of indigenous people in biodiversity conservation – the natural but often forgotten partners. The International Bank for Reconstruction and Development/The World Bank

Stefanova V, Hart K, Znaor D, Kratovalieva S, Damjanovski D, Andonov P, Mukaetov D, Petkovski S, Sidorovska V (2012) Farmers and nature together: high nature value farming and agri-environment payments for the Republic of Macedonia. Avalon, The Netherlands, pp 1–58

Străjeru S, Ibănescu M, Constantinovici D (2009) Landrace Inventory for Romania, in European landraces: on-farm conservation, management and use. Biovers Tech Bull 15:137–142

Teklu Y, Hammer K (2006) Farmers perception and genetic erosion of Ethiopian tetraploid wheat landraces. Genet Res Crop Evol 53:1099–1113

Tester M, Langridge P (2010) Breeding technologies to increase crop production in a changing world. Science 327:818–822

Thomas E, Vandebroek I, Sanca S, Van Damme P (2009) Cultural significance of medicinal plant families and species among Quechua farmers in Apillapampa, Bolivia. J Ethnopharmacol 122:60–67

Thörn E (2011) Plant genetic resources a guarantee for future welfare and common concern. 5th Balkan symposium on vegetables and potatoes, book of abstracts, Tirana, p 55

Thornton TF, Maciejewski SA (2012) Collaborative engagement of local and traditional knowledge and science in marine environments: a review. Ecol Soc 17(3):8

Trakansuphakon P (2010) Strategy workshop on rotational farming/shifting cultivation and climate change. Indigenous Knowledge and People Foundation (IKAP), Chiang Mai

Turnbull D (2009) Working with incommensurable knowledge traditions: assemblage, diversity, emergent knowledge, narrativity, performativity, mobility and synergy. Retrieved from: ThoughtMesh, http://thoughtmesh.net/publish/279.php

Uprety Y, Asselin H, Boon EK, Yadav S, Shrestha KK (2010) Indigenous use and bio-efficacy of medicinal plants in the Rasuwa District, Central Nepal. J Ethnobiol Ethnomed 6:13

Weatherhead E, Gearheard S, Barry RG (2010) Changes in weather persistence: insight from Inuit knowledge. Global Environ Change 20:523–528

Wekundah M, Joseph M (2012) Why protect traditional knowledge? African Technology Policy Studies Network Biotechnology Trust, Africa. Special paper series, no. 44

White A, Molnar A, Khare A (2004) Who owns, who conserves and why it matters. Forest Trends, Washington, DC

Wilcox M, Bodeker G (2007) Malaria. In: Bodeker G, Burford G (eds) Traditional, complementary and alternative medicine policy and public health perspectives. Imperial College Press, London, pp 239–254

Wilson EO (1992) The diversity of life. WW Norton & Co. Inc., New York, p 287

Zachariev I, Gogushev G, Velkovski N, Gesheva M (2012) Manual for collectors of medicinal herbs and berries. Delchevo, Municipality Delchevo, pp 1–132

Zeven AC (1996) Results of activities to maintain landraces and other material in some European countries *in situ* before 1945 and what we may learn from them. Genet Resour Crop Evol 43:337–341

Zeven AC (1998) Landraces: a review of definitions and classifications. Euphytica 104:127–139

4 Plant Genetic Resources and Traditional Knowledge and Their Intellectual Property

Nguyen Van Kien and Nguyen Thi Ngoc Hue

Abstract

Today, humankind is facing greater challenges than before. Pressures of population growth, environmental pollution, impacts of climate change, growth and sustainable development, biological diversity and benefit sharing are rapidly increasing. It is crucial to humankind in cooperation of research and development of plant genetic resources for sustainable management. Under globalisation, access to plant genetic resources and relevant traditional knowledge and the benefit sharing arising out of their use are key issues for the developed as well as developing countries. These will provide a steady and fairly legal basis for whole relevant parties. Although international conventions/agreements/protocols on plant genetic resources and relevant traditional knowledge access and benefit sharing are already there, application and implementation of these conventions/agreements/protocols are not easy in fact. The differences and disputes of matters prevented global collaboration on plant genetic resources/relevant knowledge use for sustainable development goals. Then, there are acquired flexible approaches as well as social responsibility and professional morals to every party on this issue. Regarding challenges of food security in the future, a common outlook in management and use of plant genetic resources, indigenous/traditional knowledge and intellectual property should be discussed, shared and released.

4.1 Introduction

Development is a heritage, creation and innovation process. It helps increase the capacity and efficiency of labourers. However, every innovation takes time and money for creation. New innovations need protection for its products and benefit sharing to its heritages. Genetic resources (GR) in general and Plant Genetic Resources (PGR) in particular are considered as special natural resources because of their recyclability and development. They provide basic services for humankind's needs such as food, beverage, feeds and industrial products and bring huge profits for relevant parties. However, as mentioned to PGR, we always think of relevant traditional knowledge (TK). This knowledge is important to man-

N. Van Kien (✉) • N.T.N. Hue
Plant Resources Center, Hanoi, Vietnam
e-mail: kienguyenvan8@hotmail.com

R.K. Salgotra, B.B. Gupta (eds.), *Plant Genetic Resources and Traditional Knowledge for Food Security*, DOI 10.1007/978-981-10-0060-7_4

age and exploit PGR for meeting the increasing needs of man. At first stages of history of humankind, PGR are major medicinal suppliers for health care and protection. Over generations, our ancestors accumulated and withdrew invaluable experience/knowledge of using PGR for different needs. The experience/knowledge is still used until now without legal regulations, norms and standards. Many of them are applied into pharmaceuticals, food and others. Of course, the products should be registered and granted protection of intellectual properties. Consequence, whoever owns the intellectual properties (IPs) will get huge benefits and profits from users/applicants of the IP (Acharya 1992).

Today the whole world is debating on the conservation and sustainable utilisation of PGR and TK as well as sharing of benefits arising out of their use. With the advancement of technologies in various fields, PGR and TK are commercially used by the stakeholders in the developed nations. Indigenous knowledge has its own importance in developing and developed countries. But the benefits arising out of the use of TK are not being equally shared by the stakeholders. Moreover, globalisation of the use of potential PGR and TK, folklore and diverse cultures has attracted the world communities towards these resources. Subsequently, the demand of these valuable indigenous resources has also increased significantly (Tripathi 2003). With the commercialisation of PGR and TK, IP came up to provide legal rights to the innovators. IP is the recognition and reward given to the industrialised persons (Helfer 2002). The main aim of this law is to protect and give safeguard to the indigenous people of a country against the loss of genetic resources and to provide equal benefits of sharing their commercialised PGR and TK (Figs. 4.1 and 4.2).

Moreover, conventional IP laws are not very effective to protect the interests of local and indigenous people. Before the inception of the CBD in 1992, most of the PGR and TK were overexploited for commercial purposes by the developed countries. The stakeholders of developed countries have granted Intellectual Property Rights (IPRs) on most of the PGR and related sources with sharing of benefits with the local and indigenous people. Moreover, the local and indigenous farming communities do not protect their PGR because of non-availability of any strong laws (Correa 1998).

The PGR and TK should be preserved and protected against the overexploitation of these resources, and people should be aware about the protection and sustainable utilisation of genetic resources. Most of the indigenous communities are not well organised and indigenous peoples have different opinions regarding the protection of these valuable genetic resources. In most of the countries, IP laws are not so strong to protect their PGR and TK. Therefore, there is a need to strengthen the IP laws in a country for the protection of PGR and TK. Moreover, there is a need to utilise the IP laws effectively for the protection of PGR and TK and sharing of benefits. For protection of PGR and TK and their sustainable utilisation, both national and international IPs should be brought together for sustainable utilisation of these valuable resources. Before the implementation of any project, there is a need to consider the interest of the indigenous farming community, access to PGR and distribution of these resources. The benefits arising out the utilisation of PGR and TK should be equally distributed amongst the farmers for further conservation and sustain-

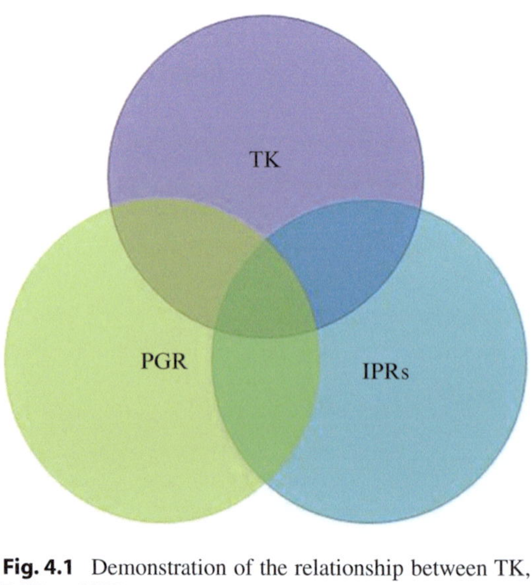

Fig. 4.1 Demonstration of the relationship between TK, PGR and IPRs

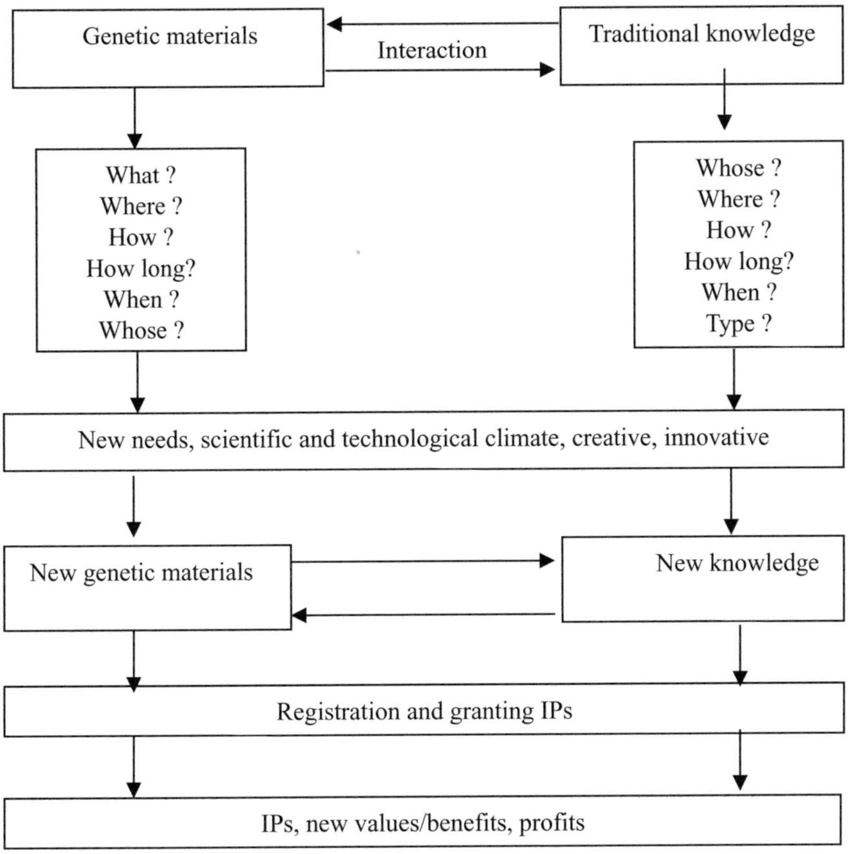

Fig. 4.2 Process of developing new plant resources and knowledge from PGR and relevant TK

able development of the biodiversity (Helfer 2002).

4.2 PGR, TK and IP: Concepts and Objectives

The genetic resources, particularly plant genetic resources for food and agriculture (PGRFA), are the important and the major components of securing the food security. These are the integral components for the sustainable production of agriculture. In addition, effects of climate change, such as drought, global warming, salinity, high temperature and infrequent floods, are the major challenges for the production of agricultural crops on a sustainable basis. Therefore, PGR and TK are also providing resilience to climate change for the production of stable food crops. The genetic diversity amongst these resources is important to avoid the genetic erosion and sustainable food production in the future (FAO 2012). All terminologies or definitions are used for legal basis for stakeholders at different levels (national, regional, international) in the area of PGR, TK and IP. General objectives of PGR, TK and IP are promoting the use of diverse genetic material and relevant TK to create new progress in materials and knowledge for meeting the demands of human beings.

Development of PGR in general and crop genetic resources in particular has been occurring for a long time. Over time, these PGR and TK have been exchanging and developing towards different purposes (Posey 1991). TK in some places in the world is rich and greatly varied and includes knowledge on conserving genetic resources, health protection, farming systems and plant and animal species' uses and management. It is a reflection of the different areas they live in and the different needs associated with the individual groups. The use of plants for medicinal treatment of diseases is especially well developed amongst the minority groups. In recent years, research institutions, private pharmaceutical companies and NGOs have recognised the importance of conserving this knowledge but also using it to create benefits, of which it has not always equitably been shared with the minority groups. The absence of a system for protection of TK in a country can easily result in the further erosion of this valuable knowledge and a permanent loss to local communities, the country and the global community at large (Greaves 1996). Nowadays, all PGR and indigenous/traditional knowledge occurred in a complex evolution and development. This process is related to multi-ethnic groups and has happened in different places and in various environments at the same time or different times (Posey 1991).

4.3 IP in Plant Genetic Resources and Indigenous Knowledge

IP is important and considered as a driving force to lead the development of developing and least developing countries. It is a tool to encourage the powerful creation and innovation for increasing the efficiency and capacity of workers/labourers (WIPO 2010). For protection of TK and genetic resources under IP laws, the IP system and farmers' system of protections overlap with each other. The World Intellectual Property Organization (WIPO) has worked with its members to clarify these issues. In current technological advances in agriculture and information technology, there should be a necessary call for constant re-evaluation of this system to equalise the benefit-sharing aspect. Moreover, to avoid the dispute on IP protection, the issues should be discussed thoroughly. IP is one of the important tools to increase the plant genetic diversity and its sustainable utilisation and equal sharing of benefits amongst the relevant stakeholders. There is no doubt that protection of PGR and TK enhances the cost of products, but it can provide equal benefit sharing amongst the different constituents. Moreover, to grant the IP protection to plant and animals, a conflict rises due to morals of the indigenous people.

The scope and nature of IP protection laws vary from one country to another and from one region to another. IP protected in one country may not be recognised in another country. Various international organisations such as CBD and WIPO formulated different international rules and regulations for protection of IP but differences are still there (Mugabe 1994). As developed countries allow IP protection for patents over GMOs, most of the developing countries are opposing this and are not willing to extend patents to such subject matter. The duration of granting the patents also varies from one country to another, and the different countries have their different terms and conditions for disclosing IP-related information. Moreover, some of the countries are not interested to disclose the whole information about the resources. The developed countries like the USA and UK have developed a strong and strict patent system compared to developing and underdeveloped nations (Goldstein 1997). In some cases, particularly in least developed nations, the IP protection system has to implement by framing various norms.

IP protection in the case of TK of indigenous farming communities has opened a new dialogue to discuss on ethics issues. Most of the debates at international levels on these issues remained controversial and unresolved. This is due to the lack of adequate information about the TK and IP laws. The basic concepts about the TK and IP should be clear. The policymakers should focus at the international levels for its effective implementation. These issues should be discussed at the international level with world organisations, stakeholders and NGOs to resolve these complex issues. The Trade-Related Aspects of Intellectual Property Rights (TRIPS) agreement of the World Trade Organization (WTO) does not deal with the protection of TK and related knowledge. However, NGOs and the international organisations like the CBD are dealing to resolve and protect these issues. The holistic approach of the

CBD with NGOs is providing an intergovernmental forum to have a discussion on these complex issues to come out with a certain measure (WTO 2000).

The debates held at the national and international levels are oscillating between the two groups. One group favours the protection of TK and related knowledge under IP laws, whereas the second group is opposing it. They are considering TK and related knowledge as a public good and these should not be protected. Moreover, the indigenous people are depending on these goods for their livelihoods. The first group has the opinion of providing IP protection to TK and related knowledge which promotes the new innovations amongst the indigenous farming communities. This will also promote their economic growth and living standards. The indigenous people will be aware with the modern agricultural production technologies and their sustainable managements and utilisation. On the other hand, bringing TK and related knowledge under IP protection could provide some incentives and income to the preservation of these resources and PGR on sustainable manner. It is also expected that the developed nations and stakeholders can provide equal sharing of benefits arising out of the use of these resources. The developed countries have moral duties to provide equal shares to the indigenous people for the commercial use of the PGR and TK (Greaves 1996).

The other group which opposes the protection of IP to TK and related knowledge has expressed their views that providing the IPRs to these resources will be overexploited by the stakeholders and the developed countries. This will also abolish the social and traditional cultural issues of the indigenous people in the rural areas. The informal system of management of PGR and TK by the indigenous farming communities will lead to a decrease in genetic diversity in the agroecosystem. Moreover, the TK and related knowledge are considered as the inherited property of nature which has been passed on from one generation to another. The IP protection to these valuable resources of indigenous people will create a monopoly for the stakeholders and developed nations for commercialisation and overexploitations. The indigenous people will lose their rights over the TK and traditional practices of production and utilisation of resources (Agrawal 1995).

4.3.1 Protection of TK and TRIPS

Providing Intellectual Property Rights (IPRs) in TK has added new dimensions in TRIPS. The minimum standards have been set for all countries that follow the IPRs. After enforcing IPRs, the maximum barriers in international trade are removed. The protection of TK under IPRs should not provide a free flow of trades amongst the countries. The legitimacy related to TK is to be framed in such a way that IPRs should not provide any hindrance in international trade. A complete IPR system should be developed for all member countries of TRIPS (Dutfield 1997).

In the present legitimacy of the TRIPS system, it is not possible to protect the TK and related knowledge. Under this system, a patent applicant must disclose all the information regarding Article 29.1 of the TRIPS agreement, and this requires that a patent applicant discloses sufficient and clear information regarding the creation and innovation as per the normal conditions for patenting any invention. At the same time, the opponents should respond quickly and give detailed information about the losses caused by patenting of TK and related knowledge of indigenous people (Dutfield 1997). Due to lack of knowledge and update information about the IPR systems, the indigenous communities and local people cannot avoid the overexploitation of genetic resources and get the sharing of benefits. The indigenous people cannot monitor and implement the IPRs in their TK and related knowledge. Moreover, TRIPS does not provide the patenting of TK and related knowledge.

It is stated that the inventions related to TK and other knowledge cannot be patented under Article 27.2 of TRIPS. Some of the indigenous knowledge is related with human morality and their commercialisation may hurt the ethics values of indigenous people. It is important to protect the moral values of the local and indigenous people. In the TRIPS, the morality and ethics values of the indigenous people have not been discussed. It is mentioned that the innovations that cause any harm to human beings, animals and plants may be excluded from the patent. But some of the member states are providing the patents which harm the environment. Most of the developed countries are providing patents to GMOs and to altered genes and a modified organism which shows the qualities of a normal organism.

Earlier, the contributions made by the local and indigenous people were not recognised anywhere. Even the contributions made by the indigenous people in the form of providing important genetic resources as well as the basic materials for crop improvement programmes have not been recognised. Indigenous people are debarred from any kind of benefit sharing out of the use of their genetic resources. Although the indigenous farming communities have been conserving and maintaining genetic diversity and using the genetic resources on a sustainable basis, the stakeholders are not providing any incentives to them. Moreover, no particular law has been framed to safeguard the interest of these people. Most of the drugs today are developed from the conserved genetic resources of indigenous people who are maintaining these resources over the years. Moreover, the knowledge of indigenous people is not considered for IP, whilst the scientific knowledge of the researchers is being treated as patentable. Most of the TK of the local and indigenous people has also been commercialised and patented by the companies in developed nations. These companies have not provided any compensation to indigenous people. Now it is very difficult to get the compensation from these private firms.

These complex issues for the protection of TK under the IP system should be discussed at an international forum. The indigenous people should benefit from institutional arrangements and be financially strong to protect their rights on these resources. The IPR-related awareness programmes should be framed to provide the update information to the indigenous people (Dutfield 1997).

4.3.2 TK and Indigenous People in the Human Rights Agenda

In general, indigenous peoples' rights include rights to land, PGR and other genetic resources and experience developed through practices, self-determination and culture. All the indigenous peoples' rights are community based and their rights on any community items are collective in nature. Most of the local people in the rural areas have collective rights on the communities' property. Similarly, TK and related knowledge could not be owned by a single individual. Moreover, in the developed countries, the concepts of collective rights of the communities are not accepted. So it is impossible to implement and utilise the developed nations' concept of patenting TK and related knowledge (Donnelly 1989).

Most of the national and international rules and regulations are not recognising IPRs on TK and related knowledge. It has been debated at the national and international forums that the rights on indigenous knowledge are not the rights from human rights. The provisions under the Universal Declaration of Human Rights (UDHR) and the International Covenant on Economic, Social and Cultural Rights (ICESCR) can be interpreted to cover rights on TK and related knowledge. However, these cannot be used as legal tools for the protection of local and indigenous people's knowledge because of certain limitations. Indigenous and local knowledge of a community is a collective knowledge and all the individuals in the community have equal rights (Donnelly 1989).

4.4 International Institutions: Linkages and Multilaterals of PGR, TK and IPs

Nowadays, there are many international institutions involved in sectors of PGR, TK/IK and IPs. Because all these sectors are premises and fundaments of growth and development of the world, the United Nations (UN) and its units are interested to cover these areas. The UN focuses on main international issues such as promotion of sustainable development, human rights protection, upholding of international law and humanitarian aid delivery. Also, various programmes, funds and specialised agencies of the UN focus on specific global issues. At present, the FAO is mainly responsible for PGRFA. At the international level, all the international institutions have different aims and agendas for the welfare of humankind and preservation of biodiversity. There is linkage amongst different international institutions with regard to PGR and TK (Table 4.1). These institutions are responsible for providing food to everyone, conserving genetic resources, using these resources on a sustainable basis and equal sharing of benefits and IP protection of indigenous peoples' TK (UNESCO 2003).

The FAO is seen as a pioneer in fighting poverty and these institutions are considering, building and establishing multilateral mechanisms for accessing and sharing the benefits of PGR and relevant indigenous knowledge/TK to increase the capacity and productivity of global food systems as well as human rights in accessing food (FAO 2012).

Today PGR and TK as well as IP are cross-cutting topics related to many international institutions. Different global institutions are trying to cover their duties and mandate those that are regulated by functions and missions in the UN house. Under globalisation, these topics are the best emerging ones in the UN house and are key driving forces to monitor the interaction amongst units and specialised agencies of the UN. A consistent policy of units and specialised agencies in the UN house is needed to address the topics for accessing and sharing of benefits. PGR and TK could contribute and promote the development of both developing countries and developed countries (Posey and Dutfield 1996). Under the UN house, its units and agencies are cooperating to build mechanisms on accessing and sharing of benefits of PGR and indigenous knowledge/TK for sustainable development goals, including global food security and living in the planet.

4.4.1 International and Regional Perspectives

Historically, many indigenous people communities survived on their TK which is being used for their livelihood. Indigenous people developed different products by using their experience and traditional practices for their livelihood. In most of the Asian countries, local and indigenous communities meet up their basic needs from the products they developed and earn some money for their livelihood by using their knowledge. Even now, indigenous people maintain their health by using traditional medicines derived from herbs and trees. Due to vast utilisation of PGR and TK for the development of new products, a number of indigenous communities are facing challenges for their livelihood and survival. Most of the private companies are commercialising their products to earn money by overexploitation of PGR and TK. They are using these valuable resources without seeking the permission and sharing the benefits with the local and farming communities. Today most of the companies have developed their products by using the TK of the indigenous knowledge and traditional practices with the help of modern techniques. Moreover, the scientists are developing and modifying new products of indigenous knowledge in such a way that it is very difficult to provide the recognition and

Table 4.1 Linkage of international institutions with PGR and TK

FAO	UNESCO	WTO	WIPO	UNCEF (CBD)
Food and nutrient	Culture, science, education	Global trade	Intellectual property	Benefit sharing
ITPGR	Heritage	TRIPS	IPR	ABS
			PVP	
			UPOV	

Source: UNESCO (2003)

benefits to the indigenous communities from where such knowledge has been taken for the development of products (Tripathi 2003). The industries are protecting these newly developed products under IP protection without giving any benefits to the indigenous people.

The globalisation and commercialisation of PGR and TK are the more threatening and challenging concerns to important genetic resources. In order to protect these, policies should be framed at the national and international levels. These policies must concern with the IPR and benefit sharing of TK, and its related knowledge and benefits should be given to the local and indigenous communities. The international organisations related to IPRs have cooperated on different occasions to make some attempts to protect the TK and folklore. The WIPO developed the 'model provisions' for the protection of indigenous knowledge at the national level. These provisions were developed to address the concerns regarding illicit exploitation of expressions of folklore, and several countries utilised these model provisions to enact a national legal framework for providing such protection. Subsequently, there is a need to protect these resources at the international level with regard to the rapid increase and uncontrolled use of such expressions by means of modern technology. In 1997, it recognised the need of new international standards for the legal protection of folklore (Tripathi 2003). Pursuant to the recommendations of the World Forum, four regional consultations were organised by WIPO-UNESCO in 1999 to identify needs and issues, as well as proposals for future work, relating to expressions of folklore.

TK and its related knowledge are rarely protected under TRIPS and these resources can be protected in association with Geographical Indications (GI). Therefore, the TK can be protected through GI if it is associated with the sale of goods (Tripathi 2003).

4.5 Challenges to PGR, TK and IP

In the changing world, humankind not only uses sustainable natural genetic resources to meet the increasing needs but also ensures access and equitable and fair benefit sharing amongst stakeholders. However, these seem to be not carried out in fact. The following are the reasons to explain these causes.

4.5.1 Borderless PGR and TK

The world is known through continents, seas, nations, religious and ethnics. Plant genetic materials and relevant indigenous/traditional knowledge today are happening in complexity exchanges, evolution and development. They could be products of many communities and ethnic groups in many nations and continents. But nowadays, genetic materials and indigenous/traditional knowledge are known as properties of several communities and ethnic groups in nations. Thereby, genetic materials and indigenous/traditional knowledge are used until now in various communities and ethnic groups.

4.5.2 Ownership and Rights to PGR and TK

For plant genetic materials, ownership and rights to genetic materials will be established to all stakeholders owning these genetic materials, but rights will be divided in different levels to owners of genetic materials. Rights to stakeholders will be based on agreements amongst parties or be based on relevant international norms if agreements were built before developed international protocols/agreements.

4.5.3 Ease Lost when TK Was Not Fully Documented

Communities in many countries have their own storehouse of TK about genetic resources. These genetic resources are passed on from one generation to another one. Few people outside these communities are aware of such knowledge or how they use them, leading to threats to the community and the TK.

4.5.4 IP of Indigenous/Traditional Knowledge

WIPO has made a tremendous contribution to the development in the areas of PGR, TK and folklore and their bioprospecting in a sustainable manner. At the same time, the contribution of local people to the preservation and development of PGR for food and health cannot be ignored. To state that the entire array of issues relating to TK, genetic resources and folklore has economic implications is obvious. The developing and least developed countries have possessed rich biological diversity, PGR, associated TK and folklore, whereas the developed countries have the technological resources and the wherewithal to exploit these natural resources commercially. The issue thus becomes one of the sharing of benefits accruing out of such exploitation. The common good of humankind should, therefore, act as the catalyst for achieving an equitable distribution of benefits across the world. In this endeavour, WIPO is an ideal and a neutral platform and has both the credibility and the ability to harmonise the very complex issues involved to ensure justice for all (Tripathi 2003).

The twentieth century is the century of knowledge, shaped by science and technology, which has been able to penetrate the deepest secrets of the hitherto uncharted biological and genetic resources. In fact, it has been so incisive that even the mystery of the human genome has been unravelled. Practically, there is no aspect of our day-to-day life which is untouched by technology. The nuptials of Research and Development (R&D) efforts in science and technology on one hand and the untapped biological and genetic resources on the other give birth to new knowledge which could be productive, as it could be explosive. It is explosive in the sense that it might throw up many debatable issues (Tripathi 2003).

To get the rights, favours and profits of IP and indigenous/traditional knowledge, relevant stakeholders also carry out procedures/processes to get a grant of IP as instructions from local/central authorities in different nations. These are not easy for indigenous/traditional knowledge and indigenous communities, and normally, these terminologies and concepts have cultural value to the UNESCO network. Especially, indigenous/traditional knowledge related to agriculture in the rural areas is under the modules of the FAO. But to get the grant of IP and relevant benefits, community knowledge must be registered in the WIPO network. Therefore, it is essential to have a closer collaboration between the UNESCO and WIPO on indigenous/traditional knowledge (UNESCO 2003).

Despite such provisions, benefits of suppliers of this knowledge are almost not guaranteed by items on Access and Benefit Sharing (ABS) as analysed above. Sub-item c, item 2 and Article 60 of the Biodiversity Law stipulates that sharing of benefits with related parties, including the distribution of IPRs over invention, results in their TK copyrights. Therefore, the benefits are only shared if there are indigenous knowledge copyrights. However, the current legal regime does not stipulate about indigenous knowledge copyrights even if there is a general provision in Article 64 on indigenous knowledge copyrights. The member countries protect the TK. These member states also encourage the organisations to document the TK copyrights. The governments should help in guiding the methods of registration of copyrights of TK and its related knowledge. However, the governments have not issued any regulation guiding the registration of indigenous knowledge copyrights on PGR. In particular, there is no provision for determining the rights and interests of indigenous communities. Meanwhile, ministries and state agencies need to play a key role in protecting the rights and interests of indigenous communities who keep indigenous knowledge (Mugabe 1994).

4.6 Trends in PGR, TK and IP

In India, various suggestions like documentation of TK and its related knowledge, patenting of innovations and development of a sui generis system have been done prepared. To check the biopiracy of TK and associated knowledge, proper registration and documentation are required in order to provide benefits to the indigenous communities who are maintaining the resources for a long time. The registration and documentation of TK and associated knowledge could make the patent examinations as prior art to scrutinise the inventions based on TK. It is presumed that such registrations and documentations would help in tracing different communities with whom the benefits have been shared (Tripathi 2003).

The rapid development of modern biotechnology in the last decades creates the ability to use PGR to develop new products and acknowledges their contribution to human life, from essential medicines to methods to increase food security. Cooperation, exchange, exploitation and use of PGRFA as well as biological diversity in the future quickly increased amongst multilateral and bilateral institutions under globalisation. These are proofed throughout bilateral or multilateral agreements on PGR conservation and use for sustainable development goals amongst donors, international/regional research and development institutions and other relevant nations. A series of agreements on PGR conservation and use for food and agriculture are achieved for hunger eradication and poverty reduction but to reach IP in TRIPS is difficult to apply in the future. It is a complex of legal systems in countries globally. It is further difficult as disputes of IP in indigenous/traditional knowledge are occurring amongst parties (Juma 1989).

Although TK and cultural link to the PGR help in the conservation of the resource, nowadays TK related to genetic resources is being eroded. In some localities, the younger generation has less interest in the TK of their locality and ethnicity. If there is no protection method, many of the TK of ethnic people in use and in conservation will be lost.

4.7 Success Stories

The plant gene banks at international, regional and national levels are the clearest evidences of cooperation on PGR conservation and use for food and agriculture. The plant gene banks of CGIAR and the national plant gene banks in the USA, China and India or gardens such as the Royal Botanic Gardens in Kew maintained hundreds of millions of accessions of crop species and wild plant species. These plant gene banks provide genetic materials to create new plant genetic materials to meet the increasing needs of yield, quality, resistance and tolerance to various abiotic stresses. New genetic materials are contributed to sustainable hunger eradication and poverty reduction in least developed and developing countries. Furthermore, indigenous/traditional knowledge is applied to discover new pharmaceuticals or new materials.

For example, the Traditional Knowledge Digital Library (TKDL) has already documented more than 35,000 Ayurvedic systems of medicines. Whilst documenting the detailed information with regard to description, preparation methods and bibliography and their uses have been given in the patent form. The retrieval of these resources will be based on the Traditional Knowledge Resource Classification (TKRC) and International Patent Classification (IPC). The local names of the plant species are converted into scientific names and traditional Ayurvedic descriptions of diseases into modern medical terminology. TK has been documented in the digitised format which would help in preventing the use of patents which are already in the public domain.

4.8 Approaches and Solutions

4.8.1 International Standards/Norms

This approach should be considered and encouraged to be applied to issues of PGR, indigenous knowledge/traditional knowledge and IP at a global level with affiliation of multilaterals and stakeholders. International standards/norms are built through a multilateral mechanism in which involvement of key stakeholders to identify shortcomings and obstacles, concerns and disputes occurred in the future. From that, solutions are considered to develop suitable international standards/norms in the area.

4.8.2 Professional Morals and Social Responsibility for Sustainable Development

An approach is awareness raising of stakeholders in the area through education and training. The education and training will help individuals, organisations and institutions having activity in the area to understand clearly the rights and obligations in joining the area. This also helps the society and community to understand and act right to any stakeholders running out of regulation. These are just professional morals but also social responsibilities from both multiparties in policy, science, enterprises and society. Exchange and cooperation on the use of PGR and indigenous/traditional knowledge for sustainable development and for human rights should be continued to stimulate the global scientific and social community.

4.8.3 GMOs and Stories

GMOs provide prominent traits for agricultural, commercial and health sectors. GMOs reduce risks in these sectors. GMOs changed the view in food, agriculture and pharmacy. In this stage, GMOs such as golden rice, cotton, maize, soybean, etc., delivered the best solution to resistance/tolerance to help humankind in overcoming biotic/abiotic stresses or improving health. However, there are some concerns of GMOs that threaten the ecosystem and its diversity.

4.9 Futuristic Need and Importance of PGR and TK

Biodiversity in the world today is known as resilience of humankind to various climates. Under this topic, diversities of PGR and TK are specific products of different communities in various specific environments. They express adaptation, evolution and development of human communities.

4.9.1 Increasing Needs

PGR and TK are seen as a solution to meet the increasing needs of humankind in the present and future. Diversity gives multiple options to humankind and the community. Diversity found in the genetic resources is also known as a replaced/supplementary solution to resources.

4.9.2 A Multicoloured World

Under a changing world, biodiversity is a crucial factor to humankind which could adapt to climate changes in the future. These changes could cause from natural and social climates. At different developmental levels, ethnics and nationals will have various approaches and solutions to integrate and embed into globalisation and try to keep its independence which are distinguished features of culture and knowledge. These will create a multicoloured world under globalisation. At the PGR area as well as relevant indigenous/traditional knowledge and its IP, biodiversity plays a crucial feature under globalisation and the changing world.

4.9.3 Preparation and Response to an Unpredictable Future

With an unpredictable future, warning of risks and disasters is done by relevant international organisations. Before warning of any natural disasters and political risks, ABS in genetic resources and TK as well as implementing rights and obligation of IP protection will promote humankind and overcome threats as well as warnings of risk and disasters in the future.

4.9.4 Future Generations and Our Planet

PGR and TK are invaluable properties of vicar generations for us and future generations. They are the biggest and the most productive biological firms. They join all important substance cycles of living. The diversity of PGR also contributes to the diversity and existence of other kingdoms and the planet. Indigenous knowledge and IP are scoping terms in a changing world. These terms will contribute to developing a suitable mechanism for stakeholders which could help in accessing and sharing the benefits of PGR and TK and associated knowledge to manage and use them for sustainable development of humankind. Besides preserving and developing the TK value, documenting that knowledge to determine the reuse and effectiveness is imperative. The information collected from farmers, combined with on-site investigation, is used to analyse and assess the strategic direction for the conservation and efficient use of genetic resources for today and for tomorrow.

4.10 Conclusion

There is no gainsaying fact that PGR and TK are economic assets of a developing country. It is required to be recognised as economic assets of the holders such as local and indigenous communities and that they need to enjoy the same level of protection rights which are available to science- and technology-based innovations. The biopiracy of genetic resources and indigenous knowledge can be prevented by documentation of genetic resources and associated TK. The objective of getting positive rights for these resources will need a wider and complex consensus evolving process; therefore, it is necessary that different forums continue to work actively. At the same time, also we need to recognise the urgency of implementing the IP protection directions given by the international firms. For effective realisation of the objectives of preservation and protection, there is a need to take collaborative projects on setting up databases/registries. However, dissemination strategies need to be worked out carefully with adequate inbuilt safeguards so that possibilities of misuse and/or misappropriation are minimised. Acceptable and agreed international sui generis systems will have to be created, and there is a need to take a holistic view of the entire IP system which provides equity to the creator of knowledge.

References

Acharya R (1992) Intellectual property, biotechnology and trade, vol 4, Biopolicy International. ACTS Press, Nairobi

Agrawal A (1995) Indigenous and scientific knowledge: some critical comments. Indig Knowl Dev Monit 3:413–439

Correa CM (1998) Implementing the TRIPs agreement. General context and implications for developing countries. Third World Network, Penang

Donnelly J (1989) Universal human rights in theory and practice. Cornell University Press, New York

Dutfield G (1997) Can the TRIPS Agreement protect biological and cultural diversity? vol 19, Biopolicy International. ACTS Press, Nairobi

FAO (2012) Food and Agriculture Organization. The state of food and agriculture. ftp://ftp.fao.org/docrep/fao/009/a0800e/a0800e.pdf

Goldstein P (1997) Selected statutes and international agreements on unfair competition, trademark, copyright and patent. The Foundation Press, New York

Greaves (1996) Tribal rights. In: Brush, Stabinsky (eds) Valuing local knowledge: indigenous people and intellectual property rights. Island Press, Covelo

Helfer LR (2002) Intellectual property rights in plant varieties: an overview with options for national governments. www.fao.org/Legal/prs-ol/years/2002/list02.htm

Juma C (1989) The gene hunters: biotechnology and the scramble for seeds. Zed Books/Princeton University Press, London/Princeton

Mugabe J (1994) Technological capability for environmental management: the case of biodiversity conservation in Kenya. Ph.D. dissertation submitted to the University of Amsterdam, The Netherlands

Posey D (1991) Intellectual property rights for native people: challenges to science, business, and international law. International symposium on property rights, biotechnology and genetic resources, Nairobi, Kenya

Posey D, Dutfield G (1996) Beyond intellectual property. International Development Research Center, Ottawa

Tripathi SK (2003) Intellectual property and genetic resources, traditional knowledge and folklore:

international, regional and national perspectives, trends and strategies. J Intellect Prop Rights 8:468–477

UNESCO (2003) Language vitality and endangerment. Document submitted by *Ad Hoc* Expert Group on Endangered Languages to the International Expert Meeting on UNESCO Programme Safeguarding of Endangered Languages, UNESCO Paris

WIPO (2010) World Intellectual Property Organization, http://www.wipo.int

WTO (2000) International trade statistics. https://www.wto.org/english/res_e/statis_e/stat_toc_e.htm

Farmers' Rights to Plant Genetic Resources and Traditional Knowledge for Livelihood

5

Vijai Pal Bhadana, Shiv Datt, and Pawan Kumar Sharma

Abstract

Farmers' Rights are basically about enabling farmers to conserve, develop and utilise plant genetic resources and traditional knowledge and recognising and rewarding them for their contribution to the global genetic pool. The Convention on Biological Diversity recognised the sovereign rights of countries for use of these resources. It affirmed that the owners have rights of their innovations, associated traditional knowledge and traditional practices whether or not it can be protected by intellectual property rights. It is an obligation for the governments to protect the ownership of the resources and associated knowledge through legal policy or law. Subsequently, the International Treaty on Plant Genetic Resources for Food and Agriculture ensured the conservation and sustainable use of plant genetic resources and the equitable sharing of benefits from their use. Farmers' Rights are a foundation stone of this Treaty, as their realisation is a precondition for the conservation and sustainable use of the world's vital plant genetic resources and traditional knowledge. The Treaty recognises the enormous contribution that the farmers of all regions have made, and will continue to make, for the conservation and development of these resources throughout the world. It further stipulates that governments to protect and promote Farmers' Rights and choose the measures to do so according to their needs and priorities.

V.P. Bhadana, Ph.D. (✉)
Crop Improvement Section, ICAR,
Indian Institute of Rice Research (IIRR),
Rajendranagar, Hyderabad 500030,
Andhra Pradesh, India
e-mail: vp_pb@yahoo.co.in

S. Datt, Ph.D.
Indian Council of Agriculture Research (ICAR),
IP & TM Unit, Krishi Anusandhan Bhavan-I, Pusa,
New Delhi 110012, India
e-mail: shivdatts@rediffmail.com

P.K. Sharma, Ph.D.
ICAR-National Bureau of Agriculturally Important
Microorganisms, Mau, Uttar Pradesh 275101, India
e-mail: pawan112000@gmail.com

R.K. Salgotra, B.B. Gupta (eds.), *Plant Genetic Resources and Traditional Knowledge for Food Security*, DOI 10.1007/978-981-10-0060-7_5

5.1 Introduction

Traditional Knowledge (TK) and Plant Genetic Resources (PGR) have contributed significantly to technological advancement of society and improvement of human livelihoods.

PGR and TK also play a fundamental role in many contexts. In agriculture, the conservation and development of genetic resources are crucial for the maintenance and the improvement of crop varieties to meet the food security (FAO 1996). Moreover, the TK of indigenous and farming communities has played an important role in conservation and development of genetic resources. Genetic diversity of agricultural crops is the very basis of farming system.

Plant genetic diversity is probably more important factor for the improvement and development of farming than any other single environmental factor. The plant genetic diversity particularly for smallholder farmers has the vital role for preventing the spread diseases and insect pests. The genetic diversity also helps in securing the food security, livelihoods and poverty eradication. Ever since, the farmers have worked to develop plant genetic diversity particularly PGR and TK through saving, selecting, exchanging, selling and improving seeds and propagating material. However, these routine practices are being restricted all over the world (Andersen 2008). Recently, Plant Breeders' Rights (PBR) serve to restrict the use of PGR in the form of seeds and plant materials protected by such rights. In many countries, traditional crop varieties are excluded from the market by prohibiting exchange or sales of such varieties by seed laws and seed certification rules. As a result, the genetic diversity that existed in agriculture 100 years ago has been eroded, and genetic erosion is still continuing at a rapid pace (FAO 1998).

Farmers' Rights are recognised as the rights arising from the past, present and future contributions of farmers in conserving, improving and making available plant genetic resources for food and agriculture (PGRFA) (FAO 1998). To formulate these rights, it took years of debates and negotiations in the context of a legally binding international agreement. In 2001, it was concluded with the adoption of the International Treaty on Plant Genetic Resources for Food and Agriculture (ITPGRFA), which underscores the need to promote and protect Farmers' Rights at both national and international levels. Certain measures are suggested in Article 9 of the Treaty, covering the protection of TK, benefit sharing and participation in decision-making. As per this Treaty, farmers have a right to save, use, exchange and sell farm-saved seeds and propagating material, but particular directions for implementing these rights are not mentioned (Andersen 2005). The responsibility for implementing Farmers' Rights at the country level lies with the government according to their needs and priorities, as appropriate and subject to national norms (Fig. 5.1).

The nature and terms under which PGR can be accessed have changed dramatically in recent years. The free flow and trade of PGR have been replaced by a wide range of national and international legal instruments that aim at regulating access, control and use of genetic biodiversity and to protect TK. Of late, the attempts aimed at extending Intellectual Property Rights (IPRs) to PGR are of particular concern. The possibilities for appropriation of genetic resources that support food and agriculture production have made PGR valuable assets for their potential commercial use (Bala Ravi 2004). The exchange of these resources under commercial terms could potentially benefit the many developing countries that are rich in biodiversity. The traditional communities also reap the gains from the use of genetic resources. However, misuse of PGR and associated knowledge may take place if access to these resources is not managed properly at the international level (Helfer 2002). The intellectual property (IP) system is of particular concern for developing countries with regard to IPRs, PGR and TK as it allows for rights to be claimed over indigenous knowledge even when changes made to it are minor, while at the same time, local communities are unable to protect their TK or offensive mechanisms to claim such indigenous knowledge.

Fig. 5.1 Traditional method of producing quality paddy seeds in mid hills of Himalayas

5.2 Farmers' Rights

The purpose of Farmers' Rights is to ensure full benefits to farmers and support the continuation of their contributions. These rights are in terms of species or varieties of plants cultivated by the farmers. Their purpose is to encourage farmers and their communities to conserve, utilise and improve PGR (FAO 1996). Farmers' Rights are basically enabling the farmers to contribute their work as innovators of PGR and rewarding them for their contribution in conserving genetic resources. Thus, these Rights are critical to ensure the conservation and sustainable use of PGRFA and consequently for food security. Farmers Rights also play a vital role to fight against rural poverty in developing and under developing countries.

The concept of Farmers' Rights provides a measure of compensation to formal IPRs and patents. They compensate for the latest innovations with little consideration of the fact that in many of these innovations are based on the recently accumulated knowledge and inventions that have been made over millennia by generations of men and women in different parts of the world (FAO 1996). All modern plant varieties have genes from farmers' traditional varieties, landraces or wild crop relatives. These are the women and men who have historically provided the germ plasm upon which scientific plant breeding is contributed. All the germ plasm is provided by small-scale subsistence farmers to plant breeders whose varieties are then developed and adapted for medium- to large-scale commercial farmers (Spillane 1996).

In its Article 9, the ITPGRFA Treaty recognises the enormous contribution that the indigenous communities and farmers of all regions of the world have made and will continue to make for the conservation and development of PGR which constitute the basis of food and agriculture production throughout the world. It gives governments the responsibility for enforcing Farmers' Rights and lists measures that could be taken to protect and promote these rights.

The protection of TK relevant to PGRFA:

(a) The right to equitably participate in sharing of benefits arising from the utilisation of PGRFA and agriculture
(b) The right to participate in making decisions relating to the conservation and sustainable use of PGRFA at the country level

The International Treaty also recognises the value of giving support to the farmers and local and indigenous communities for their efforts in the conservation and sustainable use of PGR, including through a funding strategy. As per this strategy, in developing countries, priority will be given to the implementation of plans and programmes that have been agreed upon for farmers, especially in least developed countries and in countries having economies in transition, who are conserving and sustainably utilising PGRFA (Sahai 2003).

The contributions of indigenous communities to the sustainable conservation and regeneration of PGR and TK have been substantial, and it has been widely accepted that their contributions should be recognised. However, a number of key questions remain such as how to recognise and attribute a true value to these contributions. Furthermore, there has been much debate at the national and international levels on how to activate or implement 'Farmers Rights'. The resource poor rural people are dependent on small-scale farming in developing countries (Spillane 1996). Both men and women farmers have contributed to domesticating, developing, conserving and making available PGR and other natural resources to which they have access to earn their livelihood and ensure the wellbeing of their family through the provision of basic requirements such as food, fuel and water and income from the land (Spillane 1996).

5.3 Farmers' Rights to Use and Exchange of Farm-Saved Seed

The International Treaty is unclear on Farmers' Rights to save, use, exchange and sells farm-saved seed. Section 9.3 of the Treaty does not give much direction on Rights (Article 9 on Farmers' Rights) that farmers have to save, use, exchange and sell farm-saved seed in accordance with national law and as appropriate. The preamble sets out that the rights of farmers to save, use, exchange and sell farm-saved seed and other propagating material are very important. Since no specific rights are mentioned in the Treaty, the preamble is not quite clear on this point. It is important that farmers be granted rights; each country has the right to define these rights legally as per its laws. International commitments restrict the freedom that countries have to define such legal space for farmers. Most countries are members of the World Trade Organization (WTO) and are, therefore, required to implement the WTO Agreement on Trade-Related Aspects of Intellectual Property Rights (TRIPS). WTO member countries are required by the TRIPS to protect plant varieties either by patents or by sui generis system or a combination (Correa 1998). The limits to a sui generis system and the meaning of an effective sui generis system are not explicitly defined in the text which means the countries have to provide some sort of plant breeders' rights. According to the Union for the Protection of New Varieties of Plants (UPOV), the way to comply with the provision of an effective sui generis system is to adopt the UPOV Convention model. There are several proponents of this stand. The UPOV recognises the comprehensive rights of plant breeders to save, reuse, exchange and sell seeds. It is still possible to enable small-scale farmers to save and reuse seeds within strict limits. Farmers are totally prohibited from exchanging and selling seeds. All this applies to seeds protected with PBRs and not to traditional varieties. There has been opposition to UPOV model from some countries, and many organisations fear that UPOV membership would be detrimental to the rights of their farmers to save and share propagating material. The TRIPS Agreement provides only minimum standards, and there is scope for the development of solutions that have compatibility with the demand for Farmers' Rights (Helfer 2002). WTO member countries must therefore meet their TRIPS obligations regarding PBRs while at the same time creating the necessary legal space for the realisation of Farmers' Rights under the ITPGRFA Treaty.

5.4 Protection of Farmers' Rights for PGR and Indigenous Knowledge

The main objective from the perspective of Farmers' Rights would be to grant all such rights to farmers. It means that farmers will have the right to save, use, exchange and sell farm-saved seed, whether from varieties protected with IPRs or not. Plant breeders need to be compensated for their efforts, and other issues of plant health concerns are also to be resolved. The ultimate success story would tell about a country where farmers have all these rights. India is one country which has the most extensive legislation on this topic in the world. In most of the other countries that have legislated on plant variety protection (PVP), Farmers' Rights are more limited, often circumscribed by acts of legislation, such as PVP acts and regulations concerning seeds and seed certification (Leskien and Flitner 1997). There is a need to have a regulation that is less stringent or avoids the adoption of a stricter regulation.

5.4.1 TK Related to Agro-biodiversity

TK related to agro-biodiversity is vital to understand the properties of plants, their uses and cultivation methods. One way to protect and promote Farmers' Rights, as given in Article 9.2 (a) of the International Treaty, involves the protection of TK relevant to PGRFA. The farmers' varieties and associated TK should be documented, and seeds preserved in gene banks should be shared so that these precious resources do not become extinct. However, delegates voiced their concern about the legal status of such collections (Yeshitela 2001). If readily available, the companies and other commercial players can use them without obtaining Prior Informed Consent (PIC) from the farmers or entering into benefit-sharing arrangements with them, and there is apprehension that local communities might lose control of their PGR and TK. This points the dilemma between sharing seeds and TK to avoid extinction – and protecting it against misappropriation. This fear basically hinders conservation work aimed at enhancing farmers' varieties and fortifying their seed systems – which is vital to the future of our plant genetic heritage (Andersen 2008).

Keeping in view these considerations, an ultimate objective of activities aimed at protecting TK related to agro-biodiversity would be to document such knowledge and ensure its free sharing amongst farmers. One challenge in registering and documenting traditional crop varieties lies in the genetic heterogeneity of these varieties (Shiva and Jafri 2003). Problem arises due to difficulty in describing them as varieties and there is fear of misappropriation. For a plant breeder to be granted PVP, it is sufficient to discover a variety and develop it in terms of genetic purification. In such situations, farmers will not be in position to challenge such a right, if the prior existence of the variety is not documented. For that reason, developing improved methods of documenting traditional varieties can represent important achievements for protecting TK against misappropriation as well as against extinction.

5.4.2 TK Related to Farmers' Rights

Most of the countries in South Asia such as Bangladesh, Bhutan, Chile, India and Vietnam have taken account of the protection of farmers' TK relevant to PGRFA in their legislation. Several countries have provisions pertaining to the protection of farmers' TK in their regulations on access to genetic resources such as Ethiopia and the Philippines. Some countries have included TK protection in legislation pertaining to indigenous peoples, as in Peru (Tamiru and Manda 2002). India also has provisions in its PPV&FR legislation on PVP. However, results with regard to implementing such legislation are not very encouraging, and thus there are few successes to highlight in this report. The legislation is often aimed at protection against misappropriation of TK, and there are number of measures that cannot be legislated but which aim at protecting TK from extinction. In the Philippines, a local farmers' association has documented and facilitated the sharing of the rice varieties developed by the community. Many of these varieties have been developed through participatory plant breeding and are further improved every season. Farmers' knowledge concerns not only older varieties and traditional practices but also expertise in selecting the best material for further propagation and improving varieties. Agricultural history shows that seeds and other propagation material were considered part of the 'common heritage of mankind', and this idea has dominated to this day. This open access to PGR was a central factor in the spread and development of crops globally and has played a key role in laying the foundations for scientific research and the development of modern high-yielding varieties (Tamiru and Manda 2002). This led to the development of the commercial seed industry and, with it, the demands for private property rights to PGR.

5.4.3 Equitable Benefit Sharing

To protect and promote Farmers' Rights as per recommendations of the International Treaty concerns the right to participate equitably in the benefit sharing arising from the utilisation of PGRFA (Article 9.2 [b]). However, in the Treaty, in Article 18 on the multilateral system on ABS, the most important benefits are facilitated access to PGRFA, the exchange of information, access to and transfer of technology, capacity building and the sharing of monetary and other benefits arising out of commercialisation (Shiva 2005). Moreover, it is specified that under the multilateral system, the benefits accruing from the use of PGRFA should flow primarily to farmers in all countries, especially in developing countries and other countries, who conserve and sustainably utilise PGRFA.

All these provisions are about the multilateral system and do not directly relate to the provisions on Farmers' Rights in the International Treaty, and they point to benefit sharing which is relevant for interpreting Article 9.2 (b) on benefit sharing as a way to protect and enhance Farmers' Rights. Monetary benefits are a form of benefit sharing. Second, we see that benefits are not only to be shared with those farmers who grow plant varieties that are utilised by commercial breeding companies, but farmers in all countries are conserving and ensuring sustainable use of agro-biodiversity. This approach has been prevalent in the FAO ever since Farmers' Rights and benefit sharing were first recognised officially in 1989 (FAO 1998).

In seeking to operationalise the concept of benefit sharing with regard to Farmers' Rights and based on the international stakeholder survey on Farmers' Rights (Andersen 2005), the following goals could apply:

(a) Ensuring that incentive structures in agriculture are weighted towards farmers who are conserving and sustainably using PGRFA at an equal footing with, or more than, farmers are practising monoculture of genetically homogeneous plant varieties. Such incentives may include supporting farmers through extension services, provision of loans on favourable terms for the purchase of farm animals and other required inputs, help to market products from diverse varieties and other infrastructure measures.
(b) Provision of reward and support systems to enable farmers to benefit significantly from their contributions to the adding value to the crops they grow and conserve for improved livelihoods and increased income. Many small-scale programmes and projects show the enormous potentials in this regard – such as community gene banks in India, dynamic conservation coupled with participatory plant breeding in France, participatory plant breeding, farmers' field schools and activities aimed at capacity building and marketing. Today, these benefits are achieved mostly through programmes taken by non-governmental organisations (NGOs), inter-governmental organisations (IGOs) and some extension services that only have a limited number of farmers. A major challenge is to scale up these activities so that all farmers engaged in the maintenance of agro-biodiversity can participate in the sharing of these benefits.

(c) Ensuring recognition of farmers' contributions to the global genetic pool and to express that these contributions are valued by society. Proponents claim that farmers should be granted IPRs on an equal footing with breeders, as a matter of fairness. Others hold the view that such a system would discourage farmers from sharing seeds in the expectation that these could benefit them economically. Such a situation could be harmful to traditional seed systems and could negatively affect Farmers' Rights to seeds. Another way to recognise farmers' contributions could be to provide some sort of remuneration to farmers who register varieties in seed catalogues for free use amongst other farmers. Another way of recognising farmers and farming communities is through awards for innovative practices, as has been done in several countries.

5.5 International Agreements and Farmers' Rights

The relationship between agriculture, PGR, TK and IPRs is a debatable and ongoing issue in several multilateral institutions. These include the WTO, the WIPO, the CBD and the FAO. Several international agreements are dealing with abovementioned issues, and thus their treatment of PGR and TK is interrelated. These international agreements include the CBD, the TRIPS Agreement, the UPOV Agreement of new varieties, the International Undertaking on the Utilization of Plant Genetic Resources for Food and Agriculture and the ITPGRFA Agreement. The fact that several instruments touch on issues related to PGR and TK ultimately complicates their implementation and allows for different interpretations of the obligations (Rahmato 2004). Developing countries, supported by several NGOs, have highlighted the need to create rationality between the various international agreements. However, to date the negotiations are continuing, with few results with regard to introducing measures for the defensive protection that developing countries seek in order to prevent the misappropriation of genetic resources and TK.

The relationship between the TRIPS Agreement and the CBD is one of the most debatable. While the TRIPS Agreement is seen by many developing countries and NGOs to be in tension with the CBD, many developed nations hold the view that there is no conflict between the two instruments. NGOs emphasise the importance of implementing the CBD and harmonising the various international agreements that deal with PGR and TK. To ensure the sustainable use of biological resources and prevent their misappropriation, and for the protection of TK, the developing countries are calling for harmonising the TRIPS Agreement with the CBD. The USA argues that the TRIPS and CBD are sufficiently flexible to carry out their parallel implementation on the national level in a nonconflicting manner and that an explicit harmonisation is therefore unnecessary (WTO 2000).

Both TRIPS and the CBD touch on issues concerning genetic resources and intellectual property (IP). The CBD deals with access to genetic resources and creates an obligation for the fair and equitable sharing of benefits emanating from the utilisation of genetic resources on mutually agreed terms and informed consent. The TRIPS Agreement allows IPRs to be extended to genetic resources. The Agreement obliges WTO members to protect plant varieties, through a patent or through a sui generis regime or through a combination of both. In order to create rationality with the CBD, the amendment to TRIPS that developing countries and NGOs envisage would primarily aim to address concerns about biopiracy.

5.5.1 ITPGRFA and Farmers' Rights

The ITPGRFA Treaty aims at the conservation and sustainable use of crop genetic resources, their accessibility and the sharing of benefits arising from their use. The contributions of farmers in conserving and developing agro-biodiversity and ensuring its availability are recognised

(Andersen 2008). As per the Treaty, it is the responsibility of the national governments to realise Farmers' Rights, and the governments are free to choose measures according to their own needs and priorities. There are three potential measures to protect and promote Farmers' Rights that are listed: protection of TK relevant to PGRFA, equitable sharing of benefits arising from the utilisation of such resources and participation in decision-making related to the conservation and sustainable use of these resources. Besides, the rights of farmers to save, use, exchange and sell farm-saved seed are also mentioned. Creation of awareness is very important to give impetus to the realisation of Farmers' Rights as well as to deal with the obstacles and challenges.

Farmers' Rights are essential for maintaining plant genetic diversity, which is the cornerstone of all food and agricultural production in the world. Realising Farmers' Rights means enabling farmers to preserve and develop crop genetic resources and incentivising them for their important contribution to the global genetic pool. The realisation of Farmers' Rights recognised in the ITPGRFR is therefore closely linked to sustainable development. Article 9.1 of the Treaty recognises the contribution of indigenous and local communities (ILCs) and farmers towards the conservation and development of PGRs. Article 9.2 identifies three measures to protect and promote Farmers' Rights: (a) protection of TK relevant to PGRFA, (b) the right to equitably participate in sharing benefits from the use of PGRFA and (c) the right to participate in national decision-making on conservation and sustainable use of PGRFA. The overall objectives of the Treaty on the conservation and sustainable use of PGRFA and related provisions on in situ conservation and sustainable use should guide the implementation of Farmers' Rights.

Article 5.1 of the Treaty supports the farmer and community efforts to manage and conserve PGR on farm and in situ conservation of wild plants and wild relatives of crop plants for food production, including the efforts of ILCs. Article 6 promotes the sustainable use of PGRFA through appropriate policy framework and legal measures, which may include fair agricultural policies that promote diverse farming systems. The protection of Farmers Rights is essential to enable in situ conservation of PGRFA and their sustainable use. The Nagoya Protocol on access to genetic resources and benefit sharing under the Convention on Biodiversity also requires countries to take measures to ensure equitable benefit sharing from the use of TK associated with genetic resources and genetic resources held by communities. It emphasises the need for prior and informed consent of communities to access TK and genetic resources where communities have the right to grant access, based on mutually agreed terms. It also emphasises the need for customary laws, community protocols and procedures to be considered into account when implementing its provisions on TK and encourages support for the development of community protocols for access and benefit sharing relating to TK (Ayalew et al. 2005).

5.5.2 Farmers' Rights Under TRIP

The growth of agriculture, biotechnology, seed and drug industry relying primarily on biodiversity and, in many cases, associated TK systems has alerted the biodiversity and TK-rich communities and countries around the world. There is a feeling that the outside users of local and indigenous knowledge as well as genetic resources have not fulfilled their economic institutional and ethical responsibilities towards knowledge and resource providers. In the changing global scenario, most Asian countries have to deregulate and also liberalise their economy to gain access to international capital. The pressure on PGR is increasing, but the livelihood options of the local communities and marginalised people in many cases have not improved. Under such a situation, the concerns about asymmetry in sharing of benefits through the use of TK and biodiversity resource are bound to become more acute and urgent. The NGOs and other civil society continue to level charges of biopiracy against various international companies and research organisations because of lack of reciprocity in sharing of benefits. Tribal farmer communities conserve various kinds of genetic resources, and these resources provide very useful inputs into seed, biotechnology and drug industry. Following are the issues which need to be tackled while revising TRIPS:

1. A registration system is required at national and international level to protect the land races, and the local communities need to be incentivised so that they reveal the knowledge they have about these plant varieties or local herbs.
2. The communities represented by the village councils or their federation considered for the purposes of the property rights as body corporate should have the power to protect the community knowledge.
3. In cases where the land races have been documented and included in the national or international gene banks, the biodiversity users have the responsibility to share part of the benefits so that rewards for conservation are available to the communities. It has to be appreciated that ex situ gene banks contribute to the cause of conservation, but these cannot be a substitute for in situ conservation.
4. The new uses of existing diversity should be subject to registration and filing of patents. New use patents are not allowed by many countries. They should reconsider their position for the sake of empowering local communities and to take benefits from this provision.
5. In so as the right to share benefits from commercial use of land races is concerned, a longer duration of protection of more than 20 years may be considered.

The public sector breeding has depended on access to the collection in gene bank and may get affected if every user has to approach the community to take prior permission to use the germ plasm which they had originally collected. In many cases, this may not be impractical. In gene banks, the passport data sheets do not include large number of research institutions and have no information about the village or the local community from where the seeds were collected. In addition, the communities themselves exchange genetic material for their own use. The efforts to create and conserve biodiversity may suffer unless and until exchanges for local self-use public purpose are excluded from the requirement of any need to take permission from the originating community.

5.6 India's Protection of Plant Varieties and Farmers' Rights Act

The Protection of Plant Varieties and Farmers' Rights (PPV&FR) Act of 2001 in India is the most far-reaching legislation with regard to establishing Farmers' Rights to save, use, exchange and sell farm-saved seed. A unique aspect of the Act is that it confers rights on breeders, farmers and researchers. The Act recognises the farmer as cultivator, conserver and breeder. A farmer who has bred or developed a new variety shall have the right to register it and be entitled for other protection in like manner as a breeder of a variety under this Act. The farmers' variety can be registered if the application contains declaration as specified in clause (h) of subsection (1) of Section 18. Moreover, a farmer who is engaged in the conservation of landraces and wild relatives of economic plants and their improvement through selection and preservation shall be entitled for recognition and reward from the gene fund. The Act establishes nine rights for farmers (Bala Ravi 2004), of which the most important in this regard are the right to seed and the right to compensation for crop failure (Art. 39):

1. The provisions on the right to seed specify that farmers are entitled to save, use, sow, resow, exchange, share and sell farm produce, including seeds of varieties protected by plant breeders' rights. However, they cannot sell seeds of protected varieties as branded packages. This is a very liberal legislation that allows farmers all the customary rights they previously enjoyed.

2. The Act seeks to protect farmers from exaggerated claims by seed companies regarding the performance of their registered varieties. The breeder is required to reveal to farmers the performance of the variety under given conditions. If the variety fails to perform as claimed, farmers may claim compensation from the breeding company through the authority which administers the Act.

This Act not only protects the rights of farmers to save, use, exchange and sell farm-saved seed, but it ensures that these seeds are of good quality or at least that farmers are informed about the quality of seed they buy. In addition, safeguards are provided against innocent infringement by farmers (Bala Ravi 2004). Farmers who unknowingly violate the rights of a breeder are not to be punished if they can prove their lack of awareness about the existence of such a breeder's right (Article 42).

Ensuring Farmers' Rights to save, use, exchange and sell seed is a success with regard to this component of Farmers' Rights, as these rights are enforced through the Act. Whether the provision on compensation in case of crop failure can be implemented in practice is a moot point, as there have been no cases so far. On the whole, India's PPV&FR Act is the most advanced in terms of Farmers' Rights to save use, exchange and sell seed to date. It applies to all farmers in India and to all crop species (Ramanna 2006). So far, India is the only country where a law has been passed establishing and securing Farmers' Rights to save, use, exchange and sell seeds.

The most important lesson for others is that it is possible to uphold Farmers' Rights to save, use, exchange and sell farm-saved seed within the framework of legislation on plant variety protection. India is a member of WTO and TRIPS and thus obliged to provide for the protection of plant varieties. With its 2001 legislation, the country has become TRIPS compliant on the protection of plant varieties. Other countries should pass similar laws without neglecting their obligations towards the TRIPS Agreement (Pant 2008). It should be mentioned, however, that India has applied for UPOV membership on the basis of its 2001 Act. The application was made in 2002, and the country has so far not been granted such membership, as its 2001 Act does not comply with the strict requirements of UPOV. Furthermore, a massive and enduring advocacy is needed to succeed with demands for Farmers' Rights in the context of the development of laws for plant variety protection. In India, MS Swaminathan Research Foundation (MSSRF) has initiated a programme for capacity building amongst farmers, grass-roots democratic institutions and non-governmental and community organisations with the assistance of the government, to enhance the implementation of Farmers' Rights as provided for in the Protection of Plant Varieties and Farmers' Rights Act of 2001 (Swaminathan 1994).

5.7 Equitable Benefit Sharing of PGR and TK

Equitable benefit sharing at community level is very important to ensure that conflicts over benefits are minimised amongst communities and that benefits and conservation incentives are widely distributed, rather than captured by local elites.

5.7.1 Recognising Collective Rights

Knowledge is believed to come from God and so is always considered to be collectively held in the community, even if it can be attributed to a particular inventor or provider in the community, in which case both collective and individual rights should be recognised. Knowledge and related genetic resources are widely shared within and between communities, and this sharing is important to sustain traditional subsistence economies in often difficult environments, and no individual can survive based on their knowledge and resources alone. Sharing allows farmers to innovate further and add to the stock of knowledge and genetic resources. In this context, recognising individual or exclusive rights would not only threaten livelihoods but also the innovation systems which sustain and enrich PGR and TK (Feyissa 1999).

5.7.2 Understanding the Drivers of Traditional Innovation

The meausres to protect the indigenous knowledge rights aim at providing incentives for this innovation. This is the original purpose of IPRs and is particularly important for protection of this knowledge given its rapid loss. It is estimated that 50–90 % of all indigenous knowledge will extinct or threatened by 2100 (UNESCO 2003). Indigenous knowledge relating to PGRFA is a subset of this knowledge and needs to be sustained as part of wider TK and livelihood systems. In developed countries, IPRs' protection and incentivise innovations are through financial benefits. These financial benefits and incentives alone are unlikely to be sufficient to promote traditional innovation which could undermine the traditional cultural values, collective/sharing practices and dependence on natural resources that sustains this knowledge.

Land is an essential resource for traditional subsistence economies that depend directly and substantially on natural resources for meeting basic needs for livelihood (food, agriculture, healthcare, income and cultural/religious needs). Landscapes provide access to wild gene pools and wild plants for food production and healthcare and to sacred wilderness areas for sustaining spiritual beliefs. Landscapes not only have economic value but are tied to cultural identity and spiritual beliefs. They also provide the physical space for sharing and conservation practices based on customary laws. The wider the area for sharing and exchange between farmers and communities, the richer the genetic and intellectual basis for further innovation.

5.7.3 Protecting 'Bundles of Rights'

The policies and laws to protect indigenous knowledge should not only protect this knowledge but also rights to the associated genetic resources, landscapes, cultural and spiritual values and customary laws in order to protect TK from loss as well as misappropriation. Most policies only protect the intellectual component of knowledge systems but not the biological, cultural and landscape components that sustain TK and innovation systems. They separate rights over TK which are vested in communities and rights over genetic resources, which vested in governments. The knowledge forms part of genetic resources which have been domesticated, improved and conserved by farmers/communities, and these two are used in traditional practices and transmitted together. In the holistic indigenous worldview, knowledge and genetic resources are inextricably linked and cannot be separated (Brush 2003). Cultural and spiritual values and customary laws also play an important role in sustaining TK and PGRFA but are being eroded by various processes and policies.

5.7.4 Recognising Customary Laws

The procedures to protect TK should be based on the customary laws and practices of indigenous and local communities for protecting and sustaining TK, rather than existing IPR standards. Community knowledge related to seeds, farming and everyday healthcare is openly shared, while specialised or sacred knowledge is restricted to elders, family or clan members. The obligation to share is particularly strong in relation to seeds, and the principle of reciprocity means that the more seeds are shared, the more seeds are received. Hence, reciprocity promotes diversity of resources, and the wider the space for sharing and exchange, the greater the genetic diversity.

5.7.5 Supporting Local As Well As National Measures

National measures alone are unlikely to be enough to effectively protect TK or farmers rights and need to be complemented by measures at local and community level, particularly if Farmers' Rights are to be a tool for in situ conservation. The threats to TK and PGRs and drivers of change affect communities at local level and hence require local as well as national responses to effectively address them. At the same time, protecting indigenous knowledge in situ may also the best way to protect the rights of communities over their TK, because it enables communities to strengthen control and management of TK and customary laws.

5.7.6 Promoting Reciprocal Exchange of PGRs

As explained above, financial benefits are not the only or even the best type of benefits to contribute to the conservation and sustainable use of PGRFA by subsistence farmers. Communities have shared many genetic resources with outsiders over the years but received nothing or little in return. According to the customary law of reciprocity, they expect genetic resources in return for those shared. Genetic resources have been eroded and are increasingly critical to enable farmers to adapt to climate change.

5.7.7 The State of Support to Farmers in the Country

The bulk of genetic diversity of agricultural crops is still on the farms of local farmers, who continue to conserve, manage, improve and maintain the genetic diversity in its dynamic state. These resources and farmers' efforts to manage them are important to ensure the availability of the resources for use. The value of these resources is not limited to their use within the country as shown by the contributions to world agriculture of early collections by Harlan and Vavilov. Sustainable conservation and improvement of the genetic diversity of the resources are required for their development and use, which cannot be achieved without practical efforts of local farmers and indigenous communities. The farmers are supported in various ways. There are governmental rural development programmes such as healthcare services, water supply, infrastructure and school building, seed and input supplies as well as bilateral and NGO projects that fulfil the same goals. NGOs implement important grassroots activities such as environmental and socioeconomic development programmes to promote reforestation and agroforestry, soil protection, on-farm conservation, strengthening of seed supply systems and market promotion for farmers' products. These projects have restored lost optional crops and degraded ecologies. Dissemination of appropriate agricultural technologies for increasing productivity and improving sources of animal fodder and energy is amongst the areas that still need more strengthened support from which farmers can benefit (Buckland 2004).

Such programmes, whether implemented by governmental organisations or by NGOs, all help farmers and can be considered as contributions towards Farmers' Rights. Although these projects contribute to improve farmers' livelihoods, they are not that directly framed within the context of promoting the concept of Farmers' Rights. However, even if these programmes are implemented within the context of the promotion of Farmers' Rights concept, much of what can be done is what these programmes are doing now. It would be useful if forums are created to raise awareness on the concept of Farmers' Rights within NGO circles, in order to enable them to observe and promote of Farmers' Rights within the framework of the implementation of relevant projects.

5.8 Barriers to the Realisation of Farmers' Rights

Efforts made to address community and Farmers' Rights can be considered as a progressive step although there are some gaps in content of their rights. The general understanding is that most of communities are farming communities. These are involved in mixed crop-livestock agriculture, and groups with special knowledge and traditional practices such as traditional medicine practitioners are also included in it. Farmers and Farmers' Rights within the context of the local community are not clearly defined. These are small-scale farmers, living on agricultural practices within diverse farming systems throughout the country. They have small land holdings for survival rather than high economic gain (Padmavathi et al. 2001). Local farmers constitute local communities, so it may not be practical to draw dividing lines between communities and farmers. Nevertheless, local communities and farmers have been perceived as different entities, and separate proclamations for community and Farmers' Rights have been drafted. It has been a challenge to define farmers and communities, and the result has been inconsistencies in different policies.

There are also some limitations in the recent proclamations in setting the scope of rights and roles of communities and farmers. Farmers' Rights as noted in the PBR Proclamation are restricted to farmers' use of the seed of formally bred varieties; there is no mention of the role of farmers in conserving and developing crop genetic diversity. A major problem has been poor awareness of the conceptual frameworks, issues, trends and challenges of Farmers' Rights amongst the relevant stakeholders including policymakers and farmers themselves.

5.8.1 Lack of Awareness

Leaving aside the individuals and institutions taking part in the process of negotiations, most stakeholders are not informed about the objectives and concept of Farmers' Rights. Farmers are totally unaware of the issues. As the current perception of Farmers' Rights is protected in the recently approved PBRs Proclamation is limited. In other words, the concept of genetic resources is reduced to its supposed essence as a commodity, presumably separable from its complex relationships with farmers' activities and agro-ecological systems, and valued only in terms of its immediate consumption. This perception restricts farmers' roles and tasks in the process of conserving and developing diversity to simply growing of varieties of immediate commercial use (Gubo and Andrey 2001).

5.8.2 Lack of Understanding of Local Structures

Over 85 % population of groups within the community depends on the practice and knowledge of medicinal plants. But such groups are still small-scale farmers that have varying and immediate interactions with the environment they live in. They are members of 'local communities' as per Article 8j of the CBD. They do not depend on only a single crop or crop variety for their subsistence but grow several varieties of crops/plants, harvest and still domesticate semiwild and wild plant species (Yemane 2003). They do not only depend on crop cultivation but practise mixed farming systems. There is lack of awareness about farmers and communities, and this has become a limitation in understanding the issues, trends, the extent of the rights and challenges of implementing farmers' and community rights.

5.8.3 Lack of Strategic Institutional Arrangements

Due to overlapping of the policies and mandates of institutions, there is lack of integration of interlinked programmes into a system. There is a need to facilitate division of tasks amongst institutions. Institutional focal point systems to coordinate relevant policies and programmes of various institutions are not efficient. For instance, various biodiversity, seed and environment policies recognise the role of farmers and local communities and their right to equitable shares of benefits from the use of resources they maintain and develop. But there have been no strategies or integration of institutional approaches to fulfil the commitments made in these policies (Bayu and Estrella 2004).

5.9 Lessons for the International Implementation of Farmers' Rights

New agricultural technologies have been widely used with little or no regard for genetic diversity and are now being realised. In some cases, these technologies have brought about changes in agro-ecological requirements and, thereby, adversely affecting farmers' own long-term productive potential leading to large number of farmers remaining below the poverty line. Similarly, the impact of national laws and policies to increase production and improve food security in a lop-sided manner has contributed to the increase of genetic erosion in crops. The aftermath of such practices was amongst the reasons why measures for conservation and recognition of farmers' contributions in preventing genetic erosion became issues of concern. Recognition of farmers' contributions in the past, present and future has ethical and ecological dimensions, i.e. the imperative for equal recognition of creativity and the economic imperative that enables the provision of health and nutrition services. The interest in PGR that farmers of the world have maintained and developed through generations cuts across national boundaries, public and private sectors and rural and urban communities of the world. These resources need to be further maintained and developed to satisfy human needs now and in the future (Nagarajan et al. 2008).

Farming communities live and work in environs where local social norms have an important influence on their lives and the way they manage and use the resources around them. This shows that subsistence farmers do practise for conservation and not just for the sake of conservation as such. They practise it to meet their immediate and future needs and as a security for their livelihoods in the case of unpredictable situations such as drought and disease epiphytotics. It is important to recognise that the rights of farmers and plant breeders are closely linked with the seed supply system (Hardon 2004). In most cases, variety protection and licensing are privileges in the formal systems that protect plant varieties. If farmers are allowed to enjoy the same privileges, they may get incentives for the conservation of traditional varieties. As such, it may counterbalance the current trend that farmers discard diversity and grow crop high-yielding varieties. Therefore, there is reason to believe that such a system will not adversely impact traditional seed systems but rather strengthen them.

The international community should take into account that genetic erosion of PGRFA is a serious problem in almost all countries. Moreover, ex situ gene bank systems to conserve plant genetic resources seem beset with many weaknesses, including inadequate lack of funding for running the facilities, lack of capacity to systematise information on conserved samples, problems in maintaining some species under such conditions and isolation of genetic materials from evolving environmental changes (FAO 1996). Genetic erosion cannot be controlled by the ex situ system alone. The problems of genetic erosion can be overcome by integrating ex situ and in situ approaches. The farmers' role in managing the in situ conservation system is an essential requirement, and recognition of and support to farmers' in this regard can be linked to Farmers' Rights.

The international community needs to support farmers so that these resources continue to exist. These responsibilities should be reflected in a concerted manner and through the Governing Body of the ITPGRFA. One way to facilatate the involvement of the international community for the implementation of Farmers' Rights is through work promotion and support to governments to frame policies and legislations that enable them to implement Farmers' Rights effectively. The Governing Body should stress the obligations of both developed and developing countries to support farmers' practices of conserving and developing plant genetic resources beyond funds raised in relation to the commercialisation of improved varieties. It is important that the Governing Body develops programmes so that benefits to farmers lead to achievement of conservation goals. One strategy might include the establishment of gene funds at both national and international levels, where the national governments and various local projects on bioprospecting can be used as national sources of gene funds.

Such funds could be channelled to the support of farmers to promote in situ conservation and enhancement of farmers' varieties (Brush 1994), as well as to strengthening ex situ facilities at the local such as farmers' storage facilities and community seed banks and central levels.

The way of thinking and actions have to be rationalised in order to overcome the challenges the world is facing due to unfair way of exploiting global resources. International agricultural policies such as the World Bank, the CGIAR system, WTO and the TNCs have the responsibility in assisting the ITPGRFA and farmers in order to ensure universal food security for the present and the future.

5.10 Recommendations on Measures to Implement Farmers' Rights

In order to implement the Treaty's provisions on Farmers' Rights, governments need to take measures at four levels:

5.10.1 Developing Effective National Legislation for Protection of Farmers' Rights

National legislation to protect indigenous knowledge should recognise this knowledge as the collective heritage of indigenous and local communities. This should be developed with their active participation and leadership and be fully designed on the basis of customary laws rather than western IPR standards. They should ensure that farmers and communities have the authority to decide over access and use of their indigenous knowledge. In order to implement the right of farmers to equitably participate in sharing benefits, laws on TK protection and ABS should recognise the rights of farmers and communities over their varieties in the exercise of national sovereignty. This should include Farmers' Rights over traditional varieties held in situ, including crop wild relatives, and over traditional varieties which have been collected from their land/communities and are held ex situ. To facilitate this, a list of traditional varieties of communities and farmers could be developed with their participation, including in the development of criteria to be used for identifying these varieties. The Prior Informed Consent (PIC) of farmers and communities should be required for access to these varieties, along with equitable benefit sharing from their use based on mutually agreed terms. In order to protect TK from loss as well as misappropriation, and to support in situ conservation, legislation is also needed to protect farmers' and ILC rights to traditional landscapes, cultural values and customary laws associated with TK and genetic resources.

5.10.2 Addressing the Impacts of Other Policies and Laws on Farmers' Rights

Current laws and policies seem to favour the rights and interests of plant breeders and agribusiness over those of poor farmers. The protection of Farmers' Rights needs to be commensurate with that provided to plant breeders, and other policy constraints also need to be identified and addressed. As PBRs under UPOV 91 which extends breeders' rights on farm-saved seeds, seed laws which require registration based on uniformity and distinctiveness and prevent exchange of unregistered seeds and agricultural subsidies which flood the market with cheap modern foods, making it very difficult for small farmers to sell their products/varieties.

5.10.3 Supporting Farmer and Community Level Initiatives

Protecting Farmers' Rights requires new practical tools for PIC and equitable benefit sharing. Supporting policy pilot experiments at farmer and community level provides a way to test out and develop these new approaches and inform the design of policy and law. At the same time, this will build the capacity of farmers and others to implement Farmers' Rights and ensure that new laws are framed on farmers' needs and based on practical experience. Such experiments might include the development of agreements between farmers and plant breeders for access to genetic resources and equitable benefit sharing. Supporting community initiatives and capacity is also important to enable poor farmers to protect their rights. Moreover, national laws on Farmers' Rights alone may not be enough due to many threats that communities face at subnational and local level.

5.10.4 Supporting and Institutionalising Farmer Participation in National Decision-Making on PGR Conservation and Sustainable Use

Farmer participation in national decision-making is far from being standard practice. Laws to ensure public participation tend to be nonexistent or poorly implemented in developing countries. Legal reforms are often needed to ensure that traditional farmers can participate in decisions relating to PGRs. New institutional structures are also needed to enable farmer representatives to participate in national policy and legal processes on genetic resources and agriculture. To ensure that farmers can actually influence the outcome of decisions, they have the same voice and influence as trade and economic actors. This is also likely to require funding for farmer information, capacity building and consultations at local level to enable farmers to participate effectively.

5.11 Conclusion

Farmers' Rights are recognised as the rights arising from the past, present and future contributions of farmers in conserving, improving and making available PGRFA. It should be the responsibility of all to recognise that farmers and farm communities are very important to the world and economy at present and in the future. It is also important to recognise that farmers in many countries are facing pressures that are suppressing their livelihoods and their capacity to contribute to the world food system. The tendency to further marginalise farmers through monopolisation of food sources would certainly lead the world into serious conflicts over food sources. Farming communities should have the rights to decide over access and use of their knowledge. Recognising the values of PGRFA and in harmony with several international treaties, developing countries should come up with the initiatives such as enactment and implementation of the BDA in line with CBD and PVP&FRA Act, providing unique rights to farmers on par with qualified breeders.

References

Andersen R (2005) The history of farmers rights – a guide to central documents and literature, FNI Report Lysaker, Norway

Andersen R (2008) Governing agrobiodiversity: plant genetics and developing countries. Ashgate, Aldershot

Ayalew D, Stefan D, Madhur G (2005) Property rights in a very poor country: tenure insecurity and investment in Ethiopia. Global Poverty Research Group, GPRG-WPS-021. World Bank, Washington

Bala Ravi S (2004) Manual on farmers' rights. M.S. Swaminathan Foundation, Chennai

Bayu M, Jaime E (2004) Report of the brain-storming workshop. Genetic Resources Policy Initiative, Addis Ababa

Brush SB (1994) Providing farmers' rights through in situ conservation of crop genetic resources. CPGR FAO, Rome

Brush SB (2003) The demise of 'common heritage' and protection for traditional agricultural knowledge, paper prepared for conference on biodiversity, biotechnology and the protection of traditional knowledge. St. Louis, MO

Buckland J (2004) Ploughing up the farm: neoliberalism, modern technology and the State of the World's Farmers London

Correa CM (1998) Implementing the TRIPs agreement. General context and implications for developing countries. Third World Network, Penang

FAO (1996) The State of the World's plant genetic resources for food and agriculture. FAO, Rome

FAO (1998) The State of the World's Plant Genetic Resources for Food and Agriculture FAO, Rome

Feyissa R (1999) Mainstreaming biodiversity conservation towards sustainable agricultural development: an Ethiopian perspective. In: Food security through sustainable land use: policy on institutional, land tenure and extension issues in Ethiopia. Proceedings of the first national workshop of NOVIP Partners Forum on Sustainable Land Use, Ethiopia

Gubo Q, Andrey R (2001) The shift to cereal monocropping, a threat or a blessing': toward sustainable agricultural production in the highlands of southeast Oromia, Ethiopia

Hardon J (2004) Plant patents beyond control. Biotechnology, farmers' seed systems and intellectual property rights, Agro special no. 2. Agromisa Foundation, Wageningen

Helfer LR (2002) Intellectual property rights in plant varieties: an overview with options for National Governments. www.fao.org/Legal/prs-ol/years/2002/list02.htm

Leskien D, Flitner M (1997) Intellectual property rights and plant genetic resources: options for a sui generis system, Issues in genetic resources. IPGRI, Rome

Nagarajan S, Yadav SP, Singh AK (2008) Farmers' variety in the context of the protection for plant varieties and Farmers Rights Act, 2001. Curr Sci 94(6):7092

Padmavathi, Chintalapati, Kefyalew D, Kassahun T, Gubo Q, Andrey R (2001) The shift to cereal mono-cropping, a threat or a blessing': toward sustainable agricultural production in the highlands of southeast Oromia, Ethiopia

Pant R (2008) Protection and empowerment of indigenous plant breeder community in India. Ph D. thesis, Jawaharlal Nehru University, New Delhi

Rahmato D (2004) A review of Ethiopia's land tenure system and new policy initiatives: report prepared for the World Bank, Addis Ababa

Ramanna A (2006) Farmers' rights in India – a case study. The farmers' rights project, the Fridtj of Nansen Institute, Lysaker, Norway

Sahai S (2003) India's protection of plant varieties and Farmers' Rights Act, 2001. In: State of the rights of farmers and plant breeders in Asia. Gene Campaign, New Delhi

Shiva V (2005) Seed patents and Police States. In: Seed dictatorship and food fascism. Navdanya, New Delhi

Shiva V, Jafri AH (2003) Need for a genuine *Sui generis* law to defend farmers' rights as traditional breeders: the inadequacies of the PVP Act, 2001. Bija, p 10

Spillane C (1996) Farmers rights in the conservation and use of plant genetic resources. FAO, Rome. http://www.fao.org/docrep/x0255e/x0255e03.htm

Swaminathan MS (1994) Farmers' rights and plant genetic resources – recognition and reward: a dialogue. (eds), M.S. Swaminathan, Chennai, India

Tamiru M, Manda J (2002) Case study report – ethio-organic seed action project, Ethiopia: options for supporting on-farm conservation in Eastern and Southern Africa. Funded by the UK Darwin Initiative and BMZ/GTZ

UNESCO (2003) Language vitality and endangerment. Document submitted by Ad Hoc Expert Group on endangered languages to the international expert meeting on UNESCO programme safeguarding of endangered languages, UNESCO

WTO (2000) International trade statistics. https://www.wto.org/english/res_e/statis_e/stst_toe.htm

Yemane B (2003) Food security situation in the pastoral areas of Ethiopia, GB, Oxfam

Yeshitela K (2001) Loss of forest biodiversity associated with changes in land use: the case of Chewaka-Utto tea plantation. In: Imperative problems associated with forestry in Ethiopia. Proceedings of a workshop organized by Biological Society of Ethiopia. Faculty of Science, Addis Ababa University

Plant Genetic Resources and Traditional Knowledge: Emerging Needs for Conservation

6

Mehfuz Hasan and Hasan M. Abdullah

Abstract

Plant genetic resources and traditional knowledge comprise an inimitable universal heritage, and their conservation and utilization are of instantaneous concern. As it is the basic source of all types of agricultural activity, the conservation and protection of these precious materials are of immense potential. Plant genetic resources conserved by the farmers constitute our invaluable assets to meet the growing demands to increase crop production and productivity. The Convention on Biological Diversity is engaged with the genetic erosion and waning use of agrobiodiversity in modern-day agriculture. This is perhaps the most comprehensive intergovernmental agreement concerning for conservation, proper utilization of genetic resources, and giving out the benefits arising out of exploitation in an equitable way. Concern about the looming accessibility of agricultural production, food security, and environmental stability has encouraged the conservation of plant genetic resources and indigenous knowledge to the pinnacle of the international development strategies. Plant genetic resource and traditional knowledge conservation and utilization have been the source of dramatic scientific changes over the course of the last few decades. Precise evaluation and documentation of plant genetic resources and traditional knowledge are a prerequisite for their sustainable utilization to secure the food security.

M. Hasan, Ph.D. (✉)
Department of Genetics and Plant Breeding, Bangabandhu Sheikh Mujibur Rahman Agricultural University, Gazipur 1706, Bangladesh
e-mail: saikathu@yahoo.com

H.M. Abdullah, Ph.D.
Department of Agroforestry and Environment, Bangabandhu Sheikh Mujibur Rahman Agricultural University, Gazipur 1706, Bangladesh
e-mail: hasan.abdullah@bsmrau.edu.bd

6.1 Introduction

This is really tough to know how many species have been developed in 3000 million years since the Earth was created. We really don't know how many of them disappeared or how many are being generated. However, it is assumed that approximately 10 % of the species that came into

R.K. Salgotra, B.B. Gupta (eds.), *Plant Genetic Resources and Traditional Knowledge for Food Security*, DOI 10.1007/978-981-10-0060-7_6

existence received scientific names which is about 250,000–300,000 species. Of these 20 % species have supplied 90 % of our food requirements; nine species provided 75 % of our main food necessities, and only four species, i.e., rice, wheat, maize, and potatoes, have supplied 60 % (FAO 2012). We can now easily guess the reasons which have led to the erosion of many of our genetic resources.

Plant diversity is characterized by an estimated 300,000 species of higher plants. Only about 7000 species have been domesticated till now and cultivated by humans over the millennia for various proposes. Our nutrition is actually supplied by mere 30 plant species. It provides 95 % of dietary energy or protein (http://www.nbpgr.ernet.in/Why_Conserve_PGR.aspx). Every year a small number of new varieties are released around the world according to the need for a particular terrain. But nature has given us a tremendous amount of diversified plants along with their wild relatives. These diversified plants provide the sources for breeding new materials which can withstand biotic and abiotic stress environment and can be grown in different regions of a particular territory.

6.2 Importance of Genetic Diversity in Plant Genetic Resources

Plant Genetic Resources (PGR) can be considered as any type of reproductive or vegetative propagating material of cultivated varieties and newly developed varieties, obsolete cultivars, landraces, wild and weedy species, near relatives of modern cultivars, and particular genetic stocks inclusive of elite and current breeders' lines and mutants. In other words, it can be considered as any living material of present and plausible value for mankind. PGR include all plants that possess valuable traits including all of our agricultural crops along with their wild relatives. The genes, DNA fragments and RNA, and other genomic resources are also included as PGR, therefore, are nowadays conserved in gene banks for their specific hereditary functions.

Genetic diversity of plants ensues principally as variation in the nucleotide sequence of DNA of any species. A small portion changes due to a mutation which is observed into protein variation and into marker polymorphisms, characters, and physiological and morphological variation in agronomic traits that results into varieties given different names by the respective authority (http://www.fao.org/docrep/013/i1500e/i1500e20.pdf). In sexually producing organisms like higher plants and animals, the offspring is produced through the union of reproductive cells from two genetically different parents. So, the offspring produced from this cross-pollination are genetically dissimilar from their parents.

Sexual reproduction is very crucial for upholding genetic diversity contained by a species since it coalesces the parental genetic material, ensuing genetically diverse offspring different from their parents. When a population of an organism is characterized by a large gene pool, which means the population has a broad genetic base, the organisms have a better opportunity of existing and flourishing than a population which has a narrow genetic base. In this case, some of the plants of this particular population may have some traits which would make them resistant to any kind of biotic or abiotic stress or they may have some traits which would give them better chance for survival. According to "natural selection," the fittest individuals survive and go on to imitate. So, if there is an epidemic condition which threatens a particular species, the more genetic variability there is contained by that species, the higher the likelihood that at least some of the individuals will show resistance and will endure. In the lab, through recent advancements, there are a lot of techniques which can be useful for plant breeders to improve the accessible varieties and also to generate a new genetic variant. Through conventional breeding, scientists do their utmost effort to breed for biotic and abiotic stress-resistant varieties, for superior fruit production, or for other desirable characteristics.

Genetic diversity also decreases the danger of transmitting undesirable traits. In a narrow genetic base, plants may not have much choice to breed with an array of other individuals of the population. They are then forced to breed with their close relatives resulting in inbreeding depression. The genetic makeup of the individuals in the population turns out to be more and more identical. Genetic error becomes an issue. Things may turn out very badly when closely related organisms interbreed. The hidden lethal genes in the parents can be multiplied which may show the deleterious effect phenotypically in the offspring. For instance, any organism can be a carrier of a particular disease the symptom of which may not be seen phenotypically. Closely related individuals are more likely to have the identical mutations, and when they mate the offspring may divulge the symptoms of that meticulous disease. In a population with a narrow genetic base, it is more likely that the carriers will interbreed, and after a certain period of time, the total population will be destabilized.

So, genetic diversity reinforces a population by escalating the possibility that measurably some individuals will be able to endure foremost turbulences and by building the individuals less vulnerable to inherited turmoil. Biological diversity is the disparity present in any type of organism, their hereditary material, and the environment in which they transpire. Based on this, the diversity can be classified into three levels: genetic diversity, species diversity, and ecosystem diversity. The significance of biodiversity for humanity has been evidently established in the recent time, and some may dispute that diversity is crucial for the sustainable development of diverse individual events. Biological diversity is the source of poor people to congregate their food and nutritional requirements and hold the edifying diversity of countries all over the world thus maintaining social and economic balance of the country (Shiva 1994). When we talk about the importance of biological diversity, we should understand the term "genetic erosion."

6.3 Genetic Erosion

The genetic erosion is the main threat to genetic diversity since an enormous number of individuals and their habitat can be eliminated quite quickly. The main cause of genetic erosion is the substitute of landraces by modern varieties, and the obsolete varieties in farming communities are replaced by newly developed ones. Additionally, the absolute number of local varieties is repeatedly reduced when profitable varieties are incorporated into conventional farming systems. Genetic erosion can also be caused by the appearance of new races of diseases, insect-pests, weeds, ecological degradation, and urbanization.

As it is already mentioned, genetic diversity is very important for the changing environmental condition. Diversified plant species play a vital role in adopting stress environment. For long-term viability and species' fitness, genetic diversity is important. Plant populations that are having narrow genetic bases may be more susceptible to biotic and abiotic stresses or other environmental stresses.

Genetic erosion can have tumble effects all over the ecosystem. Due to natural selection and genetic drift, some trouncing of genetic diversity can be observed under natural environment. However, these losses are usually not disastrous, since it can often be balanced by mutation and gene flow. Usually, thrashing of genetic diversity is a more severe danger to species that were previously more pervasive and have lost habitat. The genetic erosion influences the local plant species and the environment such as damages of habitat and the local crop populations, raising plants from a narrow genetic base, etc. The threat of genetic erosion in local plant species can be reduced in precise revegetation.

Seeds and planting material should be collected in such a way that genetic diversity of the geographic area would be preserved. While collecting the materials, the following things should be considered:

1. The geographic source or provenance
   The number of parental material
2. Total number of seeds (or propagules) per plant
3. Their plant-to-plant distance
4. Biography of a plant species.
   If the vegetative parts are collected from a dioecious plant, the ratio of male and female should be in equilibrium.

5. In the asexually reproducing organisms, collection should be done from different clones.

Even while purchasing plants from a nursery, the geographic source information should be obtained. This information will facilitate to know if the purchased materials are genetically diversified or not. With the geographic source information, materials' growing condition should also be mentioned. Care should be given while planting local species. Ample disparity is present in plants which are marked as cultivars subject to the original collection and the procedure it has gone through preceding its release. So, utilizing cultivars is not necessarily responsible for the genetic erosion. However, the cultivar originating from a narrow genetic base when used broadly to suppress the older plant population is more likely to cause genetic erosion of the local population.

Such nursery management activities should be encouraged where the goal is to make the best use of seeds which grow to be the vigorous seedlings. High-quality nursery management established on wakefulness of potential genetic variation in seed distinctiveness, germination necessities, and development system helps to circumvent unplanned selection and reduce the effect of the genetic erosion on the original collection.

6.4 Genetic Vulnerability

Genetic vulnerability occurs from the shape of use or indigence of genetic diversity. Populations are genetically vulnerable when they are in short of the required diversity to combat with the biotic and abiotic stresses. The perception of vulnerability entails a scarcity or low intensity of genetic diversity, most clearly recognized when enormous parts of an area are occupied with a sole cultivar. In this case, if one individual is attacked by a recently occurring disease, biotype, or any other climatic stress, the total field of the area will react correspondingly. This is because of their communal genetic makeup mainly for the genes concerned in the host plant's susceptible (or compatible) attitude (Marshall 1977; Wolfe and Barrett 1977).

6.4.1 Kinds of Genetic Vulnerability

Generally, genetic vulnerability is of the following four kinds:

1. *Genetic homogeneity*. The crop population is of a sole genotype. It may also consist of a few varieties or genotypes.
2. *Mutational vulnerability*. A single mutation can damage the entire crop population.
3. *Migrational vulnerability*. The plants are resistant to locally available biotic stress. When a new pathogen or pest migrates from another place, they become susceptible.
4. *Environmental vulnerability*. The plants in the population are resistant to the current abiotic stresses, but it lacks the adaptive mechanisms for any type of environmental stresses that may arise over time.

6.4.2 Causes of Genetic Vulnerability

6.4.2.1 Narrow Genetic Base of Crop Varieties

The genetically uniform cultivars have narrow genetic base and are thought to be the main reason of genetic vulnerability. The local plant populations may experience natural disasters, but they are genetically capable of tolerating the stresses because of their broad genetic base, while modern varieties are genetically identical that their population is rigid enough to avoid the genetic vulnerability (de Boef et al. 1996; Simmonds 1979). Local plant population may have either low genotype or environment interaction, facilitating it to stand under both stress and non-stress situations, in a mixed population (de Boef et al. 1996). It could be speculated that, in local population, the tolerant plants to the present race of the pathogen generates more offspring than the vulnerable ones because of the comparative reproductive effectiveness. Additional genetic improvement along with improved adaptation is achieved after every round of generation advancement. On the other hand, in a genetically identical population, comparative reproductive advantages cannot be achieved as the population is genetically uniform, unless there is any mutation alteration in the pathogenic race (de Boef et al. 1996).

The development of hybrids through crossing genetically highly uniform inbred lines has decreased the genetic diversity which cannot be observed in the open-pollinated varieties (Simmonds 1979). Additionally, numerous high-yielding and stress-resistant varieties are repeatedly developed through crossing with the locally adapted materials, and these works significantly shrinkage the genetic bases of the varieties.

Genetically diversified plants are more stable and easy for crop production as compared to the uniform population. The following reasons can be mentioned here.

1. It is hard for the pathogen to develop matching genes for a big number of resistance genes present in genetically diversified genotypes. So, when the population is having good number of resistance genes, the pathogen cannot match all the corresponding resistance genes simultaneously (Sharma 2001).
2. The use of mixture cultivars is advantageous due to either spatial or temporal complementarities. Spatial and temporal complementarities may happen when the allied crops are capable of using available resources over space and time.
3. The buffering effects of the components of a mixed cultivar decrease danger of natural hassle as all components of a mixed cultivar would not be at risk to a particular stress at a time.
4. If any component of a mixed cultivar becomes vulnerable to a specific stress, other components are there to control the harm.
5. The reproductive gain of the mixed cultivar produces more resistant offspring after each successive generation resulting more adaptive progenies.
6. Communal environmental alteration can be observed for the component plants.

6.4.2.2 Wide Spread of Dominant Varieties

One or a few genetically identical varieties when widely spread in a large area cause the genetic vulnerability. It creates the perfect state for the pathogens and insect-pests as well. So, the narrow genetic base of the modern cultivars is not the only reason of genetic vulnerability. It is deceptive that identical varieties from a narrow genetic base may overcome the stresses only for a short period of time after its release and be cultivated in a large land and then may suffer from serious thrashing such as unpredicted disease epidemics. The unremitting production of a sole variety year after year will also ease disease epidemics especially when host plant resistance is beaten by the mutation of disease and insect-pests.

6.4.2.3 Failure of Vertical Resistance

Varietal resistance may stop working in a shorter period of time than it takes for the improvement of a modern variety. Vertical resistance is the single gene resistance, and its collapse is considered as another cause of genetic vulnerability related with a narrow genetic base. The vertical resistance is considered as monogenic or oligogenic; the horizontal resistance is controlled by many genes (polygenic). It is, thus, exceedingly specific and accountable to alterations in races of pathogens. It is easier for a pathogen or an insect-pest to interact with the host plant having vertical resistance as the pathogen or the insect-pest needs to beat a single or a few genes of the host.

6.4.2.4 Unplanned Breeding for Susceptibility

Unplanned breeding for susceptibility followed by the introduction to disease of a variety not recognized in advance may also result in vulnerability to unpredicted diseases. One instance of unplanned breeding for vulnerability was seen in the 1970s when a new corn genotype (Texas male sterile, TMS) was released in the USA. The TMS hybrids had a lot of desirable characteristics which include resistance to the most common corn diseases. Conversely, they were not resistant to a previously insignificant strain of a fungal disease, the southern corn leaf blight (*Helminthosporium maydis*). The TMS trait was covered in 90 % of the corn sown in the USA in the 1970s and became susceptible to this pathogen. The fungus colonized all the field of susceptible host and wiped out corn crop. If the corn field hadn't been such a monoculture, the fungus wouldn't have been capable to extend as speedily, as it would have faced obstacles of genetically resistant varieties (Smolders 2006).

6.4.3 Corrective Measure against Genetic Vulnerability

Modern plant breeding applications practically direct toward breeding high-yielding varieties (HYVs) based on narrow genetic bases of different crops. It is not only the ambition of private-owned companies' Research and Development (R&D), but farmers' choice is also one of the reasons behind the substitution of landraces with genetically identical varieties (Agrios 1978). Moreover, the farming is getting more competitive, and farmers prefer to cultivate the most excellent and highest-yielding cultivars accessible to them. The extensive exercise of one or a few genetically identical varieties over large fields sometimes appears to be inevitable, which offers ideal situations for diseases, insect-pests, and other natural calamities. Such technological danger will be very stern especially to marginal farmers in areas where not only environmental variations are enormous but also the economic potential of the farmers is unlikely to permit the exploitation of procured inputs like chemicals that concede crop protection. The extent of genetic diversity dwindles with the decrease in the total number of varieties being cultivated and added concentration of fields planted to a few desirable varieties. Hence, the effect of genetic vulnerability should be reduced by utilization of appropriate breeding methods and by the efficient gene exploitation (Safeeulla 1977).

6.4.3.1 Breeding for Specific Adaptation

The rational yields with less danger are desirable than high yields with high uncertainty due to any type of stress condition for a resource-poor farmer existing under extremely susceptible condition. In such areas varieties are acquainted to fit the environmental condition rather than to change the environment to suite for the varieties. In the trivial fields where dispute of diseases, insects, and environmental fluctuations and threats are enormous, broad genetic-based varieties are preferred to grow rather than narrow genetic-based varieties. The area-specific and region-wise cultivars should be developed and grow a land area with diverse varieties to avoid any outburst of diseases and insect-pests. Precise adaptation to local situations, rather than wider adaptation and giving priorities to varieties that are more strongly placed to the producer's desires, increases genetic diversity in a specific area. It is also apparent that the accessibility of alternate varieties facilitates to change old varieties in case of any collapse.

The practice of growing narrow genetic-based varieties over a large area under marginal situations is not advisable, and the excessive homogeneity by current plant breeding has been dispraised all over the world. In such cases, it is fairly assumed that uniformity is not biologically essential or even preferred but diversity can, at least from time to time, improve effectiveness and stability (Simmonds 1979). Preferences of resource-poor farmers in the trivial areas include yield stability, resistance to biotic and abiotic stresses, and less reliance on external inputs. Farmers attain these by consciously generating genetic diversity at intra- and interspecific levels. Breeding exercises to deal with resource-poor farmers should be focused on farmers' activities and priorities to complement them and not to substitute their exercises. The physical environment and price proportions between external inputs and production outputs do not permit the exploitation of the big amount of procured inputs, mainly agrochemicals.

6.4.3.2 Systematic Gene Deployment

The objective of plant breeding programs in the prospective areas may be to boost production and efficiency through the utilization of yield escalating technologies like high-yielding varieties (HYVs) and agrochemicals for the control of diseases and insect-pests. Genetic diversity can still be preserved through efficient gene exploitation under prospective situations in various ways without preventing the necessity for mechanization to decrease the threat of genetic vulnerability. It is advisable that we should enhance the preference of germplasm accessible to farmers rather than to protect the adoption of a single or a few varieties over huge areas (Marshall 1977; Wolfe and Barrett 1977; Simmonds 1979; Rubenstein et al. 2005). The farming of closely located fields into varieties with different resistance genes (spatial gene deployment) may disperse the effects of diseases and insect-pests. Similarly, different varieties having different vertical resistance genes may be utilized in alternate years (temporal gene deployment) so that the mutation of the pathogen will be controlled (Rubenstein et al. 2005). It is established that growing of different multiline varieties gave more resistance compared to the single or few genes of deployed varieties. Few information is available that when places vertical to the way of the wind are sown with alternative varieties with a different genetic history (spatial gene deployment), each variety may perform as a barrier to the pathogen.

6.4.3.3 Maintaining Broad Genetic Base in Varieties

Extensive breeding programs may be applied to preserve variability within crop varieties and reduce the consequent threats of genetic vulnerability. Variability may be managed by ceasing homogeneity while there is still some remaining heterogeneity left or by combining late generation lines selected after purification. Mass selection can be considered as a good selection method for maintaining genetic variability in the population. Varieties developed through this method would have substantial genetic variation because numerous similar-looking plants which are variable for quantitative traits are preferred for selection and bulked. For example, the improvement of synthetic varieties has been efficient in escalating yield and stability in faba bean (*Vicia faba* L.) because heterozygosity is improved, while the threats of genetic vulnerability from varietal homogeneity is decreased (Suso et al. 2005).

Utilization of multiline varieties, i.e., mixtures of numerous similar pure lines with different genes for biotic and abiotic stress resistance in self-pollinated crops, may be anticipated to endure diseases and insect-pest aggression better than the pure lines. Mixtures with suitable resistance genes could also facilitate to decrease stringencies of various pathogens. It should be realized during the development of multiline varieties that morphologically similar-looking plants can be genetically fairly dissimilar. Successful instances of multiline varieties have been recorded in different countries. If any component of multiline varieties becomes susceptible to pathogenic races, it should be withdrawn and replaced with new resistant lines.

The risks of genetic homogeneity can be escaped if plant breeders utilize diverse sources of genes in their breeding objects (Russel 1978). The crosses between parents of distant relatives would not only be more perceptive to development, they are expected to generate higher heterosis and suitable genetic recombination and segregation in the offspring. This will also facilitate to create varieties having broad genetic bases that are not prone to genetic susceptibility. There is evidence that, like many other characters including grain yield, heterosis may be evident through biotic and abiotic stress resistance.

The perception of protecting the diversified materials is receiving little acknowledgement in various developing countries, but works have already been initiated to breed in larger diversity in some cereal crops (Rubenstein et al. 2005). Current studies in cotton also exhibited a significant decrease in levels of genetic homogeneity due to the reduction in the ratio of area planted to well-known varieties and buffering with the releases of substitute varieties by the rising number of seed companies utilizing diverse genetic makeup.

6.4.3.4 Utilization of Interspecific Varietal Mixtures

In tropical and subtropical countries, farmers have been applying intentionally interspecific varietal mixtures not only to maximize the yield and diversification but also to minimize the production risks since remote ages. It has been observed and established that the utilization of interspecific varietal mixtures decreased disease injury both on major crops and intercrops. Mixing of faba bean with field pea and maize with haricot bean or sweet potato decelerated the speed of disease spread and helped producing higher yields of diverse crops as compared to the corresponding individual cultures (Frey et al. 1977; Rajaram and Dubin 1977). It is usually assumed that interspecific diversity assists the components flee breakdown by decreasing susceptibility to particular biotic and abiotic stresses. It was experienced in practice that escalating interspecific diversity may reduce disease and insect-pest problems in each species. It is, consequently, desirable to identify a commonly favorable set of crop species. An incidence with the productivity of released varieties of faba bean and field pea under single and mixed cultures displayed that the occurrence of chocolate spot (*Botrytis fabae*) was greater in all varieties of faba bean under mixed cultures than the individual ones, while, defiantly, the occurrence of *Ascochyta* blight (*Mycosphaerella pinodes*) in field pea was constantly inferior in all the cultivars of field pea under mixed cultures than the single ones. As faba bean has naturally a rigid physique with enhanced air movement when mono-cropped, the insertion of field pea as an intercrop would certainly generate overmuch below the canopy and then decrease open air flow, build more humid state as anticipated, and worsen the intensity of chocolate spot occurrence. Furthermore, field pea covers its all or part of the biomass on the soil and when mono-cropped and based relatively in an erect situation when mixed cropped with faba bean. Hence, it is rational to anticipate that mixed cropping could rather generate a reasonably drier microenvironment disadvantageous for the growth of *Ascochyta* blight in field pea in comparison to mono-cropping.

6.4.3.5 Incorporation of Horizontal and Vertical Resistances

The horizontal resistance is more stable and less responsible to changes in pathogenic races in comparison to vertical resistance. Vertical resistance is recognized for its race specificity, while horizontal resistance is well known for nonspecificity (Higgins et al. 1998). In order to get rid of the drawbacks of vertical resistance, currently, the integration of vertical (monogenic) resistance with polygenic resistance and pyramiding of genes for vertical resistance against a numerous pathogenic races were recommended to protect hinder races of diseases (Sharma 2001; Asfaw 2004).

6.4.3.6 Escalation of Local Breeding and Informal Seed System

In trivial areas where resource-poor farmers prevail and circumstances are extremely vulnerable, the improvement of local breeding should be considered as one of the best approaches to conquer troubles linked with genetic susceptibility of modern crop varieties. Farmers have been breeding the crop varieties since remote ages, and the key point in local crop improvement is its protection of genetic diversity, both between and within species (de Boef et al. 1996). Framers' or local breeding implies the protection of local cultivars and their further development through enhancement with exotic materials and selection and the supply of seed system (production, selection, treatment, storage, and exchange).

6.4.3.7 Utilization of Molecular Techniques

The modern technologies are growing at extremely rapid speed and opening up new potentials almost quicker than the potentials are envisaged. Full genomic sequencing of cultivated plant species and their several wild relatives is now possible. Further, it is achievable to collect a huge number of sequence data on a rising number of diverse genes and on a comparatively immense number of individuals. This technical aptitude impacts on all fields of biological studies and signals for genetic diversity, genetic erosion, and genetic weakness. Although there is incredible growth in the capability of these modern techniques, they are still very expensive in financial resources. These rising molecular techniques help in analyzing, resolving, or developing indexes for adequate management of PGR (Brown and Brubaker 2002).

Molecular techniques offer the command to check genetic variation at the DNA sequence position. It is currently possible to evaluate organisms from the genomic level (FISH) and genomic in situ hybridization (GISH) down to the point of single nucleotides (DNA sequencing and single nucleotide polymorphisms (SNPs). Such knowledge would augment the legitimacy and reliability of indicators by escalating the clearness of analysis. The instant and apparent advantage is the suppleness and accuracy by which genetic diversity can be evaluated. Molecular systems can be tailored to particular organisms to accommodate diversity in breeding systems and comparative levels of genetic diversity (Hasan and Raihan 2015) and can be scaled based on the number of accessions to be checked, how many loci are required, and which genomic sequences are to be evaluated. In addition, numerous random DNA markers such as restriction fragment length polymorphisms (RFLPs) and amplified fragment length polymorphisms (AFLPs) can bc employed in mapping the genome. With sequence-tagged sites (STSs) established from expressed sequence tags (ESTs), it is even possible to utilize expressed genes definite to life history stages rather than random sequence variation to assess the genetic dissimilarities among accessions. There is a basic achievement in genetic understanding, and it is possible to confirm that two organisms differ and they can be positioned in a phylogenetic ladder depending on their shared progenitor (Hasan and Raihan 2014). When this is accomplished, the phylogenetic diversity of the materials can be predicted. This phylogenetic diversity of the materials permits the expansion and development of main collections of the gene pools of wild relatives. The measurement of phylogenetic diversity can also be applied to classify a division of related wild species that enlarges the genetic information distribution of the collection.

It is clear that the biological resources are very important for food security of a country. The rising consciousness of the significance of maintaining a holistic view of agricultural biodiversity and of concerning conservation with feasible use and improvement is a must. Collaboration between countries is required for efficient conservation and utilization of our universal biodiversity.

6.5 Participatory Plant Breeding: A Way to Empowering Farming Communities

Participatory Plant-Breeding (PPB) is the farmers' participatory breeding method which is controlled by plant breeders to various degrees of farmer involvement. It involves farmers to select advanced breeding material, on farm and according to their needs. This is a formal plant breeding approach of farmers which replaced diverse landraces or farmers' varieties with a few high-yielding homogeneous varieties and triggered monoculture farming in some areas where complex, diverse, and risk-prone (CDR) agriculture was the traditional feature. The consequences were erosion of agrobiodiversity and Traditional Knowledge (TK) and overexploitation of natural resource bases. These led to aggravating ecological problems, declining crop yields, and damaging morale of farmers as innovators. This approach has heightened the need for mutual service and education between scientists and farmers to develop sustainable production systems. Sustainable development can be achieved through participatory approaches that support local innovation and adaptation to augment diversity and enhance local capacities. In this case, the farmer is capable of performing the majority of activities in breeding programs and the breeder can take advantage of the land, the labor, and the indigenous knowledge of the farmer. Farmers' involvement would help the breeder to design breeding programs keeping in view the identified priorities of farmers.

6.5.1 TK and Participatory Plant Breeding

CDR agriculture works in small scales that differ from their surroundings such as silt-trap fields, pockets of fertile soil, flood recession zones, patches of high groundwater, etc. Such specialized, in many cases marginal, environments are generally missed by conventional soil surveys and land system studies because of their small size and dispersion. These are also overlooked by agricultural professionals. Local knowledge is often the best guide to not only where a particular wild species, crop landrace, or area of high diversity may be found but also to the understanding of local practices and preferences and how best to utilize these resources and practices (Guarino 1995). The knowledge of varieties that local men and women acquire, refine, maintain, and exchange can help the breeder in determining priorities and in making decisions how to design breeding programs.

In contrast indigenous knowledge is an interdisciplinary, holistic, and diachronic approach that farming system research and related techniques seek to emulate (Guarino 1995). The importance of indigenous knowledge in plant breeding was not fully realized in the past. The flaw has now been recognized as revealed by Article 8 of the CBD to preserve and maintain knowledge, innovations, and practices of indigenous and local inhabitants embodying traditional lifestyles relevant for their conservation and sustainable use of biological diversity.

6.6 Bringing the Farmer in the Main Stream

6.6.1 Nature of the Institutional System of Breeding

The institutional breeding starts from a broad genetic base which is rapidly narrowed down, through the selection process, to a genetically uniform variety. Such a process usually aims to satisfy a limited set of objectives as perceived by the breeder. It assumes some ability to adapt or change environments to the requirements of the variety through the use of inputs like fertilizers, water, agrochemicals, etc. This is called the "blue print" agriculture with the same crop variety(ies) and alike soil and water management system, irrespective of local peculiarities. Practical observations dictate us that "blue print" agriculture can be harmful in the long term, if not in the short term. Therefore, "blue print" agriculture cannot be the panacea for agricultural development where CDR agriculture is the fundamental feature.

6.6.2 Community System of Breeding

The community systems have maintained genetic diversity in a continuous process of selection (natural or artificial) followed by on-farm conservation of genetic resources. The diversity has been maintained as a part of the farming system to cope naturally with biotic and abiotic stresses. The farmer uses the diversity within and between species that provides sustainability and an ability to adapt to changing environmental conditions (Hardon 1995). The community approach is in harmony with the nature for sustainable utilization of resources.

6.6.3 Mutual Benefit

There are fundamental differences between the institutional system and the community or informal system of breeding. No doubt, both systems have comparative advantages. These advantages can be and need to be exploited for mutual benefit of the breeder and the farmer. The choice then would be to bring the farmer in the main stream, as a partner. At the same time, farming households, consumers, seed producers, and postharvest processors should also have their roles to play. However, farmers' involvement will be determined by:

1. Genetic diversity of potential interest to farmer/community systems in an environment
2. Development of breeding populations with specific characters like pest resistance, stress

tolerance, local market preferences, etc. for use in the development of crop varieties suited to local conditions
3. Cooperation between different forms of farmer/community systems and the institutional system, with divisions of labor and responsibilities as appropriate to a situation

It is interesting to note that, given the opportunity, the farmer is capable of performing the majority of activities involved in the breeding process. This has several implications such as:

1. Restoring the confidence of the farmer as an innovator
2. Reducing the cost of breeding through utilization of farmers' land and labor, as compared to the high costs involved in institutional breeding
3. Saving the time of the breeder from the field work which the farmer can supplement
4. Increasing biodiversity through farmers' individual selection from segregating populations in a number of locations (assuming that the selection process will take place in several locations)
5. Spontaneous adoption of lines/varieties developed and thereby spontaneous technology transfer

Local testing of breeding lines under farmers' conditions is essential to include and maintain a range of farmer and user preferences (Hardon 1995). If the local breeders and farmers access to heterogeneous segregating and early generation materials and graft them on local landraces, presumably this will lead to a plethora of region-specific and perhaps even village-specific cultivars, rather than just a few cultivars. This would imply that PPB can lead not only to generation of biodiversity but also to conservation of biodiversity (Ryan 1992).

The maintenance of plant genetic resources can be intricate to sell, but the stakes are high. There is pressing need for all those who are interested in PGR conservation and use to be more involved in all the aspects of genetic diversity such as to study, understand, enhance, conserve, and use it. To do so, we need to understand the extent and distribution of plant diverseness in species and ecosystems applying proper research, field studies, and analysis. Any conservation effort should be an effort that guides to integrated maintenance, i.e., a balance of ex situ and in situ methods. There is a need to stimulate international cooperation or joint ventures on all aspects of PGR. Genetic diversity should be understood at all the three levels, i.e., at the level of species, at the level of genus, and at the level of ecosystem. Additionally, various relations that influence allelic diverseness and variation in allelic frequencies within and between populations need to be understood. We need to survey genetic diversity using all available methods of measuring, before identifying the areas and species to be conserved ex situ and in situ. There remain many unresolved questions about the extent and distribution of genetic diversity in valuable crop species. To what dimensions and in what ways are ecological factors important for the distribution of diversity in crops and forages or for their wild relatives? It is important that these are tackled in a precise way and not by the continued addition of data in an almost random fashion that is often is the case. This will involve collaboration between scientists, research centers, and countries. In the light of increased utilization of molecular techniques for analyzing plant genetic diversity, there is also the need to link the information on molecular variation to PGR management in a more meaningful way than it is presently done, and this could be done on particular crop gene pools (Rigges 1990). The major elements that confer value on genetic diversity and its organization are:

1. The genetic principles of evolved populations and taxa or samples of these
2. The environments and ecosystems that support both the diversity and its structure
3. Its relationship with the ecosystem

The key to genetic conservation is maintaining these three elements. To attain this we need to improve access to existing knowledge as much as possible, maintain genetic continuity and integrity wherever possible, and integrate and coordinate different conservation efforts. Recently, scientists utilize a new tool named Geographic Information Systems (GIS) to carry out spatial analyses to identify diversity patterns of genetic resources. GIS can be used to interpose genetic parameters between sampled populations to apply resampling of georeferenced samples within a defined buffer zone or to develop grid-based genetic distance models. GIS is an essential tool to prioritize areas for conservation of PGR. Scientists have used spatial analysis techniques to develop conservation strategies for PGR based on molecular marker data. Moreover, GIS can be exposed in an apparent way on maps, which promotes the incorporation of these findings into the formulation of conservation strategies and the implementation of conservation action.

6.7 GIS to Augment Conservation of PGR

Geo-spatial tools are rarely utilized to conserve and manage diversity of PGR. With respect to selecting sampling locations and emphasizing the conservation of PGR, spatial analyses of diversity have been carried out mainly at the species level for crop gene pools. Only a few records have mapped intraspecific diversity to enhance the conservation of genetic resources of specific plants, grouped samples using a grid to compare diversity between geographic areas of similar size, whereas applied resampling to enable the calculation of diversity estimates with high degrees of confidence. Several diversity estimates are important to prioritize areas for conservation including recommended parameters like allelic richness and the number of locally common alleles. Since the application of molecular tools is becoming cheaper, intraspecific diversity analysis with large datasets will probably be more common in the near future. GIS tools and diversity information along with biotic and socioeconomic spatial database can identify possible drivers behind diversity and genetic erosion. This can be useful information in the development of adequate policies and measures to promote in situ conservation of PGR on farms and in natural ecosystem. GIS can also be used to link genetic data to available spatial information relevant to conservation of PGR to reveal short-term threats such as accessibility and long-term threats such as climate change. With this type of analysis, vulnerable hotspots of diversity could also be identified.

6.8 Role of International Conventions in Conservation of Genetic Resources

For conservation of genetic resources and indigenous knowledge, a number of international treaties and conventions came into force for sustainable utilization. It is obligation of all the countries to develop national strategic plans or programs for conservation and sustainable use of biological diversity. The following are some of the conventions which are dealing with sustainable utilization and conservation of biological resources.

6.8.1 Convention on Biological Diversity (CBD)

The CBD was proposed at the United Nations Conference on Environment and Development (UNCED) in June 1992. It represents a major global initiative directed toward the conservation of biodiversity (FAO 1994). This is perhaps the most comprehensive intergovernmental agreement concerning conservation, sustainable utilization of genetic resources, and sharing of the benefits arising out of such use in an equitable way. The convention provides a broad legal framework to conserve and use biodiversity. It became an international agreement on 29 December 1993 when more than 30 countries ratified it. Over 160 countries have signed the convention.

The preamble of the CBD recognizes and reaffirms (UNEP 1992):

1. The intrinsic value of biological diversity
2. The sovereign rights of states over their biological resources

3. The fundamental requirements of in situ conservation of ecosystems and national habitats
4. The supporting role of ex situ conservation
5. The decisive task of regional communities and women in the protection and feasible use of biological diversity
6. The desirability of allocating reasonably the advantages originating from the utilization of TK, skills, innovations, and practices
7. The importance of and need to promote regional and global cooperation for conservation
8. The requirement of substantial investments to conserve biological diversity

The objectives of the convention are:

1. The conservation of biological diversity
2. The sustainable use of its components and the fair and equitable sharing of the benefits arising out of the utilization of genetic resources and by proper dissemination of technologies
3. Taking into account all rights over those resources and to technologies by appropriate funding (UNEP 1992; Chauhan 1996)

At UNCED, governments reached a consensus on a global action to promote sustainable development known as Agenda 21. Both Agenda 21 and the convention emphasize the significance of improving and strengthening the capacity of countries to benefit fully from biological resources available to them. Access to new technologies and their managed use for training, information, and financial assets will facilitate developing countries to better conserve and use their biodiversity. Article 3 of CBD, which is now an international law, confers sovereign rights to nations over the biological resources that originate from a specific country ("the supreme rights to utilize their own resources pursuant to their own environmental policies"). While CBD provides the rights, it also places certain responsibilities/obligations on each country. These national responsibilities will help in caring for and preserving the biodiversity. CBD also promotes cooperation "with other contracting parties, directly or, where suitable, by expert international institutions" (Article 5).

The CBD places the obligation squarely on the countries to develop national strategic plans or programs for conservation and sustainable use of biological diversity. This obviously implies greater coordination within the existing setup of national programs to meet such needs focused on conservation and use of biodiversity including PGR. The countries that adhere to CBD require identifying different components of biodiversity and monitoring and paying attention to those that require urgent conservation measures (Article 7).

6.8.1.1 PGR Conservation Aspects in the Convention

In situ *conservation*: the CBD requires that the countries develop guidelines for selecting areas for in situ conservation, create confined areas, direct the utilization of resources so as to make a sustainable use, and protect ecosystems and the threatened species and natural habitats. It also requires that the countries promote environmentally sound development, rehabilitate degraded lands and ecosystems, and eradicate exotic species that may threaten the existence of native species, ecosystems, or habitats. It stresses on compatibility between conservation of biodiversity and sustainable use, taking equally the role of local communities, indigenous knowledge, etc.

Ex situ *conservation*: the countries are required to establish and maintain facilities for ex situ components of conservation and research on plant, animal, and microorganism biodiversity and carry out recovery and rehabilitation of threatened species. Countries also need to regulate and manage collections of biological resources without damaging the ecosystem and in situ populations following appropriate codes of conduct and policies considering cost controls, ownership determination, ex situ crop control, threat management, and building of use capabilities (MSSRF 1996).

Proper utilization of biological diversity: it is necessary for a national program to integrate the conservation and proper utilization of biological resources into nationwide decision-making policies and adopt measures to avoid or minimize adverse effects on biological diversity. CBD implies that countries should protect and encourage use of biological resources based on customary cultural traditions compatible with conservation and support local populations to develop and employ corrective action in despoiled places where biological diversity has been reduced. It also urges nations to encourage collaboration with its governmental and nongovernmental agencies, including the private sector in developing methods for sustainable use of biological resources.

6.8.2 Global Plan of Action

The GPA is part of the FAO Global System for the conservation and sustainable utilization of PGRFA which is an important element for the Commission on Genetic Resources for Food and Agriculture in fulfilling its mandate, though requiring also other important elements to complete it. The commission required the development of a rolling GPA on PGRFA with programs and activities focused on satisfying the gaps, minimizing obstacles, and managing emergency situations identified in the FAO Report on the State of the World's Plant Genetic Resources. The occasionally updated works will allow the commission to recommend priorities and to promote the rationalization and coordination of efforts. A discrete GPA for PGRFA is warranted because of their great importance to world food security and, within the greater background of biological resources, because of several features of this particular form of biodiversity.

The main aims of the GPA are (FAO 1996):

1. To ensure the conservation of PGRFA as a basis for food security
2. To promote sustainable utilization of PGRFA, to foster development, and to reduce hunger and poverty particularly in developing countries
3. To support a reasonable and unbiased distribution of the benefit arising from the use of PGRFA, recognizing the desirability of allocating equitably advantages originating from the utilization of traditional knowledge, innovations, and practices relevant to the conservation of PGRFA and their feasible use

Establishing the necessities and individual rights of farmers was recognized by national rules to have unbiased entrance to germplasm, information, technologies, financial resources, and research and marketing channels needed for them to carry on to manage and improve PGR. Developing and promoting fair and unbiased distribution of advantages originate from the use of PGRFA in their exchange between communities and within the international community was needed:

1. To assist countries and institutions responsible for conserving and using PGRFA to identify priorities for action
2. To strengthen, in particular, national programs, as well as regional and international programs, including education and training, for the conservation and utilization of PGRFA and to enhance institutional capacity

The GPA has 20 priority activity areas that are organized into four main groups. The first group deals with in situ conservation and improvement, the second with ex situ conservation, the third with utilization of PGR, and the fourth with institutions and capacity building. As the GPA is a set of integrated and intertwining activities, the placement of the activities into four groups is intended simply to help in the presentation and guide the reader to areas of particular interest. Many activities will be related and be relevant to more than one group.

6.9 Conclusion

PGR and TK constitute a unique global heritage, and their conservation and utilization are of immediate concern. A diversity of PGR and TK has long been viewed as a means of increasing both global and local food security. These genetic resources are important source of income security for livelihood to farmers in the less developed regions of the world. Genetic diversity is important both to individual farmers and farming communities and to agriculture in general. The importance of PGR and TK and threats to them have led to the creation of conservation programs to preserve these resources for future generations. Concern about the future vulnerability of agricultural production, food security, and environmental adaptability has shifted the conservation and sustainable use of genetic resources to the top of the international development agenda. To avoid the genetic vulnerability, a broad genetic base is necessary for the development of varieties. In spite of its inclusive advantage and support by the international community, in situ and ex situ conservation is still inadequate. The CBD emphasizes the conservation and sustainable use of biodiversity and advocates the equitable sharing of the benefits arising from such use. The CBD provides a broad legal framework and comprehensive intergovernmental agreement for conserving, sustainable utilization of genetic resources, and sharing of the benefits arising out of such use in an equitable way. It further obliges the wider application of such indigenous knowledge with the approval and involvement of the farmers and local people and encourages the equitable sharing of the benefits arising from the utilization of these resources.

References

Agrios GN (1978) Plant pathology, 2nd edn. Academic, London

Asfaw A (2004) Alternate approaches in deploying genes for diseases resistance in crop plants. Asian J Plant Sci 3(5):618–623

Brown AHD, Brubaker CL (2002) Indicators for sustainable management of plant genetic resources: how well are we doing? In: Engels JMM, Rao RV, Brown AHD, Jackson MT (eds) Managing plant genetic diversity. CABI Publishing, Oxon, pp 249–262

Chauhan KPS (1996) Implications of the convention on biological diversity: Indian approach. Indian J Plant Genet Resour 9(1):1–10

de Boef WS, Berg T, Haverkort B (1996) Crop genetic resources. In: Bunders J, Haverkort B, Hiemstra W (eds) Biotechnology; building on farmers' knowledge. Macmillan, London, pp 103–128

FAO (1994) Genebank standards. Food and Agriculture Organisation, Rome

FAO (1996) Global plan of action for the conservation and sustainable utilization of plant genetic resources for food and agriculture. Food and Agriculture Organisation, Rome

FAO (2012) World's plant genetic resources for food and agriculture. FAO, Rome

Frey KJ, Browning JA, Simons MD (1977) Management systems for host genes to control diseases loss. In: Day PR (ed) The genetic basis of epidemics in agriculture. The New York Academy of Sciences, New York, pp 255–274

Guarino N (1995) Formal ontology, conceptual analysis and knowledge representation. Int J Hum Comput Stud 43(5/6):625–640

Hardon J (1995) Participatory plant breeding. Issues in genetic resources. No. 3, IPGRI, Rome

Hasan M, Raihan MS (2014) Genetic polymorphism in Bangladeshi aromatic rice through RAPD analysis. Open Sci J Biosci Bioeng 1(2):23–27

Hasan M, Raihan MS (2015) Genetic variability in Bangladeshi aromatic rice through RAPD analysis. Turkish J Agric Food Sci Technol 3(3):107–111

Higgins VJ, Lu H, Xing T, Gelli A, Blumwald (1998) The gene-for-gene concept and beyond: interactions and signals. Can J Plant Pathol 20:150–157

Marshall DR (1977) The advantages and hazards of genetic homogeneity. In: Day PR (ed) The genetic basis of epidemics in agriculture. The New York Academy of Sciences, New York, pp 1–20

MSSRF (1996) Agrobiodiversity and farmers' rights. In: Proceedings of a technical consultation on the implementation framework for farmers' rights, MS Swaminathan Research Foundation, Madras, India

Rajaram S, Dubin HJ (1977) Avoiding genetic vulnerability in semidwarf wheats. In: Day PR (ed) The genetic basis of epidemics in agriculture. The New York Academy of Sciences, New York, pp 243–254

Rigges LA (1990) Conserving genetic resources on-site in forest ecosystems. For Ecol Manage 35:45–68

Rubenstein DK, Heisey P, Shoemaker R, Sullivan J, Frisvold G (2005) Crop genetic resources: an economic appraisal. United States Department of Agriculture (USDA). Econ Information Bulletin No. 2

Russell GE (1978) Plant breeding for pest and diseases resistance. Butterworths, London

Ryan JG (1992) Agricultural productivity and sustainability in the semi-arid tropics: is there such a thing as free environmental lunch? Paper for panel discussion on the occasion of ICRISAT's 20th anniversary: sustainability and environment in the semi-arid tropics: food production in the future, ICRISAT Centre, Pantacheru, India, 28 August 1992

Safeeulla KM (1977) Genetic vulnerability: the basis of recent epidemics in India. In: Day PR (ed) The genetic basis of epidemics in agriculture. The New York Academy of Sciences, New York, pp 72–85

Sharma JR (2001) Principles and practice of plant breeding. Tata McGraw-Hill Publishing Company Limited, New Delhi

Shiva V (1994) Agriculture and food production. UNESCO/Environmental Education Dossiers 9, pp 2–3

Simmonds NW (1979) Principles of crop improvement. Longman, London
Smolders H (2006) Enhancing farmers' role in crop development: framework information for participatory plant breeding in farmer field schools. Center for Genetic Resources, Wageningen University and Research Center, Wageningen
Suso MJ, Harder L, Moreno MT, Maalouf F (2005) New strategies for increasing heterozygosity in crops: *Vicia faba* mating system as a study case. Euphytica 143:51–65
UNEP (1992) Convention on biological diversity
Wolfe MS, Barrett JA (1977) Population genetics of powdery mildew epidemics. In: Day PR (ed) The genetic basis of epidemics in agriculture. The New York Academy of Sciences, New York, pp 151–163

7 Biotechnological Approaches for Conservation of Plant Genetic Resources and Traditional Knowledge

Pankaj Pandotra and Sulpha Gupta

Abstract

The conservation and sustainable use of the diversity present in plant genetic resources and traditional knowledge of germplasm within and among plant species represent economic, scientific, and societal values that have the potential to solve the food security problems arising from our expanding global population. Advances in biotechnology fields such as in vitro culture technology, cryopreservation, and molecular markers have generated significant improvements in methods of conservation of rare and endangered plant genetic resources and the traditional knowledge of germplasm, and their valuable management, in effective ways. A strategic and forward vision for conservation of plant genetic resources and traditional knowledge of germplasm and sustainable use of plant resources in the twenty-first century is of far-reaching significance for the Earth's sustainable development.

7.1 Introduction

Plant genetic resources (PGR), a vital segment of biodiversity, constitute the basic genetic material of plants having valuable resources to meet human requirements for present and future generations (Dhillon and Saxena 2003; Leisa 2004). This genetic information exists in various components of plant genetic resources such as landraces and farmers' seed varieties, obsolete cultivars, modern cultivars, breeding lines and genetic stocks, wild relatives, weedy races, potential domesticated species, and both exotic and indigenous species that can be useful for exploitation in both traditional and modern different breeding

P. Pandotra, Ph.D. (✉)
School of Biotechnology, Sher-e-Kashmir University of Agricultural Sciences and Technology of Jammu, Chatha, Jammu 180009, Jammu and Kashmir, India
e-mail: pankajpandotra@gmail.com

S. Gupta, Ph.D.
Plant Biotechnology Division, Council for Scientific and Industrial Research (CSIR), Indian Institute of Integrated Medicine (IIIM), Canal Road, Jammu 180001, Jammu and Kashmir, India
e-mail: suphlabg@gmail.com

R.K. Salgotra, B.B. Gupta (eds.), *Plant Genetic Resources and Traditional Knowledge for Food Security*, DOI 10.1007/978-981-10-0060-7_7

programs. Goals to be targeted include increasing the yield of crops and making plants disease resistant to counter various biological and environmental stresses (Ogbu et al. 2010; FAO 1996; Engels and Visser 2006; Sharma 2007).

The biological diversity that exists in plant genetic resources that we can see today, which has resulted from billions of years of various modes of evolutionary forces, constitutes the biological basis for world food security and human life support. To date only 1.7 million species have been characterized of the total estimated 5 to 50 million species of the world's biota (Groombridge and Jenkins 2000), of which 44 % are vascular plant species. Spread across the surface of the earth are 25 hotspots (Myers et al. 2000), of which India is lucky with 2 hotspots: the Western Ghats and the Indo-Burma region covering the Eastern Himalayas, endowed with a great wealth of rich biodiversity. Known as "megadiversity" countries, 17 countries account for more than 80 % of the planet's biodiversity. The majority of these 17 countries are in the developing world, such as the Andean region (Venezuela, Colombia, Ecuador, Peru), the Amazon basin (Brazil), and South Asia (Malaysia, Philippines, Indonesia, India, China) (Fig. 7.1).

The rich heritage of biodiversity wealth in India is spread over ten biogeographic zones consisting of the trans-Himalayan, the Himalayan, the Indian desert, the semiarid zones, the Western Ghats, the Deccan Peninsula, the Gangetic Plain, Northeast India, and the islands and coasts (Rodgers et al. 2002). India is considered to be the centre of origin of 30,000 to 50,000 varieties of crop plants and ranks seventh in terms of contribution to the world's agriculture. It is also the homeland of many cultivated species and wild relatives of crop plants (Tandon and Kumaria 2005), which include around 15,000 medicinal plants: 7,000 plants used in Ayurveda, 700 in Unani, 600 in Siddha, 450 in homeopathy, and 30 in modern medicine (Das 2008). The origination and utilization of traditionally based innovations, creations, and practices by indigenous and local communities have been transferred from one generation to the next, usually in oral form or written records called Traditional Knowledge (TK). The creation of such knowledge is not profit driven and is intended to support the livelihood of local communities, is held in common by the community, and no individual has sole right to claim this as property. Traditionally based knowledge on the use of medicinal plants as natural resources has been considered as one of the important assets for treatment of various human diseases.

7.2 Present Problem of PGR and TK

During the past decade, the most burning environmental issue facing our planet is the high rate of biodiversity loss. The synergetic effects of all activities such as anthropogenic pressure, introduction of domesticated species and new varieties, weed problems, urbanization, pollution, habitat destruction, deforestation, spread of invasive alien species, drastic climate change resulting from global warming, changing lifestyles, eutrophication, globalization, market economies, overgrazing, and changes in land use patterns have dramatic impacts on plant diversity and the growing increase in the number of threatened species (Kaviani 2011). More than 15 million hectares of forest cover in tropical areas perish every year (Rao 2004), and many plant species are gradually becoming extinct (Panis and Lambardi 2005). The consistent exploitation of plant wealth for medicinal purposes tends to cause these plant species to become extinct: for example, salicylic acid, the active ingredient of aspirin, is isolated from the bark of the willow tree; digoxin is the active chemical present in the leaves of foxglove; quinine occurs naturally in the bark of the cinchona tree; and the analgesic and narcotic drug morphine is obtained from the opium poppy—these are the most widely used medications in the world's health system (Vickers and Zollman 1999). The colchicine content extracted from the perennial plant named *Gloriosa superba* L., growing profusely in Africa and Southeast Asia, has caused the plant to be characterized as a threatened species because of its extensive use in the medicinal industry. The

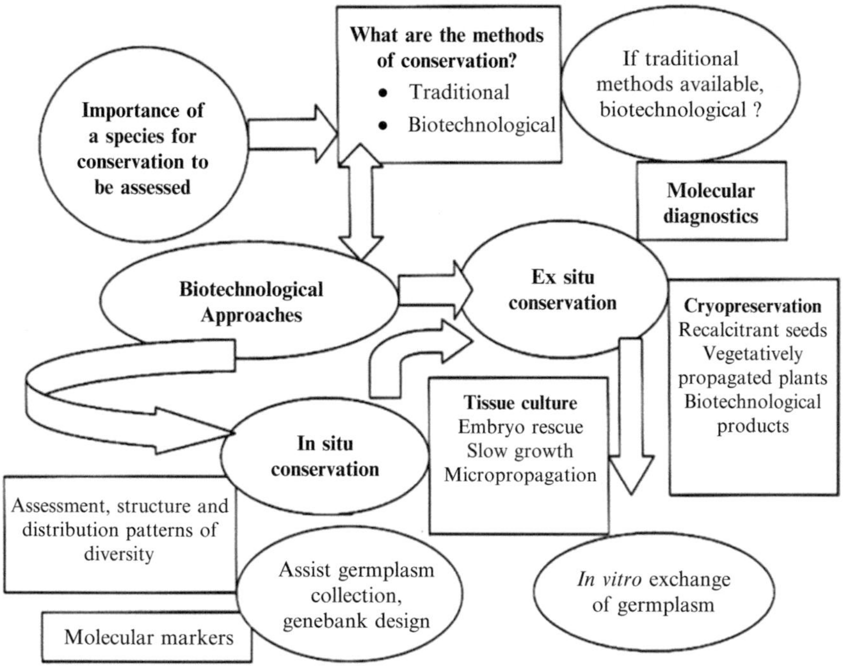

Fig. 7.1 Methodology performed for the conservation of PGR and TK of germplasm. (From Kasagana and Karumuri 2011)

extract known as pygeum collected from the bark of *Prunus africana* has been extensively used as a traditional medicine for benign prostatic hyperplasia (BPH), leading to its vulnerable status (Jimu 2011). During the year 1990, an estimated 35,000 debarked trees were being processed annually. Therefore, the International Union for Conservation of Nature (IUCN) evaluated this species grown across the globe and decided to mark them in the Red List category. Recently, the extensive harvest of some medicinal species such as *Taxus wallichiana*, *Aconitum heterophyllum*, and *Picrorrhiza kurrooa*, and from an ecological aspect that of the important forest-grown tree species such as *Quercus* spp., *Shorea robusta*, *Pinus roxburghii*, and *Cedrus deodara*, raise alarming concerns about their extinction.

Apart from the overexploitation of PGR, TK of germplasm has now regained importance from the discovery of new drugs and novel biochemical constituents from phytoresources (Alves and Rosa 2007; Pandey 2011). More than 80 % of the population living in developing nations report belief in traditional cures for many diseases by using these tradition-based medicinal plants considering the minor level of side effects in their consumption; thus, there has been a spurt in herbal industries. To earn more profits in their business growth, nowadays many pharmaceutical/nutraceutical industries are focussing on the traditionally grown plants that have been used for treatment of various diseases since long ago. These raw materials of PGR are procured in unregulated fashion to extract the medicinally important bioactive compounds. These activities become a matter of alarming concern as such unfortunately leads to overexploitation, loss of habitat, and unregulated use of this natural resource. Many medicinal plants with traditional knowledge uses have become invisible, threatened, or endangered species and are also highly affected by climate change, especially global warming effects; for example, the increased concentration of carbon dioxide gas favours C_3 plants over C_4 plants, and the occurrence of various form of diseases, pests, and high salt content in soil, etc., lead to decreased crop production (Rai 2010).

7.3 Conservation of TK

Recently many scientists in Western countries and established industries have showing increased interest in TK resources as these have a key role in saving the lives of many people by ensuring food security and the healthcare system. There has always been interest in TK systems and innovations associated with TK systems for use in various sectors such as agriculture, water management, and town planning. Rai (2010) highlighted some excellent examples of using traditionally based methods for various pest controls, and indigenous tank irrigation systems, and traditional techniques were utilized to make building and housing coverings earthquake resistant.

The holders of TK may use biotechnology-based innovation to further increase their economic position and ultimately the prosperity of the nation. Recent progress in modern biotechnology, especially in genetic engineering, has considerably broadened the limits for using different genetic resources. To address the issue of climate change from global warming caused by burning carbon fuel, an alternative solution is a combustible fuel called biofuel with the help of TK and modern biotechnology tools. For example, in Indonesia, fishermen commonly use traditional knowledge as the basis to extract oil from jatropha or jarak fruits to fuel their motorboats without any emitted gases or suspended particles in the air. The efficiency and yield of the oil extracted from these plants can be further improved with the help of techniques associated with modern biotechnology so that we can make a more eco-friendly, alternative type of fuel for combustion purposes. Recently, biotechnology-based industries have utilized corn crops to produce plastics that can also be used to produce items such as golf balls and cups. In addition, herbals can be used to produce herbal cosmetics and further increase their quality to avoid side effects through biotechnology tools. Fruits may be used as a source of various types of enzymes, and these enzyme products are utilized for many commercial purposes and for upscaling the production of enzymes and flavonoids products by fermentation technology.

The protection of TK is based on the importance of such knowledge as essential for conservation purposes. Thus, maintenance of biological diversity in farming systems generates valuable resources for the global community (Swanson et al. 1994). If a farmer's community boycotts the uses and breeding of traditional local varieties and is attracted towards planting the better-yielding modern varieties to obtain a higher income, then a tremendous loss of biodiversity wealth could occur (Swanson et al. 1994). Under this approach, the protection of TK by Intellectual Property Rights (IPRs) helps meet society's broader objectives for conservation of the environment, sustainable agriculture, and food security. Traditional knowledge has entered international debate for a number of reasons such as the importance of the rights of indigenous peoples, including the right to self-determination; the loss of TK germplasm and the apparent need for its protection; the destruction of the environment and the reduction of biological diversity; the north–south relationship; and the effects of modern biotechnology, particularly genetic engineering, and the protection of the resulting innovations by IPRs.

7.4 Different Approaches for the Conservation of PGR and TK of Germplasm

The explosive growth rate of the human population worldwide and global warming issues have become the subjects of front headline talks during meetings of all political parties across the world. Therefore, the seriousness of the loss of vanishing valuable genetic resources invoked international action to collect, maintain, and conserve PGR and TK and their diversity. The landmark signing of the Convention on Biological Diversity (CBD) event held at Rio de Janeiro in 1992 greatly increased the importance and visibility of biodiversity conservation as a crucial international issue. The CBD highlights three objectives: (1) conservation of biological diversity, (2) sustainable use of its components, and (3) equitable sharing of benefits arising from the use of genetic resources. Therefore, to meet the demand for food scarcity worldwide resulting from the explosion of population size, and to address human health concerns raised by the greater side effects associated with synthetic drugs, effective, efficient, and within time bound measures must be carried out to draw a streamline for conserving and maintaining the existing diversity of PGR and TK of germplasm so that continuing availability of the germplasm is assured for future generations. Keeping all these parameters in mind, the following are the different approaches engaged towards conserving the PGR and TK of germplasm:

1. In situ conservation
2. Ex situ conservation
3. Biotechnological approaches

7.4.1 In Situ Conservation

The process of cultivating those overexploited and endangered medicinal plants outside their natural habitat, because habitat degradation and destruction have taken place, has become an alternative approach for conservation (Long et al. 2003). It provides the opportunity to study their biology for research purposes, to understand the threats to endangered species, to continuously supply plant materials for multiplication, and to improve the quality of key agronomic traits (Temitope 2013). The only remarkable drawback associated with this type of conservation is that it creates a narrower range of genetic variation in the target species than that occurring in the wild as there is no evolutionary process taking place. It involves three methods for conservation: to conserve namely by seed storage in seed banks, field gene banks, and botanical gardens (Withers and Engelmann 1998). Also, an adequate amount of sample is required for the conservation of genetic diversity, requiring a trained and skilled person, a herculean task to manage and more prone to genetic erosion of the material.

7.4.2 Ex Situ Conservation

This approach is considered highly recommendable under the CBD in which plant species are promoted to grow in their natural habitats, that is, wilderness areas, reserves, protected areas, and within traditional farming systems, where evolutionary processes continue to operate (Ogbu et al. 2010). This approach offers several advantages that include conserving a broad range of potentially interesting alleles, enhancing and ensuring sustainable use of genetic variation for present and future human needs, assuring protection of associated species, and facilitating research on species in their natural habitats, thus allowing better evaluation and utilization. This method is greatly significant for the conservation of wild relatives of crop plants, especially tree crops and forest species, and novel genetic resources may be conserved even in home gardens (Rathore et al. 2005; Tao 2003). The numerous disadvantages associated with these methodologies are time-consuming processes, costs, covering only a small portion of the total diversity of a crop species, political interference, and social issues concerns. The danger always remains that the genetic material can be wiped out if natural disasters happen.

To stop the illegal malpractice and overutilization of important medicinal plants grown across the globe, researchers have been introducing a new promising approach in which part of the plant is used or a substitution of plant species can quench the requirements of the medicinal industry. In this concern, species of *Cryptocarya* (Lauraceae) grown in South Africa countries are commonly used as substitute plant for *Ocotea bullata* (Burch.), which is in serious shortage, to address the problems of threatened medicinal plants as both have similar pharmacological values (Zschocke and Van 2000). Some internationally based research institutes have an important mandate for particular crops to conserve and maintain germplasm, as mentioned in Table 7.1.

7.4.3 Biotechnological Approaches

Biotechnology has an important role in international plant conservation programs and in preservation of the PGR (Bajaj 1995; Benson 1999; Ramanatha and Riley 1994), which impacts growth, development, and prosperity of the various nations around the world. The innovative tools of biotechnological approaches such as cell fusion techniques, proteomics, next-generation sequencing, recombinant DNA technology, protein engineering, structural biology, and micropropagation methods have contributing significantly in the respective research areas and in generating high-quality, genetically superior planting materials for the nation. When the biological material (organs, seeds) cannot be stored and propagated feasibly in a traditional or classical way for reasons of a reproductive barrier problem or extremely reduced populations, the various approaches of modern biotechnology tools are taking place to conserve important endangered or threatened plant species and can have a supportive role in ex situ conservation efforts. Recent progress made in plant biotechnology fields such as in vitro culture, cryopreservation, and molecular biology provide us alternative approaches for conservation of PGR and their quality as evaluated in laboratory conditions (Paunesca 2009). At present, the following biotechnological methods have been used to rapid multiply and conserve endangered, rare, crop ornamental, medicinal, and forest species, allowing the conservation of pathogen-free material, elite plants, and genetic diversity for short-, medium-, and long-term duration:

1. In vitro propagation
2. Mycorrhization
3. Genetic transformation
4. DNA banks
5. Cryopreservation

Table 7.1 List of important international research institutes focus to conserve the range of germplasm material

Name	International Research Institute	Area of interest
IRRI	International Rice Research Institute, Los Banos, Philippines	Tropical rice
CIMMYT	Centre International de-Mejoramients de Maize Trigo, El Baton, Mexico	Maize and wheat (Triticale, barely, sorghum)
CIAT	Center International de-agricultural Tropical Palmira, Columbia	Cassava and beans, (also maize and rice) in collaboration with CIMMYT and IRRI
IITA	International Institute of Tropical Agriculture, Ibadan, Nigeria	Grain legumes, roots, and tubers, farming systems
CIP	Centre International de-papa-Lima. Peru	Potato
ICRISAT	International Crops Research Institute, for Semi-Arid Tropics, Hyderabad, India	Sorghum, groundnut, pearl millet, bengal gram, red gram
WARDA	West African Rice Development Association, Monrovia, Liberia	Regional Cooperative Rice Research in Collaboration with IITA and IRRI
IPGRI	International Plant Genetic Research Institute, Rome Italy	Genetic conservation
AVRDC	The Asian Vegetable Research and Development Centre, Taiwan	Tomato, onion, peppers Chinese Cabbage

7.4.3.1 In Vitro Propagation

This approach is used for the generation and multiplication (micropropagation) of a large number of plants in a mass scale way throughout the year irrespective of seasonal dependence. The callus production from explants such as leaves, seeds, nodes, and tubers, thereafter the growth of shoots, then followed by rooting, finally leads to the development of whole plantelet under controlled conditions, which after an acclimatization process is able to grow in the open field area (Rai 2010). It is tension-free methodology because a stock of pathogen-free germplasm can be raised, made available, and maintained economically for many years.

Explants used for in vitro conservation must be of the right type and in good physiological stage condition.

In many cases, species generating orthodox seeds, or their inadequate seed development process does not fulfil the aim of seed collection among the plants, then excised zygotic embryos or vegetative tissues, such as budwoods, shoots, apices or leaves, or tubers or corms can be an alternative way for the conservation of germplasm. Apical and auxiliary meristems of very small size are the preferred explants for in vitro storage. In fact, organized explants have proved better than unorganized tissues, in terms of genetic stability of the germplasm (Mandal 2003; Chaudhury and Vasil 1993; Kameswara 2004; Ogbu et al. 2010). The advantages associated with in vitro propagation include clonal propagation, production of insect- and disease-free material, zero genetic erosion, short time required for obtaining new plants, less space- and less labour intensive, allows safe mode to internationally exchange plant material, and leads to genetic modifications such as somaclonal variations including stress factor resistance and production of useful compounds in the target germplasm.

Regeneration and successful propagation of genetically stable seedlings from cultures are prerequisites for any in vitro conservation effort. Generally, organized cultures such as shoots are used for slow growth storage because undifferentiated tissues such as callus are more vulnerable to somaclonal variation. Development of the slow growth technique has been optimized successfully for medium-term conservation of plant species (Engelmann and Drew 1998; Sarkar and Naik 1999; Ogbu et al. 2010) in which the target germplasm is stored as an aseptic plant under tissue culture conditions and maintained on nutrient gels for 1 to 15 years with intermittent subculturing, depending on the genotype of the species. Several approaches have been optimized for maintenance of the slow growth technique, but in most cases, conditions such as lower temperature combined with low light intensity or some cases in darkness parameters are easily approachable. There are some cold-tolerant plant species preferably grown in the temperate region of the globe; in that case temperatures require for maintaining slow growth should be employed within the range of 0 to 5 °C, but tropical plant species need a temperature between 15 and 20 °C as they are a heat-tolerant crop. There are other aspects feasible to slow down the growth of maintained plantlets: in vitro propagation technique by manipulating the composition of culture medium, use of growth retardants, media engineering, and cutting the supply of oxygen level available to cultures by covering with a layer of liquid medium or mineral oil (Withers and Engelmann 1997). There are a few species, such as *Musa* spp., potato, sweet potato, cassava, yam, *Allium* spp., and some temperate tree species, in which the developed slow growth procedures have been executed in a routine mode for the conservation of plant genetic resources.

Withers (2002) reported the first-time standardization of protocol and propagation for cocoa (*Theobroma cacao* L.) and coconut (*Cocos nucifera* L.) in vitro collecting methods. Rai (2010) reported a list of plants such as *Aquilaria malaccensis*, *Dioscorea deltoidea*, *Guaicum officinale*, *Hydrastis canadensis*, *Nardostachys grandiflora*, *Panax quinquefolius*, *Picrorhiza kurroa*, *Podophyllum hexandrum*, *Buchanania lanzan*, *Prunus africana*, *Pterocarpus santalinus*, *Rauwolfia serpentina*, *Saussurea costus*, *Gloriosa superba*, and *Taxus wallichiana* that have been categorized under 'critically endangered.' To save the remains of these endangered species, they should be conserved through in vitro propagation.

7.4.3.2 Mycorrhization

Mycorrhization is the name of the process of infestation of mycorrhizal fungi into the roots of target plants. Generally, these fungi develop a symbiotic relationship with 90 % of the plants (Williams et al. 1994), which helps in the uptake of nutrients, particularly phosphorus, from the soil system (Vestberg and Estaun 1994). This process induces various biochemical and metabolic changes present in the regenerate raised endangered medicinal plants co-cultured with beneficial microbes that help to improve their better survival and enhance their resistance against various forms of encountered biotic and abiotic stress.

7.4.3.3 Gene Transformation

Agrobacterium tumifaciens and *Agrobacterium rhizogenes* are potent beneficial strains used to transform endangered medicinal plants into developed resistant varieties against various stress conditions and a tendency to generate the overproduction of secondary metabolites in hormone-free culture conditions. These transformation processes performed under laboratory conditions help to improve the quantitative traits of plant genetic resources such as yield improvement, and the development of insect-, pest-, drought-, and salinity-resistant plants, so to bridge the gap of concerned consumers' demand for food security problem globally and to maintain these refined plant genetic resources in the conservation programme. To date more than 120 species belonging to 35 families have been transformed successfully transgenically by *Agrobacterium*-based techniques (Birch 1997), and the data are still increasing.

7.4.3.4 DNA Banks

The conservation of genome fragments or individual genes or entire genotypes in the form of library of "DNA samples" is similar to the "Gene Library," which acts as a source of important information to scientists worldwide for their future use. There are many species that do not produce seeds or are difficult to maintain in vegetative propagation and highly threatened in the wild population; in that situation, DNA banks may provide an alternative way to conserve these types of species for short-term duration until another method could be generated (Dulloo et al. 2006). The stored genetic material in the DNA banks can be maintained at −20 °C for up to 2 years for short- and midterm duration; for long-term storage, it should be maintained at −70 °C with the help of liquid nitrogen gas. There are some important DNA banks such as The Royal Botanic Garden, the US Missouri Botanical Garden, the Australian Plant DNA Bank of Southern Cross University, and Leslie Hill Molecular Systematics Laboratory of the National Botanical Institute, established to procure and conserve DNA specimens representative of all plant families. Of these, The Royal Botanic Garden (UK) is considered the world's largest and the most comprehensive PGR DNA bank, encompassing more than 20,000 DNA specimens representative of all plant families.

7.4.3.5 Cryopreservation

Cryopreservation refers to the nonlethal storage of biological tissues at ultra-low temperatures, mostly at −196 °C by liquid nitrogen (LN). At this temperature, all activities in the vegetative propagated and recalcitrant seed species such as cell division and cellular metabolism come a halt, and thus plant material can be stored for long-term conservation of germplasm. These methods do not require subculture steps for long-term tissue culture storage, and a reduced level of somaclonal variation is observed (Martin et al. 1998). In cryopreservation, only shoot apices or meristem cultures are highly recommended as we know these are virus-free plant material and easily recovered when we revive the culture after a prolonged time (Scowcroft 1984). This method causes no change in viability, vigour, and genetic makeup of the conserved materials (Ogbu et al. 2010) and provides a good method for conservation of the species, especially woody plant germplasm (Panis and Lambardi 2005).

Using the cryogenic approach, very limited space and limited maintenance are required, and the stored plant material is protected from exogenous contamination. The excised zygotic embryo cultures are considered good for cryopreservation in species producing recalcitrant seeds, which has been successfully employed for the cryopreservation of coconut (Assy-Bah and Engelmann 1992), cocoa (Chandel et al. 1995), walnut (De Boucaud et al. 1991), jack fruit (Chandel et al. 1995), and neem (Berjak and Dumet 1996).

The common steps involved in the cryopreservation technique are the removal of all freezable water content from tissues by physical or osmotic dehydration, followed by ultra-rapid freezing (Kaviani 2011). The various techniques currently in use include classical and new cryopreservation techniques (Kameswara 2004; Withers and Engelmann 1998). Classical techniques involve freeze-induced dehydration, whereas the new technique is based on vitrification. Vitrification is defined as the transition of water content directly from the liquid phase into an amorphous phase or glass to avoid the formation of crystalline ice (Fahy et al.1984).

The classical technique of freezing procedures involved the following successive steps: pregrowth of samples, cryoprotection, slow cooling (0.5–2.0 °C/min) to a determined pre-freezing temperature (usually around −40 °C), rapid immersion of samples in liquid nitrogen, storage, rapid thawing, and recovery. To maintain the integrity of the cell membrane and increase osmotic potential of the external medium, cryoprotectants should be added to the freezing mixtures. This approach requires sophisticated and expensive programmable freezers that are not recommended for cryopreservation.

In vitrification-based procedures, cell dehydration is performed before freezing by exposure of samples to concentrated cryoprotective media and/or air desiccation. This stage is followed by rapid cooling; as a result, there will be no intracellular ice formation to take place. It offers more advantages for the cryopreservation of complex organs such as shoot tips and embryos as it consists of different types of cell with different requirements under conditions of freeze-induced dehydration as compared with classical freezing techniques. The vitrification-based procedures are operationally less complex than classical ones as they do not require the use of controlled and programmable freezers. It has greater potential for broad applicability; requiring only minor modifications for different cell types (Engelmann 1997). The seven vitrification-based procedures for cryopreservation (Engelmann 2000) include the following.

1. *Pregrowth*

   The pregrowth technique consists of cultivating samples in the presence of cryoprotectants and then freezing them rapidly by direct immersion in liquid nitrogen. This technique has been developed and standardized successfully for *Musa* meristematic cultures (Panis 2007).

2. *Dehydration*

   This is the simplest vitrification-based procedure, in which explant material should be dehydrated properly and subsequently frozen swiftly with the help of liquid nitrogen. It has been applied to the excised embryos of a large number of recalcitrant and intermediate species (Engelmann 2000). The dehydration process is usually carried out by the air current of a laminar airflow cabinet and by using a flow of sterile compressed air or silica gel. Ultra-rapid drying in a stream of compressed dry air allows freezing samples with relatively higher water content, thus reducing the level of desiccation injury. However, optimal survival is generally obtained when samples are frozen with water content between 10 and 20 % (fresh weight basis).

3. *Pregrowth dehydration*

   This process involves pregrowth of explants in the presence of cryoprotectants, then under the laminar airflow cabinet or with silica gel; the dehydration process is then afterwards followed by rapid freezing. This method has been successfully standardized in asparagus stem segments, polyembryonic cultures in oil palm, and coconut zygotic embryos (Uragami et al. 1990; Dumet et al. 1993).

4. *Encapsulation dehydration*

   The encapsulation dehydration procedure comprises the preculture of alginate-coated samples in liquid medium having a high sucrose concentration, called encapsulation; with the subsequent desiccation process, followed by rapid cooling and finally slow rewarming, this method has been very effective for cryopreservation of different plant species (Gonzalez-Arnao and Engelmann 2006). To obtain maximum recovery, the encapsulated samples should be placed on standard culture medium without extracting the explants from their alginate coating (Gonzalez-Arnao and Engelmann 2006). The survival rate of explants preserved after thawing is independent of the warming rate, for example, in carrot somatic embryos (Dereuddre et al. 1991) and orchid seeds (Wood et al. 2000). Mandal et al. (2009) reported 53.3 % recovery of growth of the embryogenic culture of *Dioscorea bulbifera* using these procedures.

5. *Vitrification*

   Vitrification is a freezing procedure that comprises a pretreatment (loading treatment) at room temperature, followed by exposure to a vitrification solution at 25°° or 0 °C, rapid cooling and warming, and final removal of the vitrification solution by washing samples with an unloading solution consisting of liquid culture medium supplemented with 1.2 M sucrose (Withers and Engelmann 1998).

6. *Encapsulation vitrification*

   In a combination of the encapsulation-dehydration and vitrification process, samples are encapsulated in alginate beads, and then subjected to freezing following the vitrification approach.

7. *Droplet vitrification*

   The droplet vitrification technique is characterized by increased cooling and warming rates compared to other vitrification-based procedures, because samples are frozen in minute droplets of polyvinyl siloxane (PVS) placed on aluminium foil strips, which are plunged directly into liquid nitrogen. This protocol significantly increases the probability of obtaining a vitrified state during freezing and of avoiding devitrification during warming (Panis and Lambardi 2005).

The establishment of a National Cryobank at the National Bureau of Plant Genetic Resources (NBPGR), New Delhi, has sole responsibility to conserve desiccation-sensitive seeds, vegetative tissues, pollen, and selected orthodox seed species. Presently about six large capacity cryotanks are capable of accommodating 30,000 to 40,000 samples of varied germplasm in the form of seeds, embryo, embryonic axes, shoot apices. and pollen for conservation. The large-scale applications of cryopreservation techniques applied to different plant germplasms are presented in Table 7.2.

7.5 Genetic Diversity of Rare and Endangered PGR and TK of Germplasm

The level of genetic variation decides the fate of survival for the short- and long-term duration of plant species in their natural growth habitat. Genetic variation depends on demographic history, distributional pattern in the geographic range, the type of mating in their reproductive systems, the availability of pollen, and seed dispersal. Successional stages are among the factors that affect genetic variation patterns among plant populations (Hamrick and Godt 1989; Nybom and Bartish 2000; Nybom 2004). However, the population size of rare and endangered PGR and TK of germplasm has decreasing gradually because of indiscriminant overexploitation of these resources, leading to loss of the level of genetic diversity. Therefore, it is imperative to interpret the importance of genetic diversity and the composition and structure of populations of rare and endangered plant species in their natural habitat to develop effective strategies for their conservation for in vivo as well as in vitro methods and gene manipulation (Gitzendanner and Soltis 2000).

Molecular marker biotechnology can be used for accurate evaluation of genetic diversity present in the PGR and TK of germplasm, including rare and endangered species. The main advantages associated with this technique are assessment at the DNA level without environmental impact, requiring only a tiny amount of material for evaluation. These analyses can evaluate at any developmental stage, using any plant part, are not seasonally based, and require only a very small amount of plant material. Associated advantages are that no technical expertise is required; a good amount of polymorphism is detected; and the analysis is sensitive, reproducible, and less expensive (Karp 2002; Karp et al. 1996). Earlier the level of variability present among the collected specimens at DNA level had been detected on the basis of restriction enzymes, as these recognize and act on restriction sites consisting of short sequences of DNA randomly present in the genome; this technique is called restriction fragment length polymorphism (RFLP). Later, the innovation of the polymerase chain reaction (PCR) technique by Kary and Faloona (1982) resulted in the generation of various types of molecular techniques that depend upon using short oligonucleotide primers and the nature of their binding to the target genome. Karp reported that techniques

Table 7.2 Large-scale application of cryopreservation techniques to different plant germplasm

S. no.	Conserve germplasm	Gene bank/country
1	Endangered medicinal plants of 50 different species	The National Bureau for Plant Genetic Resources (NBPGR), New Delhi, India
2	Rare or threatened species of more than 110 accessions	Kings Park and Botanic Garden, Perth, Australia
3	Collection of the seeds of 450 accessions of coffee germplasm	IRD Montpellier, France
4	2,200 accessions of dormant buds of apple	National Centre for Genetic Resources (CNGR), Fort Collins, USA
5	420 accessions of dormant buds of mulberry tree	National Institute of Agrobiological Resources (NIAR), Yamagata, Japan
6	630 accessions of shoot-tips of banana	INIBAP International Transit Centre, Catholic University of Leuven, Belgium
7	540 accessions of shoot-tips of cassava	International Centre for Tropical Agriculture (CIAT), Cali, Colombia
8	Pollen of 13 pear cultivars and 24 *Pyrus* species	National Centre for Genetic Resources (CNGR), Fort Collins, USA
9	700 accessions of traditional Chinese flower species	College of Landscape Architecture, Beijing Forestry University, Beijing, China
10	1,000 callus strains of species of pharmaceutical interest	Phytera, Sheffields, UK
11	Embryogenic cell lines of Several thousand conifer tree	Sylvagen, Vancouver, Canada
12	Embryogenic cell lines of coffee and cacao	Biotechnology Laboratory of the Nestle Company, Notre Dame d'Oé, France

such as Random Amplified Polymorphic DNA (RAPD), amplified fragment length polymorphism (AFLP), and microsatellites (SSRs), based on PCR, have proved successful in diversity studies of various plant species.

Molecular marker technology paves the way to ease the problems of taxonomy and phylogenetic relationships, which helps to devise the methodology of a breeding programme and effective conservation to improve the quality of germplasm. Molecular markers have been applied to study genetic diversity from natural populations and to formulate efficient strategies to capture maximum variation for genetic resource conservation (Nebauer et al. 1999; Huang et al. 2000). To facilitate better understanding of the extent and structure of diversity, molecular marker technology has proven extremely helpful to demarcate core collections of germplasm in population study by phylogenetic tree analysis, to identify the level of redundancy among germplasm collections, development of core collections, protection of plant species by adulteration with other varieties, the level of genetic drifts/shifts, testing accessions stability, and working as a supportive role of the conclusive development of efficacy operational strategies for ex situ and in situ conservation programmes of rare and endangered PGR and TK of germplasm (Drummond et al. 2000; Sharma et al. 2000; Sica et al. 2005; Meloni et al. 2006).

7.6 Germplasm Exchange of Rare and Endangered PGR and TK

Healthy germplasm material is prerequisite for storage and exchange for conservation and the judicious use of plant genetic resources. To determine the level of infected germplasm before conservation, the health of the seed should be determined by seed quality tests, and phytosanitary regulations should be included in the conservation programme. Back in the 1970s, earlier seed health tests performed on the basis of incubation period of pathogens present in the seed were labour intensive, generated false results, were time consuming, and required a huge amount of seed material for detection. The situation became miserable when only a limited amount of seed material was available for conservation. Some pathogens that do not show any type of symptoms, such as viruses and viroids, pose great risk associated during the exchange of germplasm as they cannot be detected under traditional methods. Some crops are propagated by a vegetative mode of reproduction that provides a conducive platform for virus and viroid transmission through these vegetative propagules. Because of these pitfalls many genebanks and plant health testing laboratories accept biotechnology approaches for the detection of pathogens from germplasm collections.

Earlier, an Enzyme-Linked Immunosorbent Assay (ELISA) was developed, using specific lettuce mosaic virus (LMV) antiserum replacing the standard seed assays for the detection of LMV in the lettuce crop (Falk and Purciful 1983). Later, the PCR-based technique was used to identify the suspected pathogens present in the germplasm as it is considered a more sensitive and specific approach for early detection. DNA primers specific to several important seedborne pathogens have been designed and used for detection by the PCR technique within a lesser timeframe. The designed specific primer of virulence-specific gene of causative agent used in the PCR-based diagnostic technique has been reported in many crop diseases, such as blackleg of crucifers (Taylor 1993), halo blight of beans (Schaad et al. 1995), bacterial canker of tomato (Santos and Cruz 1997), and bacterial fruit blotch pathogen in watermelon (Minsavage et al. 1995); also, hybridization-based use of a pathogen-specific probe consisting of an oligonucleotide tagged with fluorescent marker has been used in a Southern blotting technique to detect the coconut cadang-cadang viroid (Hodgson and Randles 1997).

To safely transport diseased free germplasm from one destination to another, it should be preserved as in vitro cultures having no need to worry about issues such as transportation charges, retaining septic conditions during the exchange process, requiring a volume space for storage, and restoring the regeneration capacity of exchange germplasm material into fully developed plantlets after receipt.

7.7 Conclusion

Conservation and sustainable use of attentive plant genetic resources and traditional knowledge of germplasm is essential so that we can fill in the gaps for the demands of food security. Advances in biotechnology, especially in the areas of in vitro culture techniques, cryopreservation, and molecular biology, have generated new opportunities for improved conservation, utilization, and operation of plant genetic resources. Biodiversity, the basis for sustaining life on the planet, considers the twenty-first century to be the period of bio-economy, consisting of bioscience and biotechnology fields. After agricultural, industrial, and information technology, bioeconomy may become the fourth emerging economy having the key potential to sustainable development in agriculture, forestry, and environmental protection, and will help to boost light industry, food supply, healthcare, and other micro-economy aspects. Besides these attempts, it would be better for the involvement of local communities in biodiversity conservation programs to make it more profitable and successful. To achieve this, an ecotourism concept has been introduced to promote sustainable, low-impact tourism over protected areas as a source of generating income for local communities, and economic incentives should be set for those local communities that are associated with conservation of natural ecosystems in a traditional way. Thus, a planned roadmap vision is needed for the conservation of both untapped and exploited plant genetic resources and traditional knowledge of germplasm, so that the forthcoming generations of human beings have available the benefits of goldmine resources and can meet their sustainable development.

References

Alves RR, Rosa IL (2007) Biodiversity, traditional medicine and public health: where do they meet? J Ethnobiol Ethnomed 3:1–9

Assy-Bah B, Engelmann F (1992) Cryopreservation of mature embryos of coconut (*Coccos nucifera* L.) and

subsequent regeneration of plantlets. Cryo Lett 13:117–126
Bajaj YPS (1995) Cryopreservation of plant cell, tissue and organ culture for the conservation of germplasm and biodiversity. In: Bajaj YPS (ed) Biotechnology in agriculture and forestry cryopreservation of plant germplasm I. Springer, New York, pp 3–18
Benson EE (1999) Cryopreservation. In: Benson EE (ed) Plant conservation biotechnology. Taylor & Francis, London, pp 83–95
Berjak P, Dumet D (1996) Cryopreservation of seeds and isolated embryonic axes of neem (*Azadirachta indica*). Cryo Lett 17:99–104
Birch RG (1997) Plant transformation: problems and strategies for practical application. Annu Rev Plant Physiol Plant Mol Biol 48:297–326
Chandel KPS, Chaudhury R, Radhamani J, Malik SK (1995) Desiccation and freezing sensitivity in recalcitrant seeds of tea, cocoa and jackfruit. Ann Bot 76:443–450
Chaudhury MKU, Vasil IK (1993) Molecular analysis of plant regenerated from embryogenic cultures of apple. Genetics 86:181–188
Das JS (2008) The largest genetic paradise of India lacks biotechnological implementation. Curr Sci 94:558–559
De Boucaud M, Brison M, Ledoux C, Germain E, Lutz A (1991) Cryoperservation of embryonic axes of recalcitrant seed: *Juglans regia* L. cv. Franquette. Cryo Lett 12:163–166
Dereuddre J, Blandin S, Hassen N (1991) Resistance of alginate-coated somatic embryos of carrot (*Daucus carota* L.) to desiccation and freezing in liquid nitrogen: effects of preculture. Cryo Lett 12:125–134
Dhillon BS, Saxena S (2003) Conservation and access to plant genetic resources. In: Mandal BB, Chaudhury R, Engelmann F, Mal B, Tao KL, Dhillon BS (eds) Conservation biotechnology of plant germplasm. NBPGR, New Delhi/IPGRI, Rome/FAO, Rome, pp 3–18
Drummond RS, Keeling DJ, Richardson TE, Gardner RC, Wright SD (2000) Genetic analysis and conservation of 31 surviving individuals of a rare New Zealand tree, *Metrosideros bartlettii* (Myrtaceae). Mol Ecol 9:1149–1157
Dulloo E, Nagamura Y, Ryder O (2006) DNA storage as a complementary conservation strategy. In: de Vicente MC, Andersson MS (eds) DNA banks-providing novel options for gene banks? Topical reviews in agricultural biodiversity. International Plant Genetic Resources Institute, Rome
Dumet D, Engelmann F, Chabrillange N, Duvall Y (1993) Cryopreservation of oil palm (*Elaeis guinensis* Jacq.) somatic embryos involving a desiccation step. Plant Cell Rep 12:352–355
Engelmann F (1997) In vitro conservation methods. In: Ford-Lloyd BV, Newburry JH, Callow JA (eds) Biotechnology and plant genetic resources: conservation and use. CABI, Wellingford, pp 119–162
Engelmann F (2000) Importance of cryopreservation for the conservation of plant genetic resources. In: Engelmann F, Takagi H (eds) Cryopreservation of tropical plant germplasm. Current research progress and application. IPGRI, Rome, pp 8–20
Engelmann F, Drew RA (1998) In vitro germplasm conservation. Acta Hortic 461:41–47
Engels J, Visser B (2006) Genebank management: effective management of germplasm collection. Training manual on conservation, management and use of plant genetic resources in food and agriculture. Wageningen University and Research, Wageningen
Fahy GM, MacFarlane DR, Angell CA, Meryman HT (1984) Vitrification as an approach to cryopreservation. Cryobiology 21:407–426
Falk BW, Purciful DE (1983) Development and application of an enzyme-linked immunosorbent assay (ELISA) test to index lettuce seeds for lettuce mosaic virus in Florida. Plant Dis 67:413–416
FAO (1996) State of world's plant genetic resources for food and agriculture. Food and Agriculture Oranization, Rome
Gitzendanner MA, Soltis S (2000) Patterns of genetic variation in rare and widespread plant congeners. Am J Bot 87:783–792
Gonzalez-Arnao MT, Engelmann F (2006) Cryopreservation of plant germplasm using the encapsulation–dehydration technique: review and case study on sugarcane. Cryo Lett 27:155–168
Groombridge B, Jenkins MD (2000) Global biodiversity: Earth's living resources in the 21st century. World Conservation Monitoring Centre, Cambridge, UK
Hamrick JL, Godt MJW (1989) Allozyme diversity in plant species. In: Brown AHD, Clegg, Kahler MT, Weirn AL (eds) Plant population genetics, breeding and genetic resources. Sinauer, Sunderland, pp 43–63
Hodgson RAJ, Randles JW (1997) Diagnostic oligonucleotide-probe (DOP) hybridization to detect coconut cadang-cadang viroid. In: Diekmann M (ed) Viroid-like sequences of coconut. ACIAR and IPGRI, Kuala Lumpur, Malaysia, Canberra
Huang HW, Layne DR, Kubisiak TL (2000) RAPD inheritance and diversity in pawpaw (*Asimina triloba*). J Am Soc Hortic Sci 125:454–459
Jimu L (2011) Threats and conservation strategies for the African cherry (*Prunus africana*) in its natural range: a review. Ecol Nat Environ 3:118–130
Kameswara RN (2004) Plant genetic resources: advancing conservation and use through biotechnology. Afr J Biotechnol 3:136–145
Karp A (2002) The new genetic era: will it help us in managing genetic diversity? In: Engels JMM, Ramanatha RV, Brown AHD, Jackson MT (eds) Managing plant genetic diversity. CAB International and IPGRI, Wallingford, pp 43–56
Karp A, Seberg O, Buiatti M (1996) Molecular techniques in the assessment of botanical diversity. Ann Bot 78:143–149

Kary BM, Faloona FA (1982) Specific synthesis g DNA in vitro via a polymerase catalysed chain reaction. Methods Enzymol 155:335–350.

Kasagana VN, Karumuri SS (2011) Conservation of medicinal plants (past, present & future trends). J Pharm Sci Res 3:1378–1386

Kaviani B (2011) Conservation of plant genetic resources by cryopreservation. Aust J Crop Sci 5:778–800

Leisa (2004) Valuing crop diversity. LEISA Mag 20:4–5

Long CL, Li H, Ouyang ZQ, Yang XY, Li Q, Trangmar B (2003) Strategies for agrobiodiversity conservation and promotion: a case from Yunnan, China. Biodivers Conserv 12:1145–1156

Mandal BB (2003) Cryopreservation techniques for plant germplasm conservation. In: Mandal BB, Chaudhury R, Engelmann F, Bhag Mal KL, Tao, Dhillon BS (eds) Conservation biotechnology of plant germplasm. NBPGR, New Delhi; IPGRI, Rome; FAO, Rome, pp 187–207

Mandal BB, Dixit SS, Srivastava PS (2009) Cryopreservation of embryogenic cultures of *Dioscorea bulbifera* L. by encapsulation-dehydration. Cryo Lett 30:440–448

Martin C, Iridono JM, Benito-Gonzales E, Perez C (1998) The use of tissue culture techniques in the conservation of plant biodiversity. Agron Food Ind Hi Tech 9:37–40

Meloni M, Perini D, Filigheddu R, Binelli G (2006) Genetic variation in five Mediterranean populations of *Juniperus phoenicea* as revealed by inter-simple sequence repeat (ISSR) markers. Ann Bot 97:299–304

Minsavage GV, Hoover RJ, Kucharek TA, Stall RE (1995) Detection of the watermelon fruit blotch pathogen on seeds with the polymerase chain reaction. Phytopathology 85:1162

Myers N, Mittermeier RA, Mittermeier CG, da Fonseca GAB, Kent J (2000) Biodiversity hotspots for conservation priorities. Nature (Lond) 403:853–858

Nebauer SG, Castillo-Agudo L, Segura J (1999) RAPD variation within and among natural populations of outcrossing willow-leaved foxglove (*Digitalis obscura* L.). Theor Appl Genet 98:985–994

Nybom H (2004) Comparison of different nuclear DNA markers for estimating intraspecific genetic diversity in plants. Mol Ecol 13:1143–1155

Nybom H, Bartish IV (2000) Effects of life history traits and sampling strategies on genetic diversity estimates obtained with RAPD markers in plants. Perspect Plant Ecol Evol Syst 3:93–114

Ogbu JU, Essien BA, Essien JB, Anaele MU (2010) Conservation and management of genetic resources of horticultural crops in Nigeria: issues and biotechnological strategies. J Hortic For 2:214–222

Pandey G (2011) Some important anticancer herbs: a review. Int J Pharm Stud Res 2:32–38

Panis B (2007) Fundamental aspects of plant cryopreservation. Training manual on in vitro and cryopreservation techniques for conservation of PGR. NBPGR and Biodiversity International, New Delhi

Panis B, Lambardi M (2005) Status of cryopreservation technologies in plants (crops and forest trees). The role of biotechnology. Villa Gualino, Turin, pp 43–54

Paunesca A (2009) Biotechnology for endangered plant conservation: a critical overview. Rom Biotechnol Lett 14:4095–4104

Rai MK (2010) Review: Biotechnological strategies for conservation of rare and endangered medicinal plants. Biodicersitas 11:157–166

Ramanatha RV, Riley R (1994) The use of biotechnology for conservation and utilization of plant genetic resources. Plant Genet Resour Newsl 97:3–20

Rao NK (2004) Plant genetic resources: advancing conservation and use through biotechnology. African J Biotechnol 3:136–145

Rathore DS, Srivastava U, Dhillon BS (2005) Management of genetic resources of horticultural crops: issues and strategies. In: Dhillon BS, Tyagi RK, Saxena S, Randhawa GJ (eds) Plant genetic resources: horticultural crops. Narosa Publishing, New Delhi, pp 1–18

Rodgers WA, Panwar HS, Mathur VB (2002) Wildlife protected areas in India: a review (executive summary). Wildlife Institute of India, Dehradun, p 44

Santos SM, Cruz L (1997) A rapid and sensitive detection of *Clavibacter michiganensis* subsp. *michiganensis* in tomato seed by polymerase chain reaction. Seed Sci Technol 25:581–584

Sarkar D, Naik PS (1999) Factors effecting minimal growth conservation of potato microplant in vitro. Euphytica 102:275–280

Schaad NW, Cheong SS, Tamaki S, Hatziloukas E, Panapoulos NJ (1995) A combined biological and enzymatic amplification (BIO-PCR) technique to detect *Pseudomonos syringe* pv. *phaseolicola* in bean seed extracts. Phytopathology 85:243–248

Scowcroft WR (1984) Genetic variability in tissue culture: impact on germplasm conservation and utilization. International Board for Plant Genetic Resources Secretariat, Rome, p 42

Sharma SK (2007) Indian plant genetic resources (PGR) system: role of NBPGR. Training manual on in vitro and cryopreservation techniques for conservation of PGR. NBPGR and Bioversity International, New Delhi

Sharma KD, Brij Singh BM, Sharma TR, Katoch M, Guleria S (2000) Molecular analysis of variability in *Podophyllum hexandrum* Royle, an endangered medicinal herb of northwestern Himalaya. Plant Genet Resour Newsl 124:57–61

Sica M, Gamba G, Montieri S, Gaudio L, Aceto S (2005) ISSR markers show differentiation among Italian populations of *Asparagus acutifolius* L. BMC Genet 6:17

Swanson T, Pearce D, Cervigni R (1994) The appropriation of the value of plant genetic resources for agriculture. Commission for Plant Genetic Resources, Washington, DC

Tandon P, Kumaria S (2005) Prospects of plant conservation biotechnology in India with special reference to northeastern region. In: Tandon P, Kumaria S (eds) Biodiversity status and prospects. Norasa Publishing House, New Delhi, pp 79–92

Tao KL (2003) Complementary conservation strategy for plant genetic resources. In: Mandal BB, Chaudhury R, Engelmann F, Mal B, Tao KL, Dhillon BS (eds) Conservation biotechnology of plant germplasm. NBPGR, New Delhi/IPGRI, Rome/FAO, Rome, p 51

Taylor JL (1993) A simple, sensitive and rapid method for detecting seed contaminated with highly virulent *Leptosphaeria maculans*. Appl Environ Microbiol 59:3681–3685

Temitope IB (2013) The state of ex situ conservation in Nigeria. Int J Conserv Sci 4:197–212

Uragami A, Sakai A, Nagai M (1990) Cryopreservation of dried axillary buds from plantlets of *Asparagus officinalis* L. grown in vitro. Plant Cell Rep 9: 328–331

Vestberg M, Estaun V (1994) Micropropagated plants, an opportunity to positively manage mycorrhizal activities. In: Gianinazzi S, Schuepp H (eds) Impact of arbuscular mycorrhizas on sustainable agriculture and natural ecosystems. Birkhauser, Basel

Vickers A, Zollman C (1999) Herbal medicine. BMJ 319:1050–1053

Williams PG, Roser DJ, Seppelt RD (1994) Mycorrhizas of hepatics in continental Antarctica. Mycol Res 98:34–36

Withers LA (2002) In vitro collecting: concept and background. In: Pence VC, Sandoval JA, Villalobos VM, Engelman F (eds) Vitro collecting techniques for germplasm conservation. International Plant Genetic Resources Institute, Rome, pp 16–25

Withers LA, Engelmann F (1997) In vitro conservation of plant genetic resources. In: Altman A (ed) Biotechnology in agriculture. Dekker, New York, pp 57–88

Withers LA, Engelmann F (1998) In vitro conservation of plant genetic resources. In: Altman A (ed) Biotechnology in agriculture. Dekker, New York, pp 57–88

Wood CB, Pritchard HW, Miller AP (2000) Simultaneous preservation of orchid seed and its fungal symbiont using encapsulation-dehydration is dependent on moisture content and storage temperature. Cryo Lett 21:125–136

Zschocke S, Van SJ (2000) *Cryptocarya* species: substitute plants for *Ocotea bullata*? A pharmacological investigation in terms of cyclooxygenase-1 and-2 inhibition. J Ethnopharmacol 71:473–478

In Situ and Ex Situ Conservation of Plant Genetic Resources and Traditional Knowledge

8

Deepak Rajpurohit and Tripta Jhang

Abstract

The efficient conservation and judicious use of plant genetic resources and related traditional knowledge is vital for food, health, and nutritional security. Plant genetic diversity available in gene banks, in the wild, and in farmer's fields is crucial resource for food, health, and nutritional security, augmenting livelihoods and environmental services. These resources could support acclimatization of primary food crops to climate change impacts. In pursuit of improved cereal, pulse, horticultural, medicinal, aromatic, and other cash crop varieties suitable for particular agroecosystem, availability of a diversity of alleles/genes in terms of yield, quality, resistance against disease and pest, and tolerance for drought, heat, salt, and water in the wild relatives, indigenous, local land races, and farmer cultivars is the prime prerequisite, which limits the options for sustainable resilient agriculture management. To prevent their genetic erosion, a well-devised strong linkage between in situ resource conservation, collection and ex situ conservation in gene banks and their utilization for genetic improvement through research and breeding is required for their subsequent availability to farmer, pharmaceutical and seed community. Conserving plant genetic resources in situ as well as ex situ ensures efficient explorations in order to implement the International Treaty and Global Plan of Action on Plant Genetic Resources for Food and Agriculture. These collections are valuable genetic resource of resistance to diseases and pests, tolerance to climatic stresses, and improved targeted "trait" through crop improvement and assurance against genetic erosion. It is

D. Rajpurohit, Ph.D. (✉)
Section of Biotechnology, Department of Post Harvest Engineering and Technology, AICRP on PHET, CTAE, Maharana Pratap University of Agriculture and Technology,
Udaipur, Rajasthan, India
e-mail: deepakrajpurohit2001@yahoo.com

T. Jhang, Ph.D.
Department of Genetic Resource Management, Genetics and Plant Breeding Division, CSIR-Central Institute of Medicinal and Aromatic Plants,
Lucknow, Uttar Pradesh, India
e-mail: t.jhang@cimap.res.in

R.K. Salgotra, B.B. Gupta (eds.), *Plant Genetic Resources and Traditional Knowledge for Food Security*, DOI 10.1007/978-981-10-0060-7_8

important to support in situ/on-farm conservation of agroecosystems because ex situ conservation cannot replicate the evolutionary processes and cultural practices but can act as crucial reservoir to safeguard them from calamities.

8.1 Introduction

Conserving the Plant Genetic Resources (PGR) and the related traditional knowledge is essential for achieving sustainable food, health, and nutritional security of the world. As per the Convention on Biological Diversity (CBD), "plant genetic resources and traditional knowledge include all components of biological diversity of relevance to food and agriculture and that constitute agro-ecosystems; the variety and variability of animals, plants and micro-organisms, at the genetic, species and ecosystem levels, which are necessary to sustain key functions of the agricultural ecosystem, its structure and processes." Plant genetic resources, i.e., land races, traditional varieties, modern cultivars, transgenic lines, and their wild species, serve as the base material for genetic improvement and breeding programs. The existing diversity of genetic resources, the efforts of about 10,000 years, today serves as the basis of modern high-yielding, disease-resistant, quality, and nutritionally rich crop varieties. Plant genetic resources have been used variously since onset of human civilization for food, clothing, shelter, energy, feed for domestic animals, fiber, and medicine, etc.

Evolutionary and human selection pressure over the years on the wild ancestors resulted into current genetic constellations of useful alleles. The genetic diversity arising from the encrypted variation in the DNA and their fine-tuned transcription regulation help a crop to adapt to climatic change and continually evolve under these stresses. Monetary gains attract farmers to accept high-yielding varieties, which results into loss of potentially useful germplasm, the landraces and traditional varieties, and affects the social organization of tribal groups because the new varieties and their associated technologies are accessible only to certain farmers most favored in terms of access to purchase, technical assistance, and markets (Altieri and Merrick 1987). Local landraces are still under cultivation by small farmers than rich farmers, and hence, to make crop landraces and indigenous variety conservation realistic, these programs should be integrated with participatory breeding, rural development, ensured livelihood, a stakeholder ship, and a right in their value addition marketing. For sustainability of these resources, genetic resource conservation, diversification, improvement, adaptation, and delivery to farmers through seed systems are required. Plant breeder acts as bridge between the seed conservation in gene banks and the seed production units that deliver seeds of the improved varieties to farmers. Sustainable use of PGRFA takes into consideration meeting the basic food needs, providing livelihood to marginal farmers, and protecting the environment. Agro-biodiversity management can be implemented through intensification of production; trait-targeted breeding; genetic, molecular, and metabolic characterization; evaluation of core collections; genetic enhancement and base-broadening; diversification and co-cultivation of different crops ensuring broader diversity; soil nutrient efficiency; biological control of pest; and pathogen dissemination. Simultaneously, the markets should be developed for "value-added" local varieties and underutilized crops and their "diversity-rich-"derived products. The erosion of genetic resources and climate change-induced calamities like floods, drought, salinity, poor soils, and extreme temperatures risk the world's food security (FAO 2009). Adoption of modern agricultural practices by the developing countries by cultivating high-yielding synchronized cultivars over larger areas has ceded indigenous varieties jeopardizing agricultural productivity to yield-limiting factors, e.g., Southern corn leaf blight epidemic in the United States in years 1969–1970 (Frankel and Bennett 1970; Frankel and Hawkes 1975; Harlan 1975; Barlett 1980). Establishment of agroecosystems away from centers of origin loses inherent multigenic genetic defenses against pathogens and insect pests, making crops prone to epidemic attack than in an undisturbed traditional agroecosystem (Rick 1973; Segal et al. 1980; Altieri 1983).

The enormous diversity of crop plants and the spectrum of varieties evolved and generated through man and environment, promise basic food needs of the world to date amidst prevalent biotic and abiotic stresses and climate change. Presently, only about 30 crops provide 95 % of human food energy needs, 4 of which (rice, wheat, maize, and potato) are responsible for more than 60 % of man's energy intake (FAO 2015). Since the global food security is based on small number of crops, it becomes mandatory to widen the genetic base of these crops. There should be extensive worldwide effort for the search of new crops of comparable nutritional value and quality from orphan or underutilized food crops. Tapping of underutilized and neglected food resources which are generally nutritionally rich and are adapted to low input agriculture, being produced on poor lands by the poor, is the promising new avenue for local food availability and access by the poor. The use of these species, whether wild, managed, or cultivated, can have immediate consequences on the food security and well-being of the poor and plays a key role in keeping cultural diversity alive. The population of the world is reaching above 7.3 billion, so the basic human right of access to nutritious food continuously requires attention. Asia in recent past has witnessed a decline in food production and supplies as a result of floods due to typhoons and tsunamis in India, the Philippines, Vietnam, Korea, Myanmar, Thailand, and other countries and strong earthquakes in Nepal, India, and Japan. These events indicate that agriculture will have to adapt to the conditions brought about by climate change and extreme weather events (Smith et al. 1996). In the Pacific an entire crop was decimated by disease, the taro leaf blight disease. New strains of pests and diseases such as wheat rust strains of *Ug99* are constantly emerging (FAO 2009). Through collections of natural variation in crops and in their wild relatives stored either, ex situ, in gene banks, or in situ, growing "better crops" in future will be secured.

Internationally, there is strong recommendation to allow equitable access and sharing of the benefits from use of plant genetic resources as principally, all countries benefit from germplasm that comes from another country. The chili pepper, maize, and tepary bean that evolved in Mexico and rice that evolved in Latin America are now the staple diet of most of the continents. Tepary bean (*Phaseolus acutifolius*), native to the US and Mexican arid areas, when crossed with common bean (*Phaseolus vulgaris*) has led to new varieties with climate resilience as well as nutritional advantages such as high iron content (Brush SB 1982; Egger K 1981). Sources of salt tolerance that may be transferred to crops are found in wild relatives of azuki bean and mung bean that grow commonly along the seashores of Asian countries. An agricultural strategy based on a diversity of plants and cropping systems can bring moderate to high levels of productivity through manipulation and exploitation of the resources internal to the farm and can be sustainable at a much lower cost and for a longer period of time. To salvage germplasm before it is lost by assuring its introduction into germplasm, both public (governments, genetic resource organizations, and plant breeders) and private (peasant, NGO) groups should cohere the efforts in a strategized manner. Strategies for plantation development should be put in place to alleviate the pressure on natural forests. Development of appropriate downstream industries is needed. Wherever possible, local communities and small landholders should be involved in conservation activities (Brown 1983).

8.2 In Situ Conservation

In situ conservation, i.e., conserving the biodiversity in their natural habitats wherein, the niche environment imparts them unique characteristics is generally achieved through conserving "on farm" and by means of "genetic reserves".

The Indian subcontinent, with dense biodiversity harbors around 49,000 species of plants,

17,500 being flowering plants species. India geographically localized between 80.4N′ and 370.6N′ experiences extreme ranges of temperature. The country topography maps vast mountainous ranges of 100 m hectare deserts and semiarid regions of 30 m hectare and rich coastal range of 7000 km with varying precipitation zones accounting for biological as well as landscape diversity. India consists of:

(a) Two realms: the Himalayan region represented by Palearctic Realm and the rest of the subcontinent represented by Malayan Realm
(b) Five biomes: (1) tropical humid forests, (2) tropical dry or deciduous forests (including monsoon forests), (3) warm deserts and semideserts, (4) coniferous forests, and (5) Alpine meadows (MoEF 2009)
(c) Ten biogeographic zones: trans-Himalayan, Himalayan, Indian Desert, semiarid, Western Ghats, Deccan Peninsula, Gangetic Plain, Northeast India, islands, and coasts with a total of 27 provinces

There are 15 agroclimatic zones with 127 subzones based on its rainfall pattern, soil type, availability of irrigation water, and existing cropping pattern (Agricultural Meteorology Division, India). The forest cover in India has been assessed as 692,027 km^2 which constitutes 21.05 % of total country geographical area (MoEF 2013). India which has only 2.4 % of the world's land area is one of the 17 identified megadiverse countries in the world, having about 7.5 % of the identified biological species. India's biodiversity accounts for 11.9 % of world flora with 4045 species of higher plants belonging to about 141 endemic genera from 47 families. The Indian gene center is among the 12 mega-gene centers of the world. It is also one of the Vavilovian centers of origin and the diversity of crop plants. The world's estimated species are more than 250,000 (Oppermann et al. 2015) of which 45,000 species are reported from India (MoEFF 2014). About 811 cultivated plants and 902 of their wild relatives have been documented so far from India. There are 650 gymnosperm species in the world and about 74 (11.38 %) are found in India. Of about 90,000 fungal species of the world, 27,500 are reported from India, of which 15,500 are terrestrial litter fungi, 327 are coprophilous fungi, and 450 are endophytic fungi. There are about 2300 lichen species belonging to 305 genera and 74 families reported from India (MoEF 2014). Four, out of the 34 global hotspots of biodiversity, namely, the Himalayas, Indo-Burma, Western Ghats/Sri Lanka, and Sundalands, are situated in India. About 166 species of crops including 25 major and minor crops and 320 wild relatives of crop plants have originated and developed in India (NBPGR 2007). Landraces, traditional cultivars, and farmer's varieties of many crop and horticultural plant species are prevalent in India, though a decreasing trend is noted in areas following progressive farming. Presently, the Indian diversity is composed of rich genetic wealth of native as well as introduced types. India is a primary as well as a secondary center of origin for several crops and has also rich regional diversity for several South/Southeast Asian crops as described below:

1. Primary center of origin: *Oryza*, *Vigna mungo*, *Vigna aconitifolia*, *Cajanus cajan*, cucurbits, *Gossypium arboreum*, *Corchorus*, *Artocarpus heterophyllus*, *Musa*, *Mangifera*, *Syzygium cumini*, *Amomum subulatum*, *Piper nigrum*, and several minor millets and medicinal plants like *Rauvolfia serpentina, Saussurea costus*, etc.
2. Secondary center: *Eleusine coracana*, *Pennisetum glaucum*, *Sorghum bicolor*, *Vigna unguiculata*, *Cyamopsis tetragonoloba*, *Abelmoschus esculentus*, sesame, *Carthamus tinctorius*, tropical American types such as maize, *Solanum lycopersicum*, muskmelon, pumpkin, chayote/chow-chow, chillies, *Amaranthus*, etc.
3. Regional (Asiatic) center of diversity for crops: maize, barley, amaranth, buckwheat, proso millet, foxtail millet, *Vigna radiata*, *Cicer arietinum*, cucumber, bitter gourd, bottle gourd, snake gourd, and some members of tribe *Brassicaceae*

Above-listed crops make India rich in these crop genetic resource as India is the their primary center of origin due to naturalization of the wild relative and tertiary gene pool, and intensive cultivation and breeding inputs in deployment selection and maintenance of landraces, hybrid cultivars, and synthetic and composite lines thus widening the available gene pool for the listed crops. Similarly for certain tuberous crops like sweet potato, taros, and yams, rich variability of *Coleus* species; sword beans; velvet beans; several minor fruits, such as berries and nuts; and several species of *Rubus*, *Ribes*, *Juglans*, *Pyrus*, and *Prunus* exists (IARI 1995).

As per area under cultivation, obviously food grains occupy a major share in gross cropped area as compared to nonfood grains, but the per capita availability of the land is on the decline, signaling permanent ceiling on the horizontal expansion of cultivable area, coupled with degradation of land due to salinity, waterlogging, acidity, and erosion. Extensive measures are needed to restore the soil quality for its productive efficiency through appropriate technological interventions. The quest for development over the years led to serious damages to natural resources, soil, water, climate, and biodiversity, challenging the very sustainability of agriculture. A continuous social, biological, and technological intervention is required to assess and secure the balance between crop evolution-genetic diversity, farmer's knowledge, and selections and exchange of crop varieties. Several new areas were explored for the first time under the World Bank (WB)-funded Mission Mode Project on Plant Biodiversity (National Agricultural Technology Project). Sixteen special missions were also executed in different areas that were under threat to lose the diversity due to natural or human disturbances. New accessions were collected and identified in *Musa acuminata* x *M. balbisiana* (Bhat Manohar, a Musa species – first known natural tetraploid of banana) and *Corchorus pseudo-olitorius* (new species). Besides, accessions of rare/endemic/endangered species, viz., *Citrus rugulosa* (Western Ghats), *Cucumis prophetarum* (Rajasthan), *Garcinia imberti* (Kerala), *Meconopsis aculeata* and *Podophyllum hexandrum* (Western Himalaya), *Luffa hermaphrodita* (Rajasthan), *Cycas beddomei*, *Rauvolfia serpentina*, *Aegilops tauschii* (Lahaul-Spiti, Kinnaur, and Pangi in Himachal Pradesh), *Fagopyrum emarginatum* (Kargil in Jammu and Kashmir and Kinnaur in Himachal Pradesh), *Citrus indica* (Arunachal Pradesh), and *Vanilla andamanica* (South Andaman and Nicobar Islands), were also collected (Sharma 2007).

By conserving the sites containing the plant populations, the set of ecosystems conditions is also conserved which allows for the continuation of genotype x environment interactions, adaptations, and evolution of the conserved populations, providing for a long-term dynamic situation, with the natural elements of intra- and interspecific competition and natural selection to drive the evolutionary process wherein the populations continue to evolve in nature, allowing further designing and implementing re-introductions and ecological restoration projects of various sorts. Maintenance of traditional agroecosystems and vegetation diversity is the need of the hour to preserve in situ repositories of crop germplasm in special reference to rural development projects, allowing reasonable modernization for agricultural development (Alcorn 1981).

8.2.1 Traditional Agro-ecosystem as Crop Germplasm Repositories

Climate change has started impacting the ecological communities and tribal subsistence, but conservation and judicious sustainable managements of plant genetic resources can impede these impacts. Conserving forests and other natural resources can ensure social, economic, and environmental prosperity. Traditional agroecosystems are integrated with wisdom of tribal farmers and logical interaction of farmer and environment inherited generation after generation through centuries, and this wealth of traditional knowledge must be conserved to continue reaping the benefits. A salient feature of traditional farming systems is their degree of plant diversity in time and space in the form of polycultures and agroforestry patterns. Tribes generally grow 80–125 useful plant species, mostly native medicinal plants in small areas around the houses (Altiers and Merrick 1987; Borokini et al. 2010). The pekarangan, the traditional home garden intercropping system of West Java, has 100 or more plant species fulfilling all daily requirements, about 42 % provide for building materials and fuel wood, 18 % are fruit trees, 14 % are vegetables, and the remainder constitutes ornamentals, medicinal plants, spices, and cash crops (Christanty et al. 1986). Usually, traditional agroecosystems are positioned in centers of diversity, sustaining populations of various landraces along with wild and weedy relatives of crops (Frankel 1973; Harlan 1975; Vavilov 1951). Landrace populations consist of locally adapted genetically different lines which differ in their tolerance reaction to different pathovars and races of diseases and insect pests (Browning 1974; Harlan 1975). For attaining durable plant resistance, increasing host resistance diversity by growing cultivar mixture or multilines is an effective strategy against serious epiphytotics (Browning and Frey 1969). Cultivation of barley mixtures that were grown on 3 m hectare in East Germany reducing powdery mildews infection by 80 %within 5 years (Wolfe 1992) is one of the successful examples. Wheat multilines in the genetic background of Kalyan Sona were widely grown in India (Gill et al. 1980). Cultivating multiple varieties leads to both interspecific and intraspecific genetic diversity which confers at least partial resistance to diseases specific to particular strains of the crop and allows farmers to exploit different microclimates and derive multiple nutritional and other uses from within-species genetic variation ensuring harvest security (Clawson 1985; Harlan 1975). Traditional cropping system generally involves crop plants along with wild or weedy relatives in larger ecological amplitudes which often coexist and coevolve over time with human cultures. This has been the source of novel resistance or adaptive alleles for natural selection by natural hybridization and introgression or unconsciously by farmers (Frankel and Bennett 1970; Harlan 1976; Prescott-Allen and Prescott-Allen 1983). Mexican farmers allowed teosinte to remain within or near maize fields, and through the wind, some natural cross-hybridization occurred, resulting in occurrence of natural maize-teosinte hybrids with improved yields, in the successive crop. Such hybrids and their descendants are phenotypically distinct and fertile and thus capable of passing on their genetic traits (Wilken 1977; Brush SB 1982). The process of natural hybridization has been both bidirectional, i.e., enhancement (e.g., increased pungency in cultivated chilies due to hybridization with wild chiltepines) or deterioration of quality (bitter, unpalatable flesh in domesticated squashes due to hybridization with wild gourds). The cocultivated weeds often represent specific pattern of distribution termed agrestals (Baker 1965), agroecotypes (de Janvry 1981; Barrett 1983), or arvenses. Their usage applications explored by the local turned out to be medicinal and culinary purposes (Datta and Banerjee 1978) or as biological insect-pest control (Altieri et al. 1987) and in enhancing organic matter accumulation and soil conservation (Chacon and Gliessman 1982). These plants are highly specialized in terms of adaptations to agricultural fields as a result of evolution in conjunction with particular crops grown under specific cultural conditions (Barrett 1983).

8.2.2 On-Farm Conservation of Genetic Diversity and Tribe Societal Development

As per international Convention on Biological Diversity (CBD), 1992, Article 8 emphasizes that conservation of agricultural biodiversity is important both on farmers' fields as well as in protected areas and ex situ in gene banks.

Vicissitudes of agriculture amidst changing agricultural practices, cropping patterns, and the repercussion of green revolution led to the erosion of genetic diversity. "On-farm conservation practices" involving provincial people is one of the strategies, successfully illustrated by Genetic Resource Ecology Energy Nutrition Foundation (GREEN). Another good example was the effort undertaken by NBPGR-NATP Mission Mode Project on "household food and nutritional security" focused on tribal areas and local communities with 53 centers, covering 509 villages with the partnership of over 15,000 farm families in 23 states of India, to identify life support crop species, horticultural species, and vegetable, gardening, domestic animal species, and fisheries in these areas and value addition of the local germplasm. The MS Swaminathan Research Foundation (MSSRF), Chennai, has successfully applied bottom-up efforts to conserve local agro-biodiversity by linking the livelihood security of villagers with the wider ecological security of the region. It has established effective community agro-biodiversity conservation and management programs aimed at integrated village development in tribal groups particularly in the Jeypore tract of Orissa. Many tribal communities of India are involved in Participatory Plant-Breeding (PPB) and the compilation and documentation of community biodiversity restoring on-farm participatory conservation and skill development and capacity building of farmers for agricultural diversity-based livelihoods.

On-farm conservation and promotion of pseudocereals (amaranth and buckwheat) and minor millets (foxtail millet, proso millet, finger millet, and barnyard millet) in participatory mode in Himachal Pradesh (NBPGR 2007) brought back finger millet into cultivation. Some famous rice landraces conserved at the gene bank of NBPGR were used for multiplication and subsequent restoration in their original growing areas in Bastar area of Chhattisgarh and Orissa by NBPGR; Indira Gandhi Agricultural University, Raipur; and International Rice Research Institute (IRRI), Philippines. Pockets of "Apatani plateau" (Lower Subansiri) and "Khamti Valley" (Lohit) of Arunachal Pradesh are an ideal example of in situ conservation practice for rice landraces. Many endemic plant species, their occurrence, population size, and frequencies are a matter of concern by the State Forest Department. Appropriate measure and restrictions have been imposed in favor of their natural regeneration sites. For promotion of in situ conservation of plant growing in the wild and their wild species, efforts have been made for documentation and protection of indigenous technical knowledge and germplasm of the local landraces, farmers' varieties, biodiversity conservation associated with livelihood approach particularly in the states of Gujarat and Rajasthan, by the Society for Research and Initiatives for Sustainable Technologies and Institutions, Ahmedabad, National Innovation Foundation, Ahmedabad and Ashoka Trust for Research in Ecology and the Environment. Yet comparing the size and diversity present in India, the number of crops cultivated, and the amount of genetic diversity available in different crops, the "on-farm" conservation efforts have been very limited. Therefore, for promotion of on-farm management and improvement of PGRFA, there is a need for policy support at national level. It can be achieved through a networking of existing infrastructure by involving farmers, local bodies, and government and nongovernment agencies including those involved in value addition and market research. This strategy can only work if corresponding benefits are ensured to the farmers and tribal communities of the country.

The changing agriculture scenario calls for an emphasis to be laid on in situ and on-farm conservation of PGRFA to keep the process of evolution of genetic diversity in response to changing environment. In this regard, the scope of home gardens needs to be extended to medicinal, food, and nutritional supplements at village level to ensure nutritional security, health, and improved income to the farmers. In order to maintain farm and crop diversity, critical and detailed evaluation of landraces is required. There have been examples where landraces have outyielded the released varieties. Similarly, traditional crops or crops of regional importance need value addition so that their cultivation is encouraged. It will cater the options available in context to changing food habits. To have effective conservation boost of valuable crop genetic resource, farming programs should be integrated with rural self-sustenance schemes. Native crops are the domesticated plants cultivated prehistorically within a region by its indigenous cultures. Landraces of tepary beans and cushaw squash, the native crops of Southwestern Mexico, outproduce commercial cultivars of related species as they can tolerate low soil moisture, high soil pH, salt solute concentration, and high air temperatures. Such native crop cultivation is also important for the effective utilization of limiting natural resources such as water and nitrogen. Sometimes, they have stayed long in cultivation as a symbol of cultural identity as indigenous community pride and their persistence (Nabhan 1984; Nabhan 1985).

8.2.3 Genetic Reserve Conservation: Protected Areas

Conserving the entire natural ecosystem in situ along with all its life forms as a biosphere reserve helps in recording and evaluating the undergoing environmental and evolutionary changes. Establishment of forest area reserves, national parks, sanctuaries, conservation reserves, conservation areas, mangrove conservation, community reserves, biosphere reserves, hotspot RAMSAR wetland areas, as heritage sites, costal and marine biodiversity protected areas, promotes in facilitation of in situ conservation.

8.2.3.1 National Parks

It is a protected area, by reason of its ecological, fauna, flora, geomorphological, or zoological association or importance, needed for the purpose of protecting and propagating or developing wildlife therein or its environment whether it lies within a sanctuary or not, to be constituted as a National Park, notified by the state government. Human activity is not permitted inside the national park except for the ones permitted by the Chief Wildlife Warden of the state under specified conditions as described in Wildlife Protection Act, 1972. There are 6,555 national parks throughout the world (http://www.basicplanet.com/national-parks. Accessed on September, 2015). There are 103 existing national parks in India covering an area of 40,500 km^2, which is 1.23 % of the geographical area of the country (http://www.wiienvis.nic.in. Accessed on September, 2015).

8.2.3.2 Wildlife Sanctuaries

It is a protected area of substantial ecological, fauna, flora, geomorphological, natural, or zoological significance for the purpose of protecting, propagating, or developing wildlife or its environment notified by the state government. Some restricted human activities are allowed inside the sanctuary area as described in Wildlife Protection Act, 1972. There are 531 existing wildlife sanctuaries in India covering about 3.58 % of the geographical area of the country (http://www.wiienvis.nic.in. Accessed on September, 2015).

8.2.3.3 Conservation Reserves and Community Reserves

Under Wildlife (Protection) Amendment Act of 2002, amendment to existing Wildlife Protection Act, 1972, certain protected area categories, like conservation reserves and community reserves were introduced. In India they denote protected areas which act as buffer zones or connectors and migration corridors between established national parks, wildlife sanctuaries, and reserved and protected forests of India. When such areas are uninhabited and completely owned by the Government of India, they are designated conservation areas, but if they are used for subsistence by communities and if parts of the lands are privately owned, they are designated community areas. These categories were introduced because of reduced protection in and around existing or proposed protected areas due to private ownership of land and land use. Currently, there are 65 in number with an area of 2,344.51 km^2. There are four community reserves in India, namely, Keshopur Chhamb (Gurdaspur), Lalwan (Hoshiarpur), Kadalundi (Mallapurum), and Kokkare Bellur (Mandya) with an area of 20.69 km^2 (http://www.wiienvis.nic.in. Accessed on September, 2015).

8.2.3.4 Mangrove Conservation Programme

The Mangrove Conservation Programme was launched in 1987 by the Ministry of Environment and Forest (MoEF), Government of India, under which 39 mangrove areas have been identified on the recommendation of the National Committee on Wetlands, Mangroves, and Coral Reefs on the basis of their unique ecosystems and biodiversity for intensive conservation and management. They harbor some 39 unique plant species. Activities under Management Action Plan of the central government include raising mangrove plantations, protection, catchment area, treatment, siltation control, pollution abatement, biodiversity conservation, sustainable resource utilization, survey and demarcation, education, and awareness further supplemented by research and developmental scientific inputs.

8.2.3.5 Medicinal Plant Conservation Areas (MPCAs)

India has 6,000–7,000 species estimated to have medicinal usage in folk and documented systems of medicine, like Ayurveda, Siddha, Unani, and Homoeopathy. About 960 species of medicinal plants are estimated to be in trade of which, 178 species have annual consumption levels in excess of 100 metric tones. With the financial support of national and international agencies, several civil societies are engaged in in situ conservation of medicinal plant species. For example, the Foundation for Revitalisation of Local Health Traditions (FRLHT) starting in 1993, in collaboration with the State Forest Departments and the financial support of Danish International Development Aid (DANIDA), Netherlands, and United Nations Development Programme (UNDP), has established forest gene bank sites called Medicinal Plant Conservation Areas (MPCAs) conserving regions with rich biodiversity of medicinal and aromatic plants (FRLHT 2006). They represent all forest types with large bioclimatic harboring 45 % of recorded populations of flowering and medicinal plants of Peninsular India, including 70 % of those listed in *Red Data Book*. Presently, there are 108 MPCAs in India established in various states like Karnataka (3), Kerala (9), Tamil Nadu (12), Andhra Pradesh (8), Maharashtra (13), Rajasthan (7), Orissa (5), West Bengal (7), Madhya Pradesh (13), Arunachal Pradesh (7), Uttarakhand (7), and Chhattisgarh (7). The intraspecific diversity that is conserved in the MPCA network in consultation with a dedicated scientific research organization working on medicinal plants identification and genetic improvement of medicinal plants can be utilized to provide authenticated quality planting material for commercial cultivation. MPCAs can also support recovery of threatened species and human resource development for forestry staff, local communities, and researchers in the conservation of medicinal plants for sustainable use and prospects of equal benefit-sharing.

In addition, many active projects and the organizations addressing issues of conservation of medicinal plants, traditional medicine and health care, commercialization, and their contribution to sustainable livelihoods are working in India like International Herbal Cross Society, Bhubaneswar, working on "Documentation of Traditional Health Practices in Mayurbhanj District, Orissa"; the Society for Himalayan Environmental Research working on "Conservation and Cultivation of Medicinal Plants in Mountain Areas of Garhwal Region"; Indian Institute of Forest Management working on "Community-Based Sustainable Management of Medicinal Plants in Betul District, Madhya Pradesh"; Peoples Clinic Trust, Chittoor, working on "Strengthening the Traditional Health Practices and Training in Cultivation of Medicinal Plants to the Women and Herbal Healers of District Chittoor in Andhra Pradesh"; Indian Environment Society working on "Development of Strategies for Production and Improvement of MAPs growing in the Tribal Belts of Southern Rajasthan"; Arya Vaidya Sala, Kottakkal, Kerala, working on "Development of Production to Consumption and Marketing Systems-based Strategies for the Sustainable use of Medicinal Plants in the Western Ghats, Kerala"; UTTHAN for Sustainable Development, Allahabad, Uttar Pradesh, working on "Exploring the Potentials of Medicinal and Aromatic Plants in Bundelkhand Region of Uttar Pradesh"; Heart Foundation and Yathong Foundation, Leh, Ladakh, working on "Development of Strategies and Methods to Support Traditional Tibetan System of Medicine for the Improvement of Primary health care in Ladakh Region, Jammu & Kashmir; Ladakh "; The Society for Agriculture Research Training and Health Education, Sirmour, Himachal Pradesh, working on "Promotion of on-farm cultivation of traditional hill crops and farmer's varieties"; GREEN Foundation, a community-based organization working toward the conservation of agrobiodiversity and the promotion of sustainable agriculture in the semiarid regions of South India working on "Cultivation of organic medicinal rice(s) of Kerala, The Navara Eco Farm etc."

8.2.4 Genetic Reserve Conservation: Sites of Conservation

A worldwide participatory approach by joint forest management programs, integrating the ecological and socio-economic parameters for sustainable conservation of ecosystems, has been undertaken by the International Union for Conservation of Nature (IUCN). Under these programs, the local communities and the governments have defined roles and responsibilities with regard to forest protection and development. While ensuring forest conservation, the local communities are provided with alternative sources of livelihood and assured access to forest and related produce.

8.2.4.1 Biosphere Reserves

The designation "Biosphere Reserve" for "Conservation of natural areas and of the genetic material they contain" was accorded in Man and Biosphere project by the International Co-ordinating Council (ICC) of UNESCO in its first meeting in 1971, held at Paris Biosphere Reserve network, and was formally launched in 1976. The Indian Biosphere Reserve Programme was initiated in 1986 with the objectives:

1. To conserve the diversity and integrity of plants and animals within natural ecosystems
2. To safeguard genetic diversity of species on which their continuing evolution depends
3. To provide areas for multifaceted research and monitoring
4. To provide facilities for education and training
5. To ensure sustainable use of natural resources through most appropriate technology for improvement of economic well-being of the local people

Biosphere reserves are differentiated into three interrelated zones:

(a) Natural or core zone: An undisturbed zone of suitable habitat for dwelling endemic plant and animal species; their wild relatives, i.e., all primary, secondary, and tertiary or genetic reservoirs including higher-order predators endemic to that geographical region, hence are of utmost scientific interest. A core zone secures legal protection and management, and research activities that do not affect natural processes and wildlife are allowed. Strict nature reserves and wilderness portions of the area are designated as core areas of biosphere reserve which is kept free from all human pressures external to the system.
(b) Manipulation or buffer zone: Area surrounds core zone with limited activities which do not adversely affect the ecological diversity like restoration, demonstration sites for enhancing value addition to the resources, limited recreation, tourism, fishing, and grazing, which are permitted to reduce its effect on core zone, thus safeguarding the core zone. The zone is used for research and educational activities.
(c) Transition zone: The outermost part of a biosphere reserve which is not delimited. The skills of conservation, knowledge, and management are applied for the purpose of the biosphere reserve. The activities allowed are settlements, crop lands, managed forests, and area for intensive recreation, and other economic macro-management practices are used.

The areas are used for scientific understanding of the patterns and processes in the ecosystem. Landscapes which are modified or degraded are used for rehabilitation for ecological restoration aiming sustainable productivity. It generally includes one or more national parks and/or preserves and protected larger areas. Flora and fauna are protected in regions, along with tribal communities inhabiting these regions (Ministry of Environment, Forest and Climate Change 2007). The first biosphere reserve of the world was established in 1979; since then the network of biosphere reserves has increased to 631 in 119 countries across the world with the support of the World Bank, with 18 biosphere reserves established in India (http://www.wiienvis.nic.in. Accessed on September, 2015).

8.2.4.2 Biodiversity Hotspots

These are the high biodiversity regions under risks from humans. As per Myers et al. (2000), a biodiversity hotspot region must meet two strict criteria: it must contain at least 0.5 % or 1500 species of vascular plants as endemics, and it has to have lost at least 70 % of its primary vegetation. There are 34 global biodiversity hotspots of which four are present in India i.e. Himalayas (Western and Eastern Himalayas), Western Ghats and parts of Sri Lanka (harbouring 5584 indigenous species; 377 exotic, naturalized species; and 1427 cultivated species), the Northeast and Indo-Burma, and the Nicobar Sundalands (MoEF 2014).

8.2.4.3 World Heritage Sites

These are the sites protected for outstanding universal value of cultural, natural, or with mixed properties, recognized by the World Heritage Committee (UNESCO). At present, there are 911 such sites in the world, which cover 711 cultural sites, 180 natural sites, and 27 mixed properties encompassing 152 countries. There are 39 World Heritage Sites in India, located in the states of Kerala, Karnataka, Tamil Nadu, and Maharashtra (http://whc.unesco.org/en/list/. Accessed on September, 2015).

8.2.4.4 RAMSAR Wetland Areas

The RAMSAR Convention establishes that "wetlands should be selected for the list on account of their international significance in terms of ecology, botany, zoology, limnology or hydrology." The contracting parties adopted stringent criteria of information sheet on RAMSAR wetlands and a classification system of wetland type. RAMSAR wetlands are accorded a national-level status and internationally recognized as being of significant value not only for the country, in which they are located, but for humanity as a whole. Currently, there are 168 contracting parties involved with 2,208 sites covering an area of 21.07 crore hectares and with 26 sites in India covering 6.89 lakh hectares (http://www.ramsar.org/. Accessed on September, 2015).

8.2.4.5 Important Coastal and Marine Biodiversity Area (ICMBA)

India has taken several steps to achieve the National Biodiversity Target no. 6 and Aichi Biodiversity Target no. 11 which aim to conserve a substantial portion of the coastal and marine areas in the country and the world, respectively. Toward achieving these two targets, 106 coastal and marine sites have been identified and prioritized as Important Coastal and Marine Biodiversity Areas (ICMBAs) by the Wildlife Institute of India. Sixty-two ICMBAs have been identified along the west coast of India, and 44 have been identified along the east coast. Of these, 22 ICMBAs have been prioritized for immediate conservation actions and proposed to be upgraded as protected areas under categories such as conservation or community reserve to increase participation of the local communities in governance, e.g., Gahirmatha marine sanctuary. A Marine Protected Area (MPA) is a space in the ocean where human activities are strictly regulated and special protections accorded for natural or historic marine resources by local, state, territorial, native, regional, or national authorities (http://www.wiienvis.nic.in. Accessed on September, 2015).

8.2.5 Recent IUCN Protected Area Management Categories and Governance Type

Globally protected areas differ in size and governance types and management targets. As per IUCN a protected area is defined as "geographical space, recognised, dedicated and managed, through legal or other effective means, to achieve the long-term conservation of nature with associated ecosystem services and cultural values." The IUCN protected area management categories classify protected areas based on their primary management objectives, while the IUCN governance types classify protected areas according to authority, responsibility, and accountability for them. Both IUCN and the CBD recognize these categories (Juffe Bignoli et al. 2014). At present, about 65 % of protected areas in the WDPA have an IUCN management category, and 88 % have a governance type. Protected area governance types are (11 subtypes) listed.

(A) *Governance by government*
 1. Federal or national ministry or agency in charge
 2. Subnational ministry or agency in charge
 3. Government-delegated management (e.g., to an NGO)

(B) *Shared governance*
 4. Transboundary management
 5. Collaborative management (various forms of pluralist influence)
 6. Joint management (pluralist management board)

(C) *Private governance*
 7. Declared and run by individual landowner
 8. By nonprofit organizations (e.g., NGOs, universities, cooperatives)
 9. By for-profit organizations (e.g., individual or corporate landowners)

(D) *Governance by indigenous peoples and local communities*
 10. Indigenous peoples' conserved areas and territories, established and run by indigenous peoples
 11. Community conserved areas – declared and run by local community

Protected Area Categories

I. *Strict nature reserve/wilderness area*: A protected area managed mainly for science or wilderness protection
II. *National park*: A protected area managed mainly for ecosystem protection and recreation
III. *Natural monument*: A protected area managed mainly for conservation of specific, unique natural features
IV. *Habitat/species management*: A protected area managed mainly to preserve specific species and to maintain habitats
V. *Protected landscape/seascape*: A protected area managed mainly for landscape/seascape protection and recreation
VI. *Managed resource protected*: A protected area managed mainly for the sustainable use of natural ecosystems

Till August 2014, the WDPA had around 2.09 lakh (1.97 lakh terrestrial and 12,076 marine) designated protected areas from more than 193 countries and territories. There has been significant growth of protected areas in the past decades. In 2014, 15.4 % of the terrestrial and inland water areas and 8.4 % of the marine area within national jurisdiction (0–200 nautical miles) were covered by protected areas (UNEP-WCMC 2014) (Table 8.1).

8.3 Ex Situ Conservation

Ex situ conservation entails, conservation of biological diversity components (seed, clone, live saplings, pollen, or their DNA) outside their natural habitats in storage infrastructures termed gene banks. There are about 1,750 gene bank collections in the world, of which about 625 collections are maintained in Europe comprising more than 2 million accessions from 650 institutions scattered over about 43 European countries for conservation and utilization purposes.

Gene banks play a vital role in the exploration, collection, conservation, and scientific dissemination of plant genetic diversity, essentially needed for crop improvement for food, health, and nutritional security. Gene banks support by ensuring the continued availability of limited but important genetic resources for research, breed-

Table 8.1 Global status of protected areas

Class	IUCN category	Number	Area (km^2)
IA	Strict nature reserve	11,260	130,755
IB	Wilderness area	2940	118,918
II	National park	5251	642,539
III	Natural monument	25,248	480,275
IV	Habitat/species management	58,764	300,347
V	Protected landscape/seascape	32,721	775,551
VI	Managed resource protected area	7359	899,210

Source: UNEP-WCMC (2014)

ing, and improved seed delivery for a sustainable and resilient agricultural system. Majority of the world gene bank accessions (7.5 million) are of the staple crops on which humans and livestock rely for food, including their wild relatives, landraces, and few from the native crops of indigenous importance and underutilized species (FAO 2014). The value of conserving crop genetic resources is realized only through their effective utilization which requires strong linkages along the chain from in situ resource conservation and collection, storage in gene banks, research and breeding, and dissemination of quality planting material to farmers and seed growers and ultimately consumers. Coherent coordination of gene bank curators, breeders, and national programs is needed to ensure the efficient and sustainable conservation of the PGRFA. The protocol of gene bank operation includes the preservation of germplasm identity, maintenance of viability and genetic integrity, and the promotion of access. This requires accession passport information to facilitate use of stored plant material in accordance with relevant national and international regulatory rules. Continual conservation of these plant genetic resources depends on efficient management of gene banks following standards and protocols that ensure the continued survival maintenance and availability of plant genetic resources.

Majority of the breeding programs use narrow genetic base due to lack of documented information of targeted traits. Core collections and mini core should be developed for each crop to enhance effective use of available germplasm in breeding programs. This core collection helps to identify genetically diverse trait-specific germplasm for resistance to abiotic and biotic stresses and agronomic and quality traits for use in breeding programs to develop broad-based cultivars. Molecular characterizations of core/mini core collections have helped in understanding genetic diversity and population structures. The reference sets with respect to specific well-used genotype for genetically diverse 200–400 accessions should be developed. Seeds of the core collection should be disseminated to the global research community following the standard protocol under the International Treaty on Plant Genetic resources for Food and Agriculture.

The gene bank maintenance is usually expensive. Cryopreservation, when used for large-scale screening, costs substantially. Field gene bank monitoring is also demanding in terms of labor and cost. Proactive management of optimal man hours of available scientific personnel and skilled manpower and infrastructural and financial resources under given conditions can help in managing the gene banks. In many third world countries, the availability of trained manpower and finance resource to maintain gene bank collections in a sustainable manner remains a challenge. Long-term partnerships at national, regional, and global levels together with resources for capacity development will be a boon for the development and strengthening of ongoing ex situ conservation programs.

The major responsibility of efficient gene bank is to generate passport data, the basic information of an accession, about the collector or the donor institute, time, collection site, names, numbers, availability data and other status information, different crop classifications, existence of specimens, and images (Oppermann et al. 2015). Information about scientific names of gene bank accessions is fundamental. The taxonomic information, e.g., family, genus, species, and intraspecific taxa, should reflect in the system. The gene bank management information like site of storage, availability of seed samples or lines of each accession, germinability and viability of seeds, and any seed-borne disease screening is crucial, and information is updated and digitally maintained. Metric data on highly heritable traits, controlled by a single or oligogenic genes, are generally recorded during each regeneration cycle (phenotypic characterization). Seed regeneration is influenced by storage factors, viability, and their germination capacity. The phenomics allows identification of promising new genes/alleles for their selection and incorporation in the breeding programs. Identifying traits can constitute the descriptor's list. Distribution of seed material on request to breeders, researchers, under specified guidelines such as material transfer agreement, etc., needs documentation. The information, address of the person/institution requesting the material, the order (and the items ordered), the date of purchase, and the status of the material to be delivered need a proper preferably digitized inventory (Oppermann et al. 2015).

Current research is showing that there exists variability in seed longevity for different species being conserved under similar conditions. They have significant implications for the management of multispecies gene banks like majority of national gene banks and some of the Consultative Group on International Agricultural Research (CGIAR) gene banks (e.g., ICARDA, ICRISAT, CIIMYT, CIP, IRRI, and CIAT). The Svalbard Global Seed Vault (SGSV), Longyearbyen, Norway, a glazed cave-like structure, drilled 500 ft below permafrost, in the middle of a frozen Arctic mountain topped with snow, with automated digital monitoring system which controls temperature and humidity and provides high security, has been designed to hold 4.5 billion batches of seeds of the world's main crops. The SGSV is an insurance against natural disasters such as earthquakes and tsunamis or deliberate attacks like bomb blasts or human errors such as nuclear disasters or failure of refrigeration that may erase the seeds of any important species in the other seed banks or in the wild, in the other countries. Such seed can be reestablished using seeds from SGSV. NBPGR facilitated safe transfer of about 40,000 accessions of ICRISAT mandate crops to SSGV during the past 2 years.

Ex situ research has shown that in some families of plants, the localities where they originate and the type of seed (endospermic vs nonendospermic) and intraspecific variation may be a key factor in determining their shelf life (Probert et al. 2009; Nagel and Borner 2009) and may also affect accession longevity. These factors are not well studied and need to be investigated further in a coordinated manner. The major problems on ex situ conservation are improper storage (moisture content), failure of grow-out tests and inadequate sampling procedures during field collection, and lack of whole range of diversity representation in gene banks for a given crop and its close genetic relatives. Conservation of minor crops as potential candidates for future sources of nutrition; wild relatives as source of disease resistance, wider adaptation, and increased yield; or crops for nonfood purposes such as fuel, medicine, or industrial use needs more focused attention (Frankel and Hawkes 1975; Wilkes 1983). Seed storage ceases evolutionary processes of resistance to evolve, limiting the subsequent generations from the stored seed to face higher selective pressures of the nature (Simmonds 1962).

8.3.1 Seed Gene Bank

The Genebank Standards for PGRFA has been recently revised in view of changing global policy and advances in science and technology with special reference to availability and distribution of germplasm by the FAO Commission on Genetic Resources for Food and Agriculture, Rome, (FAO 2014). These are universally followed by all the gene banks in the world for germplasm conservation as the world's 1750 gene banks vary in the accession collection size, manpower, and financial support. These standards cover seeds in gene banks, vegetatively propagated planting material in the field gene banks, and in vitro and cryopreservation gene bank. The Genebank Standards regulation must follow the ITPGRFA and the Second Global Plan of action for PGRFA. The aim of the Genebank Standards is the conservation of PGR under defined conditions meeting recognized standards based on advanced technological and scientific knowledge of seed management techniques, molecular biology, genetics, and seed storage technology. They incorporate the developments in the field of documentation and information systems essential for PGR optimization and gene bank management. Species-specific information is taken care for plants producing non-orthodox seeds and/or vegetatively propagating, as for all species, there exist different seed storage behaviors, life forms (herbs, shrubs, trees, lianas/vines), and life cycles (annual, biennial, perennial) for which it is difficult to establish specific standards that are valid for all species. The limited capacities and inadequate infrastructure challenge developing countries to ensure long-term conservation by providing ideal maintenance of seed viability and genetic integrity, thereby ensuring access to, and use of, high-quality seeds of conserved plant genetic resources. The seed bank standards deal with the conservation of the desiccation-tolerant orthodox seeds, i.e., those that can be dehydrated on low water content and are responsive to low temperatures. Lowering moisture and temperature decreases the rate of metabolic processes, increasing seed longevity (Walters et al. 2005). Genebank Standards for orthodox seeds have been defined as germplasm acquisition, drying and storage of the seed, monitoring of viability and regeneration, characterization, evaluation, documentation, distribution, safety duplication, and security/personnel (FAO 2014).

NBPGR at India leads the planning, coordination and execution of the activities of collection, characterization, evaluation, conservation, exchange, documentation, and sustainable management of diverse germplasm of crop plants and their wild relatives with a view to ensure their availability for use over time to breeders and other researchers. Efforts have been made for documentation and protection of indigenous technical knowledge and also the germplasm of the local landraces/farmers' varieties. The Indian Council of Agricultural Research (ICAR) and National Bureau of Plant Genetic Resources (NBPGR), New Delhi, with its ten regional stations/base centers/quarantine centers over different phytogeographic zones of the country and active collaboration and linkages with over 40 National Active Germplasm Sites (NAGS) have already achieved this feat. The National Genebank of NBPGR with its three types of storage facilities – seed gene bank, cryogen bank, and in vitro gene bank – has 4,08,186 orthodox seeds germplasm accessions, of nearly 1584 species (http://www.nbpgr.ernet.in/Division_and_Units/Conservation.aspx. Accessed on November 2015). Stock inventorization and monitoring of viability are performed regularly in most of the accessions. There exists a mechanism of "single-window system of germplasm supply" to the users in the country. NBPGR is the nodal organization supplying genetic resources to different breeding programs through its network. There have been a number of success stories in crop improvement programs resulting in development/genetic improvements of a number of cultivars/hybrids that have helped in enhancing the national production demonstrating the contribution in increasing the productivity. Most of the crop species have been evaluated for morphological and agronomic traits. A total of 4618 cultivars have been released in 162 crops. However, ten crops including cereals, millets, cotton, and chickpea contribute toward 50 % of the total cultivars released so far (http://www.nbpgr.ernet.in/Research_Projects/Base_Collection_in_NGB.aspx. Accessed on September 2015). The International Crop Research Institute for Semi-Arid Tropics (ICRISAT) located in India maintains 39,234 accessions of sorghum, 22,888 accessions of pearl millet, and 6804 and 1542 accessions of finger millet and fox tail millet respectively, these being there mandate crops (http://www.icrisat.org/gene-bank-crops.htm. Accessed on Nov 2015).

The seed banking prevents genetic erosion by allowing large populations to be preserved in a limited space under optimum conditions and reducing the need for regeneration. The success of long-term conservation of seeds is dependent on continuous viability monitoring and regeneration or recollection when the viability of the sample drops below a minimum level. Utmost care is needed during seed collection, regeneration, and storage as natural selection cannot be simulated and some artificial selection will be unavoidable, which inevitably leads to unpredictable genetic changes (Cohen 1991). The long-term ex situ conservation seed storage of orthodox species recommended by IPGRI is to dry the seeds to a moisture content of below 7 % and seal the dried seeds in a moisture-proof container such as laminated foil bags, aluminum cans, or glass jars for storage at a low temperature of −18 °C (for orthodox species of crops and their relatives). For wild species, a temperature of –4 °C and a moisture content of 7–8 % are recommended. The activities in seed banks are collection, seed preparation, seed drying, packaging, storage, periodic germination tests, seed regeneration, re-storage, and documentation at each stage of activity (FAO 2014).

8.3.2 Field Gene Bank

Field gene bank which includes the maintenance of live plants in pots or trays and in greenhouses or shade houses is the ideal strategy for long-lived perennials, recalcitrant species, and vegetatively propagated species (Reed et al. 2004). Major problems are the need of large space and their vulnerability to maintenance and natural disasters like rapid spread of diseases and human neglect. Outbreeders require controlled pollination for regeneration from seed. In many circumstances, they are the only available option for the conservation of important germplasm. When displayed, the plants have an important educational value and can easily be accessed for research purposes. Standards for field gene bank recalcitrant or intermediate seeds, which are propagated vegetatively, or plants that produce very few seeds or plants that require a long life cycle to generate breeding and/or planting materials are practiced differently. Field Genebank Standards involves choice of location, acquisition of germplasm, establishment of field collections, field management, regeneration and propagation, characterization, evaluation, documentation, distribution, security, and safety duplication (FAO 2014).

8.3.3 In Vitro Storage Banks

The long-term storage of recalcitrant species and vegetatively propagated species germplasm is carried out through aseptic conditions (in vitro) at low temperature under slow growth conditions or cryopreserved in liquid nitrogen at −196 °C. The main limitation of in vitro storage is the need for special equipment, techniques, and trained staff. The method involves cultures (especially shoot tips, meristems, somatic embryos, cell suspension, or embryogenic callus) being maintained under growth-limiting conditions on artificial culture media. The growth rate of the cultures can be limited by various methods, including temperature reduction, lowering of light intensity, or manipulation of the culture medium by adding osmotic agents or growth retardants (Engelmann 1999; Reed et al. 2004).

8.3.4 Cryopreservation

Cryopreservation is the storage of biological materials (seeds, plant embryos, shoot tips/meristems, and/or pollen) at ultralow temperatures, usually that of liquid nitrogen at –196 °C (Engelmann and Takagi 2000; Reed 2010). Under these conditions, biochemical and most physical processes are halted and materials can be conserved over the long term. The cryopreservation of orthodox seeds represents an advantage over low-temperature storage at –20 °C, both economically and in terms of viability, as liquid nitrogen is a relatively inexpensive cryogen and the seeds retain the same viability as they had immediately before storage. Therefore, regeneration costs are lowered and viability testing is reduced. Cryopreserved lines can be maintained as a backup for field collections, as reference collections for available genetic diversity of a population, and as a source for new alleles in the future.

Genebank Standards for in vitro culture and cryopreservation involves acquisition of germplasm; testing for non-orthodox behavior and assessment of water content, vigor, and viability; hydrated storage for recalcitrant seeds; in vitro culture and slow-growth storage; cryopreservation; documentation; distribution and exchange; security; and safety duplication.

8.3.5 Pollen Bank

Pollen preservation may be useful for base collections of species that do not produce orthodox seeds. It requires little space but some cytoplasmic genes would be lost. Like seeds, pollen can be divided into desiccation tolerant and intolerant. However, information about storage characteristics of pollen from wild species is fragmentary, existing mainly for some crop relatives and for medicinal and forest species, e.g., *Ocimum* species.

8.3.6 DNA Bank

DNA samples have been mainly used for bioprospecting and assessment of biodiversity studies. It could be used in molecular phylogenetics and systematics of extinct taxa. Its potential remains promising and should still be investigated as DNA is a very stable form in which genetic information can be stored. The creation of a network of DNA banks (DNA Bank-Net) to complement activities already being undertaken in ex situ conservation and more precisely germplasm collections can allow large quantities of genetic resources (genes, DNA) to be stored quickly and at low cost and could act as an insurance policy against rapid loss of the world's gene pool. Some major plant DNA banks include Missouri Botanical Garden; Royal Botanic Gardens, Kew; Australian Plant DNA Bank; and Trinity College, Dublin (Rice et al. 2006; Hodkinson et al. 2007). The Global Biodiversity Information Facility (GBIF) in Germany has established a DNA Bank Network in 2007 and offers a worldwide central web portal, providing DNA samples of complementary collections (microorganisms, protists, plants, algae, fungi, and animals).

8.3.7 Botanical Gardens

There are more than 1700 botanical gardens and institutions worldwide holding plant collections that serve both conservation and educational purposes. Many gardens have the mandate to preserve rare or threatened wild plants and make the plant material available for research. In India, numerous botanical gardens managed by the Botanical Survey of India (BSI) and several other organizations help in ex situ conservation of economically important as well as endangered, threatened, and rare plant species. The Indian Botanic Garden at Calcutta was established in 1787. It now spreads over an area of 110 ha and has around 15,000 plants belonging to 2500 species. Presently, there are 150 organized botanic gardens or large parks in India, of which 33 gardens are managed by the government, 40 by universities, and the rest are in public domain. The Government of India has also recently initiated establishment of National Botanical Garden at Noida in Uttar Pradesh. In all, about 150,000 live plants belonging to nearly 4000 species (including 250 endemic species) are conserved in these botanic gardens. The Millennium Seed Bank Project (MSBP) at the Royal Botanic Garden, Kew, England, is one of the largest conservation projects. MSBP's 47 partner organizations in 17 countries intend to store 25 % of the world's plant species by 2020. The Seed Information Database (SID) at Kew is an ongoing compilation of seed characteristics and traits worldwide, targeting more than 24,000 species (www.bgci.org; www.wiienvis.nic.in).

8.3.8 Gene Banks for Medicinal and Aromatic Plants (GEBMAP)

A national Gene Bank for Medicinal and Aromatic Plants (GEBMAP) was set up in 1993 under the

G-15 GEBMAP program at CSIR-CIMAP along with two other facilities at NBPGR, New Delhi, and Tropical Botanical Garden and Research Institute (TBGRI), Thiruvananthapuram, with DBT as nodal agency. The major objectives of conserving the species that are threatened, rare, or endangered, of proven medicinal value, and difficult to propagate and species of significant Research and Development (R&D) lead to the future development of drugs, survey collection and characterization of MAP wild genetic resources, strengthening of infrastructural facilities for conservation, and development of national database. TBGRI focused on conservation for select medicinal and aromatic plants with 298 accessions from 25 species from Peninsular India through the organization of field gene bank, seed bank, and in vitro repository and a cryogen bank. NBPGR holds 6671 MAP accessions belonging to 661 species. At CSIR-CIMAP, ex situ conservation of medicinal and aromatic plants seed bank, field bank, in vitro bank, and DNA gene bank is fully functional. In seed bank 2476 accessions of MAP belong to 336 families, 152 genera, and 515 species, and in field gene bank, 868 accessions of MAP belong to 80 families of 176 genera, and 227 species are being maintained.

8.4 Traditional Knowledge of Tribes and Indigenous Habitants: The Driving Force

For a country either developed or developing, the practices related to food security of local people, local communities, and local farmers play a critical role in serving countries' biodiversity. Biodiversity acts as a buffer because when one crop fails, another crop survives (Salik and Byg 2007). The biodiversity is matched by location of fields; in case of extreme environmental conditions, harvestable crop can be obtained. Local and traditional knowledge like doing multiple processes at time will help us to fight against different problems like higher prices of different food items. Different programs can be made which can be used to promote different agroecological practices like farming, fishing, animal rearing, and breeding at the same time so that in case of extreme environmental conditions, the people can survive.

Institutes like FAO are creating a link between traditional knowledge and modern technologies. In this modern era, there is a very less genetic base among cultivated crop species, so whenever a major crop is affected by a disease or pathogen, there is a huge loss in production. Reduction of diversity due to consumer preference and market demand has resulted into release of different varieties or breeds that are not having larger genetic diversity (FAO 2009). There are some species which are on the verge of extinction so there is need of hour to use modern biotechnological tools and methods like tissue culture and molecular biology to save these traditional species. Traditional farmers and their practices have conserved different plant species and varieties over hundreds of years which have created food securities all over the world. If we want sustainable development of agriculture, it is the need of the hour to follow the traditional practices and to generate such a vast knowledge which can predict the challenges of new diseases, new insects, and new pests. Documentation of traditional knowledge is a must because without documentation, the important information could be lost. These days we can observe that environmental conditions are changing frequently when compared with the past; these sudden changes in environmental conditions increase threat to the food security of the world, but if we use traditional knowledge along with modern tools, then this problem can be solved very easily. An organization such as FAO can help the farmers save the biodiversity of an area.

A large population of this world still uses their traditional knowledge for curing different diseases and physiological ailments. The world herbal trade is expected to reach US $ 7 trillion in 2050 from US $ 120 billion. Out of 3000 species/varieties known to be in international trade (Schippman et al. 2006), approx. 960 species/varieties are under commercial cultivation. All over the world these days, medicines and health products from plants are in great demand (Schippmann et al. 2003). Today only 12 % of world biodiversity has been used for medicines. There are still 88 % of plants available which can be checked for their different potential biological active molecules which can create a wonder in the area of pharmaceutical industry.

Traditional farmers have created great variety, variability, and robustness for food and agricultural sustainability. Indigenous and local communities conserve and use domestic and wild species sustainably, which helps to ensure food security, improved livelihoods, incomes, and participation in markets. Food and agricultural sustainability are key aspects of sustainable development, but economic, environmental, and social sustainability is now challenged by energy crises, conflicts over resources, inequitable distribution, pandemics, ecosystem degradation, and the impacts of climate change. Climate change, extreme weather events, and disasters are having increasing impacts on rural poor populations, threatening lives, food security, agricultural production, and the fisheries and forest and aquatic resources on which people depend for their subsistence. Many rural communities have vast knowledge of previous variations in climate and weather and have developed mitigation and adaptation strategies for ensuring food security. Communities apply traditional knowledge in early warning systems that calculate risks or detect extreme weather events, droughts, or floods. They use it in adapting subsistence strategies for agriculture, fishing, forestry, and foraging, improving water and resource management, enhancing ecosystems, and selecting which resources to use to mitigate or adapt to climate change effects. Traditional knowledge is used to observe, monitor, and report weather-related changes in food and agricultural systems and to adjust to these climate-related impacts. Traditional knowledge must be documented or else its loss will lead to food insecurity, poverty, conflicts, low livelihoods, and biodiversity. Extreme weather events are becoming increasingly frequent and intense; responses are needed that manage and reduce the risks and that improve preparedness for climate-related disasters and emergencies affecting food and agriculture. Such responses should combine traditional and local knowledge with data and applications from cutting-edge science and technology and should involve innovative collaboration among local communities and government and nongovernmental organizations (NGOs). Communities change their settlement patterns and food and agricultural practices to adapt or mitigate such climate change impacts as droughts; floods; sea level rises; changes in temperature; flora, fauna, or water conditions; and outbreaks of pests and diseases. The traditional knowledge and networks of communities and households can be mobilized to prepare for, mitigate, and manage disasters related to climate change before, while, and after they occur. To adapt to or reduce the impacts of climate change, rural communities select and improve both traditional and introduced seeds and crop varieties. Communities can reduce food insecurity risks by complementing their traditional knowledge and practices with information and support from governments and others, including rapid response systems and capacity building for disaster preparedness, mitigation, and management. Through the Globally Important Agricultural Heritage Systems (GIAHS) global program, FAO supports innovative agricultural systems at the landscape level, as well as the traditional knowledge and practices and the bio-cultural dynamics that maintain unique agroecological systems. GIAHS makes use of the cultural dynamics and traditional institutions and practices that enhance agrobiodiversity, food security, livelihood sustainability, and water and soil management in the face of climate, environmental, and social change.

From the 17,000 to 18,000 species of flowering plants, 6000–7000 are estimated to have medicinal usage in folk and documented systems of medicine, like Ayurveda, Siddha, Unani, and Homoeopathy. More than 80 % of the world's population use traditional plant-based medicine for primary health care, according to the World Health Organization (WHO). Demand is growing worldwide for medicinal plants, as well as herbs and aromatic plants. There is global resurgence in traditional and alternative health-care systems resulting in world herbal trade which stands at US$ 120 billion and is expected to reach US$ 7 trillion by 2050. Of the 3000 or so species known to be in international trade (Schippmann et al. 2006), there are approximately 960 for which commercial cultivation is underway or in development of which 178 species have annual consumption levels in excess of 100 metric tones. This signals that 70–80 % of the medicinal plants being traded in the world originate from wild collections. There is growing demand for plant-based medicines, heath products, pharmaceuticals, food supplements, cosmetics, etc., in the national and international markets. Conservation and sustainable use of medicinal plants need immediate focus of conserving biodiversity and promoting and maintaining the health of local communities, besides generating productive employment for the poor with the objective of poverty alleviation in tribal and rural areas. As medicinal plants are the basic resource of human health, appreciation for the preventive and therapeutic values of herbal remedies, in addition to their low-cost wide accessibility and cultural relevance, and increased consumption of medicinal plants through expansion of local regional and global markets have increased pressure on a resource that is largely harvested from depleted wild population in shrinking wild habitats. Their conservation lags far behind the current global demand. The current and potential value of medicinal plants – their value to local community health, regional markets, and global health security and trade – is widely recognized as a reason to conserve tropical forest ecosystems. Due to population growth and increasing urbanization, the demand for plant-based drugs is rising very rapidly in cities and towns. Worldwide, the increased popularity of non-Western treatments ("alternative" or complementary medicine) is creating a rapidly expanding market for both crude drugs and sophisticated compound preparations. Since less than 10 % of the world's biodiversity has been evaluated for potential biological activity, many more useful natural lead compounds await discovery with the challenge being how to access this natural chemical diversity. The human population explosion coupled with the improved standard of living has led to unmanaged exploitation of these plants resulting in imminent danger of extinction for some of them. Two hundred fifty-six medicinal plants in India have been listed in the *Indian Red Data Book* as endangered. Most of these wild medicinal plants are confined to certain habitats with a restricted geographic range. Their rarity coupled with large-scale destructive collection from the wild has resulted in conservation efforts being initiated by governmental and NGOs focused on their conservation and sustainable use.

8.5 Regulatory Agencies and Their Legal Policies

The FAO, CBD, ITPGRFA, International Plant Protection Convention (IPPC), and the WTO Sanitary and Phytosanitary Measures Agreement (SPS) are the main regulatory agencies for framing legal policies. The *Cartagena Protocol on Biosafety to the Convention on Biological Diversity* is an international agreement which aims to ensure the safe handling, transport, and use of living modified organisms (LMOs) resulting from modern biotechnology that may have adverse effects on biological diversity, taking also into account risks to human health. It was adopted on January 29, 2000 and entered into force on September 11, 2003. In 2010, the CBD adopted the Nagoya Protocol on "Access to Genetic Resources and the Fair and Equitable Sharing of Benefits Arising from their Utilization to the Convention on Biological Diversity," which has potential for impact upon germplasm exchange. FAO decides the appropriate policies and strategies at the national, regional, and international levels, essential to create an enabling environment for the development of sectors responsible for sustainable use, including plant breeding and seed sectors, and promotes capacity building in plant breeding and the seed sector, particularly in developing countries, mainly through support of education and training and institutional support. It also promotes exchange of technologies, know-how, plant genetic resources, and knowledge among all stakeholders involved.

The adoption of the ITPGRFA in 2001 was a major milestone for plant genetic resources conservation act. FAO, together with Biodiversity International acting on behalf of the CGIAR, founded the Global Crop Diversity Trust in 2004, to ensure the conservation and availability of crop diversity for food security worldwide. The Global Partnership Initiative for Plant Breeding Capacity Building (GIPB) was launched in 2006, with the mission of enhancing the capacity of developing countries to improve crops through better plant breeding and delivery systems. This initiative is contributing to a more comprehensive implementation of the ITPGRFA and the GPA for the conservation and sustainable utilization of Plant Genetic Resources for Food and Agriculture (PGRFA) as well as the Global Crop Diversity Trust (GCDT).

The Second Global Plan of Action (GPA) for PGRFA is a strategic framework for the conservation and sustainable use of the plant genetic diversity on which food and agriculture depend. The second GPA reaffirms the commitment of governments to the promotion of PGR as an essential component for food security through sustainable agriculture in the face of climate change. These include new policies and international agreements that affect conservation, use and exchange of genetic resources, shift in food production trends, changing roles of public and private sector in crop improvement and delivery systems, advances in biotechnology, genomics and information technologies, new products derived from agriculture, impact of new pests, climate change, and rapid urbanization on plant genetic erosion and vulnerability. The second GPA aims to promote cost-efficient and effective global efforts to conserve and sustainably use PGRFA; link conservation with a greater use of plant germplasm; strengthen crop improvement and seed systems to foster economic development, create capacities, strengthen national programs, and widen partnerships for PGRFA management; and strengthen implementation of the ITPGRFA.

Many gene banks of the world with rich and valuable resource of crop diversity face severe fund crunch. The sole global response to this threat is the GCDT. The Trust is a unique public-private partnership raising funds from individual, corporate, and government donors to establish an endowment fund that will provide complete and continuous funding for key crop collections in eternity. In line with the ITPGR and the GPA for the conservation and sustainable utilization of PGRFA, GCDT goal is to advance an efficient and sustainable global system of ex situ conservation by promoting the rescue, understanding, use, and long-term conservation of valuable plant genetic resources.

The World Information and Early Warning System is a worldwide dynamic mechanism to foster information exchange among member countries by gathering and disseminating information on PGRFA and is an instrument for the periodic assessment of the State of the World's PGRFA. It is an important component of the FAO Global System on Plant Genetic Resources for Food and Agriculture. It contains information of structure of 190 national PGR programs and activities; more than 5 million accessions belonging to more than 18,000 species, the PGRFA, and seed laws and regulations database (70 countries), *World List of Seed Sources* database (approx. 8000 entries from 150 countries), and crop varieties database (about 65,000 varieties from 1249 cultivated crops).

International agricultural research institutions, coordinated by the CGIAR, Washington, are focused mandate crops and have extensive seed collections for staple and economic crops such as rice, maize, wheat, barley, millets, pulses, oil seeds, tuber crops, banana, tropical forage, and fruits. The collections in these seed banks are well documented, and the institutions are networked among themselves and with several other institutions. The CGRFA established a subsidiary Intergovernmental Technical Working Group on Plant Genetic Resources for Food and Agriculture (ITWG-PGRFA) at its seventh session in 1997, to address issues specific to PGRFA. The Secretariat of the ITWG-PGRFA lies with the AGP Division in FAO Headquarters.

In India important legislations have been enforced in response to international developments. These are the "Biological Diversity Act (BDA) 2002," "Protection of Plant Varieties and Farmers' Right Act 2001" (PPV & FRA), and

"Geographical Indications of Goods (Registration and Protection) Act 1999." In addition, amendments have been made in the "Patents (Amendment) Act 2005" and the "Plant Quarantine Regulation of Import into India Order 2003" which have a bearing on PGR management.

8.6 Lacunae's and Future Prospects/Action Needed

A concise assessment of the status and trends of plant genetic resources is needed to identify the most significant gaps and needs in order to provide a basis to update the rolling GPA. Agricultural ecosystems that harbor diversity of PGRFA are more in a state of flux, as compared to other natural ecosystems, since cropping patterns are changing every season. Modern agricultural practices strongly favor reduction of crop diversity by providing the subsidies for cultivating high-yielding varieties and reducing weeds/wild plants diversity by using crop protection measures. By providing positive incentives to local communities, this diversity may be maintained. PGRFA diversity should be assessed at regular intervals to record changes in species population as well as monitoring genetic erosion, if any. Assessment of loss of diversity in farming systems using continued analysis of land use patterns, in crop diversity within a crop species through genetic diversity analysis, and in wild and weedy relatives at in situ level and characterization, evaluation, multiplication, and conservation with specific emphasis on breeding "trait"-specific material need illustrative documentation of information on diversity in PGRFA, linkages with organizations involved in management of PGRFA, and awareness generation in farmers and tribal communities through free trainings and demonstrations at their place. Regular survey and upgradation to assess status of PGRFA with reference to the number of crops cultivated in an area and number of varieties of each crop being cultivated in that area in relation to diversity available and genetic erosion, to add unexplored areas as well as areas explored more than 20 years back, is recommended. Measurement of genetic erosion and assessment of concentration of diversity should be monitored through ground surveys and modern tools (GIS) for mapping of PGRFA. To better understand the roles and values of the diversity of PGRFA, emphasis should be given on the food and cultural habits of the tribal communities of the country which are highly dependent on local diversity for their survival. International linkages, project formulation, and sufficient funding for such activities need to be explored. Training on the study of genetic erosion needs to be given to at least one plant breeder from each crop-based agricultural university and is also required in developing and using early warning systems for controlling genetic erosion. An awareness campaign needs to be launched at grass root level among actual stakeholders.

Expanding agriculture leads to habitat loss and fragmentation, drainage of wetlands, and impacts on freshwater and marine ecosystems through sedimentation and pollution. The important threats include introduction of new high-yielding and improved cultivars, urbanization, deforestation, shifting cultivation, overexploitation and lack of regeneration, genetic erosion, human interference, faulty and unscrupulous collection, modernization of agriculture, biotic and abiotic stresses, natural disasters, lack of availability of seeds of local varieties, changing cropping patterns due to commercialization of agriculture, unpredicted weather, changes in food habits, lack of awareness, and less R&D thrust, including value addition on traditional crops and crops of minor importance. Although the survey and inventorization activities have been included in the draft national biodiversity action plan, there is a need for training scientists in the survey and inventorization of PGRFA in the country. The main constraints indicates lack of focused and coordinated approach by the concerned organizations, insufficient financial support, and need for strengthening skills of technical staff. Unlike ex situ conservation, in situ and on-farm conservation activity especially on PGRFA has not been given as major responsibility to any of the organization. Despite a strong national policy support for the conservation of crop's wild relatives, they are not usually considered in Environmental Impact Assessments (EIAs). The plant species that require immediate attention for in situ and on-farm conservation have been prioritized, and suitable sites based on the quantum and nature of genetic diversity have also been identified. However, a comprehensive strategy is required to be framed and put into action at national level. For development of effective conservation strategies, there is a need for prioritizing collection of threatened landraces and species. In fact, true landraces and farmer's varieties are required to be identified along with their pedigree records and other related information. In many cases, the same landrace is known by various names. Critical evaluation of true landraces, including biochemical and molecular characterization, is the need of the hour. For better management practices to reduce genetic changes or loss of genetic integrity, attention is paid to select suitable regeneration environment, adequate population size, and proper handling of regenerated material in close collaboration of crop-based institutes. Priorities for expanding ex situ conservation activities would be wild and weedy relatives of crop plants and trait- and agroecological-specific germplasm. Funding may be required in crops of regional/global importance. Priorities setting for ex situ conservation are clear in India because of its commendable strength in this area. However, this may need facilitation for targeted collections based on specific trait(s) or regional/global importance. There is a built-in duplicity of accessions in the system, wherein the accessions conserved at NAGS and the crop-based institutes as active collection are conserved as base collection in the National Genebank.

Though large germplasm has been amassed under ex situ conservation, a good number of accessions collected earlier lack passport, characterization, and/or evaluation data, which in many cases has to be extracted from archive records. The present emphasis is to collect and assemble trait-specific accessions from different agroecological zones/sources to facilitate their use in crop improvement. There is a national mechanism for registration of the potentially valuable germplasm with unique traits to facilitate their documentation and use in basic research and crop improvement. Also, there has been intense genetic erosion of primitive types and landraces. Many wild and wild-related species are difficult to maintain due to differences in their adaptability and seed biology/dormancy. Most wild species are not easily crossable with cultivated species. However, the use of new biotechnology tools has resulted in the development of few intergeneric and interspecific varieties in some of the crops. Moreover, the genes carrying resistance/tolerance to stresses in wild relatives are linked with undesirable genes that slow down the progress in breeding. Further, in many cases, sources for resistance to some of the important pest diseases are not available. Although in most of the crops a large number of accessions have been assembled, their management for effective use has become difficult. Core collections have been developed only in few crops. Therefore, there is a need for strengthening the application of techniques like development of gene pools, core collections, trait-specific core sets, etc.

8.7 Conclusion

The access to trait-specific, well-characterized plant genetic resources along with secondary and tertiary gene pool to the world breeders for trait-specific targeted crop genetic improvement will be a key to future increases in crop productivity. Presently, the era of collecting germplasm for gene banks has passed. The paradigm is now shifting to how to best use what is available in the gene banks. Hence, initiatives to characterize, evaluate, and promote the use of PGRFA have become increasingly important. FAO and partner organizations are, therefore, very much involved with the GIPB that promotes use of plant genetic resources. Conservation focuses explicitly on maintaining the diversity of the full range of genetic variation within a particular species or taxa. Conserving PGRFA ensures the future adaptability of cultivars and wild populations; preserved traits ensure sustainable agriculture to promote the use of genetic resources commercially and to conserve genetic diversity for multiple reasons. Food and agriculture production have relied collectively on worldwide efforts, like genetic resources domesticated elsewhere and subsequently developed in other countries and regions. Continued access to plant genetic resources and a fair and equitable sharing of the benefits arising from their use are therefore essential for food security. The genetic diversity, the variation in the molecular building blocks that control expression of individual traits, is at the core of a crop's ability to continually undergo these changes. The combination of current and historical genetic diversity underpins the potential to adapt crops to the changing needs of farmers and consumers. Agriculture itself is one of the greatest threats to biodiversity worldwide. Today's international breeds are based on a narrow genetic base; food security is threatened when these breeds are affected by pests and diseases and climate changes. The drastic reduction in the diversity of food systems and crop genetic resources is due to changes in consumer preferences, population growth, changes in land cover and land use, loss of biodiversity, and expansion of uniform commercial varieties and breeds that led to the demise of traditional knowledge and invite disease epidemics easily. For sustainability of these resources, the conservation, diversification, adaptation, improvement, and delivery to farmers through seed systems are needed.

References

Alcorn JB (1981) Huastec noncrop resource management. Human Ecol 9:395–417

Altieri MA (1983) Agroecology: the scientific basis of alternative agriculture. Division of Biological Control, University of California, Berkeley

Altieri MA, Merrick LC (1987) *In situ* conservation of crop genetic resources through maintenance of traditional farming systems. Econ Bot 41:86–96

Altieri MA, Anderson MK, Merrick LC (1987) Peasent agriculture and conservation of crop and wild plant resources. Conserv Biol 1:49–58

Baker HG (1965) Characteristics and modes of origin of weeds. In: Baker HG, Stebbins GL (eds) The genetics of colonizing species. Academic, New York, pp 147–168

Barlett PF (1980) Adaptation strategies in peasant agricultural production. Ann Rev Anthropol 9:545–573

Barrett SCH (1983) Crop mimicry in weeds. Econ Bot 37:255–282

Borokini TI, Okere AU, Giwa AO, Daramola BO, Odofin WT (2010) Biodiversity and conservation of plant genetic resources in field gene-bank of the National Centre for Genetic Resources and Biotechnology, Ibadan, Nigeria. Int J Biodivers Conserv 2:37–50

Brown WL (1983) Genetic diversity and genetic vulnerability-an appraisal. Econ Bot 37:4–12

Browning JA (1974) Relevance of knowledge about natural ecosystems to development of pest management programs for agro-ecosystems. Proc Am Phytopathol Soc 1:191–194

Browning JA, Frey KJ (1969) Multiline cultivars as a means of disease control. Ann Rev Phytopathol 7:355–382

Brush SB (1982) The natural and human environment of the central Andes. Mt Res Dev 2:14–38

Chacon JC, Gliessman SR (1982) Use of the "non-weed" concept in traditional tropical agroecosystems of south-eastern Mexico. Agro-Ecosyst 8:1–11

Christanty L, Abdoellah O, Iskander J (1986) Traditional agroforestry in West Java: the pekarangan (homegarden) and talun-kebun (shifting cultivation) cropping systems. In: Marten G (ed) The human ecology of traditional tropical agriculture. Westview Press, Boulder

Clawson DL (1985) Harvest security and intraspecific diversity in traditional tropical agriculture. Econ Bot 39:56–67

Cohen JI (1991) Ex situ conservation of plant genetic resources: global development and environmental concern. Science 253:866–872

Commission on Genetic Resources for Food and Agriculture (2015) Second global plan of action for plant genetic resources for food and agriculture. Food and Agriculture Organization of the United Nations (FAO), Rome, 29 November 2011

Datta SC, Banerjee AK (1978) Useful weeds of West Bengal rice fields. Econ Bot 32:297–310

de Janvry A (1981) The agrarian question and reformism in Latin America. Johns Hopkins University Press, Baltimore

Egger K (1981) Ecofarming in the tropics-characteristics and potentialities. Pl Res Dev 13:96–106

Engelmann F (1999) Management of field and in vitro germplasm collections. Proceedings of a consultation meeting, Cali, CIAT, and Rome, IPGRI

Engelmann F, Takagi H (2000) Cryopreservation of tropical plant germplasm. Current research progress and application. Japan International Research Center for Agricultural Science/IPGRI, Tsukuba/Rome

FAO (2009) FAO and traditional knowledge: the linkages with sustainability, food security and climate change impact. Food and Agricultural Organization of the United Nations, Rome

FAO (2014) Genebank standards for plant genetic resources for food and agriculture. Rev edn, Rome

FAO (2015) FAO and Post 2015 Nourishing people nurturing the planet. 100 facts in 14 themes linking people, food and the planet FAO and the Post-2015 Development Agenda FAO Corporate Task Team on Post-2015 Economic and Social Development Department Food and Agriculture Organization of the United Nations 00153 Rome

Frankel H (1973) Survey of crop genetic resources in their centres of diversity. FAO/IBP, Rome

Frankel H, Bennett E (1970) Genetic resources in plants-their exploration and conservation. International biological programme handbook. Blackwell, Oxford

Frankel H, Hawkes JG (1975) Crop genetic resources for today and tomorrow. Cambridge University Press, Cambridge

FRLHT (2006) Conservation and adaptive management of medicinal plants-a participatory model: medicinal plants conservation areas and medicinal plants development areas. FRLHT, Bangalore

Gill KS, Nanda GS, Singh G, Aujla SS (1980) Studies on multilines in wheat (*Triticum aestivum* L.). Breeding of a multiline variety by convergence of breeding lines. Euphytica 29:125–128

Harlan JR (1975) Our vanishing genetic resources. Science 188:618–622

Harlan JR (1976) Genetic resources in wild relatives of crops. Crop Sci 16:329–333

Hodkinson TR, Waldren S, Parnell JAN, Kelleher CT, Salamin K, Salamin N (2007) DNA banking for plant breeding, biotechnology and biodiversity evaluation. J Plant Res 120:17–29

Indian Agricultural Research Institute (1995) India country report to the FAO International technical conference on Plant Genetic Resource, Leipzig

Juffe-Bignoli D, Burgess ND, Bingham H, Belle EMS, de Lima MG, Deguignet M et al (2014) Protected planet report. UNEP-WCMC, Cambridge

MoEF (2007) Protection, development, maintenance and research in biosphere reserves in India. Guidelines and proformae Government of India Ministry of Environment And Forests, New Delhi

MoEF (2009) India fourth national report to convention on biological diversity. Ministry of Environment and Forests, Government of India, New Delhi

MoEF (2013) India state of forest report, Ministry of Environment and Forest, Government of India

MoEF (2014) India's fifth national report to the convention on biological diversity. Ministry of Environment & Forests, Government of India

Myers N, Mittermeier RA, Mittermeier CG, da Fonseca GAB, Kents J (2000) Biodiversity hotspots for conservation priorities. Nature 403:853–858

Nabhan (1984) Replenishing desert agriculture with native plants and their symbionts. In: Jackson W, Berry W, Colman B (eds) Meeting the expectations of the land. North Point Press, San Francisco, pp 172–182

Nabhan GP (1985) Native crop diversity in Aridoamerica: conservation of regional gene pools. Econ Bot 39:387–399

Nagel M, Borner A (2009) The longevity of crop seeds stored under ambient conditions. Seed Sci Res 20:1–12

NBPGR (2007) State of Plant Genetic Resources for Food and Agriculture in India (1996–2006): A Country Report. National Bureau of Plant Genetic Resources, ICAR, New Delhi, 70 p

Oppermann M, Weise S, Dittmann C, Knupffer H (2015) GBIS: the information system of the German gene bank. doi:10.1093/database/bav021

Pande HK, Arora S (2015) India fifth national report to convention on biological diversity. Ministry of Environment and Forests. Government of India, New Delhi

Prescott-Allen R, Prescott-Allen C (1983) Genes from the wild. Earthscan Paperback, London

Probert RJ, Daws MI, Hay FR (2009) Ecological correlates of ex situ seed longevity: a comparative study on 195 species. Ann Bot 104(1):57–69

Reed BM (2010) Plant cryopreservation. A practical guide. Springer, New York

Reed BM, Engelmann F, Dulloo ME, Engels JMM (2004) Technical guidelines for the management of field and in vitro germplasm collections. Handbooks for Gene banks No. 7. IPGRI, Rome

Rice N, Henry R, Rossetto M (2006) DNA Banks: a primary resources for conservation research. In: de Vicente MC (ed) DNA Banks-providing novel options for gene banks. Topical reviews in agricultural biodiversity. International Plant Genetic Resources Institute, Rome, pp 41–48

Rick (1973) Potential genetic resources in tomato species: clues from observations in native habitats. In: Srb AM (ed) Genes, enzymes and populations. Plenum Press, New York, pp 255–269

Salick B, Byg A (2007) Indigenous peoples and climate change. Tyndall Centre for Climate Change Research, Oxford

Schippmann U, Cunningham AB, Leaman DJ (2003) Impact of cultivation and gathering of medicinal plants on biodiversity: global trends and issues. In: Biodiversity and the ecosystem approach in agriculture, forestry and fisheries satellite event on the occasion of the Ninth regular session of the Commission on Genetic Resources for Food and Agriculture, Rome, pp 1–25

Schippmann U, Leaman D, Cunningham A (2006) Cultivation and wild collection of medicinal and aromatic plants under sustainability aspects. Medicinal and aromatic plants. Springer, Dordrecht, pp 23–45

Segal A, Manisterski J, Fischbeck G, Wahl I (1980) How plant populations defend themselves in natural ecosystems. In: Horsfall JG, Cowling EB (eds) Plant disease: an advanced treatise. Academic, New York, pp 75–102

Sharma SK (2007) Indian plant genetic resources (PGR) system: role of NBPGR. Training manual on *In vitro* and cryopreservation techniques for conservation of PGR. NBPGR and Bioversity International, New Delhi

Simmonds NW (1962) Variability in crop plants, its use and conservation. Biol Rev 37:422–465

Smith JB, Ragland SE, Pitts GJ (1996) A process for evaluating anticipatory adaptation measures for climate change. Water Air Soil Pollut 92:229–238

UNEP-WCMC (2014) Global statistics from the World Database on Protected Areas (WDPA). UNEP World Conservation Monitoring Centre, Cambridge

Vavilov NI (1951) Phytogeographic basis of plant breeding. In The origin, variation, immunity and breeding of cultivated plants, pp 13–54. Chron Bot 13: 1–366

Walters C, Wheeler LM, Grotenhuis JM (2005) Longevity of seeds stored in a gene bank: species characteristics. Seed Sci Res 15:1–20

Wilken GC (1977) Integrating forest and small-scale farm systems in middle America. Agron Ecosyst 3:291–302

Wilkes HG (1983) Current status of crop plant germplasm. CRC Crit Rev Pl Sci 1:133–181

Wolfe MS (1992) Barley diseases: maintaining the value of our varieties. In: Munk L (ed) Barley genetics VI. Munksgaard International Publishers, Copenhagen, pp 1055–1067

Access and Benefit Sharing on the Use of Indigenous Traditional Knowledge

9

Gulzar S. Sanghera, Dharminder Bhatia, and K.S. Thind

Abstract

Knowledge, innovations and practices that are generated in a unique, tradition-based context from indigenous peoples is termed as traditional knowledge that forms an integral part of indigenous peoples' social, cultural and spiritual values. Access to various biological resources, fair and equitable sharing of benefits from the use of these resources and TK has become an important agenda after the CBD came into force. The debates are still going on the requirement for institutional mechanisms to regulate the access and benefit-sharing agreements, defining ownership of biological resources and their associated TK. With the advent in recent technologies, TK associated with natural biological resources and their economic and scientific importance has attracted a wide range of private sector stakeholders for their utilization. In the past, TK has often been accessed without the consent of the owners of TK and without sharing the benefits arising out of its utilization. Various measures for the protection of TK, through either adapted intellectual property rights or sui generis systems, are in vain as long as the holders of TK and the specific context in which TK has been generated, transmitted and preserved do not enjoy appropriate protection through the recognition of indigenous peoples' rights. Indigenous rights can gain strength from progress made and legal standards set in international level elaborating measures for the protection of TK and vice versa. This chapter explores and reports the various negotiations on an international regime on access and benefit sharing in the framework of the Convention on Biological Diversity that provides a forum to push indigenous peoples' right to their TK.

G.S. Sanghera, Ph.D. (✉)
Regional Research Station, Punjab Agricultural University, Kapurthala 144601, Punjab, India
e-mail: sangheragulzar@pau.edu

D. Bhatia, Ph.D. • K.S. Thind, Ph.D.
Punjab Agricultural University, Ludhiana 144001, Punjab, India
e-mail: d.bhatia@pau.edu; ksthind-pbg@pau.edu

R.K. Salgotra, B.B. Gupta (eds.), *Plant Genetic Resources and Traditional Knowledge for Food Security*, DOI 10.1007/978-981-10-0060-7_9

9.1 Introduction

In the past, indigenous people (IPs) hold the responsibility to manage their own resources and knowledge about plant and animal use, including methods of preparation, storage and management, which is of great economic significance. Their knowledge of biological and genetic resources had already formed the basis for seed, pharmaceutical and natural product industries. Management of different natural resources, viz. soil fertility maintenance, stream and coastal conservation and forest and agricultural system models, have provided viable, time-tested options for sustainable developments adapted to microclimate variations and local sociopolitical ecosystems (Showers1996; Tandon 1981; Meredith 1997). Nowadays, IPs face increasing external pressures to share their information, contribute their knowledge and practices and endorse developments involving their lands, territories, biological and genetic resources and cultural products and performances (Howes and Chambers 1979).

The contributions that TK has made in the past to world food and medicinal sources, as well as IP's current significant contributions to agriculture, water and forest management, have done little to offset political marginalization of these peoples' recognition (Labatut and Akhtar 1992). There is overlapping between interests and concerns of IPs with those of local communities in the developing world, but is far from identical. IPs comprise 'nations within nations', and those living in developing countries often suffer human rights abuses and political marginalization from countries' governments. There is continuous degradation and expropriation of indigenous lands and resources without adequate legal protection in the developed and developing worlds (Showers 1996). International standard setting activities have addressed these problems to some extent, but much more remains to be done. Some important and useful standard setting instruments are the International Labour Organization (ILO) Convention, 169 Concerning Indigenous and Tribal People in Independent Countries, 1989, the Convention on Biological Diversity (1992) (CBD), the FAO International Treaty on Plant Genetic Resources for Food and Agriculture (ITPGRFA) and the Draft United Nations Declaration on the Rights of Indigenous People.

In this context, the UN Convention on Biological Diversity is very important for its provisions related to the protection of indigenous rights, interests in biological and genetic resources and the TK and practices associated with them. The task for protection of 'intangible' knowledge is very difficult and challenging. Recent attempts for such protections have been framed within the contexts of legal regimes and policy developments concerning bioprospecting, access to biological and genetic resources, benefit sharing and intellectual property rights. However, these developments provide limited scope for recognizing intangible knowledge yet. This chapter argues that TK relating to the conservation of biological diversity cannot be protected without international and national regimes that protect the human rights of the holders/owners of this knowledge as well as their specific indigenous integrated and comprehensive rights. It reviews a range of developments, existing and emerging, including both international and domestic that provide or have the potential to protect indigenous rights in TK and biodiversity.

9.2 Traditional Knowledge: Its Role and Value

TK has been discussed in several international forums; no consensus on an internationally acceptable definition has been reached in particular. It generally refers to tradition-based literary, artistic or scientific works; performances, inventions, scientific discoveries, designs, marks, names and symbols; undisclosed information; and all other tradition-based innovations and creations resulting from intellectual activity in the industrial, scientific, literary or artistic fields (Correa 2001). Sometimes, it is also referred to as indigenous knowledge that can be defined as native, born or growing or produced naturally in a specific area (Fig. 9.1). It is not introduced from

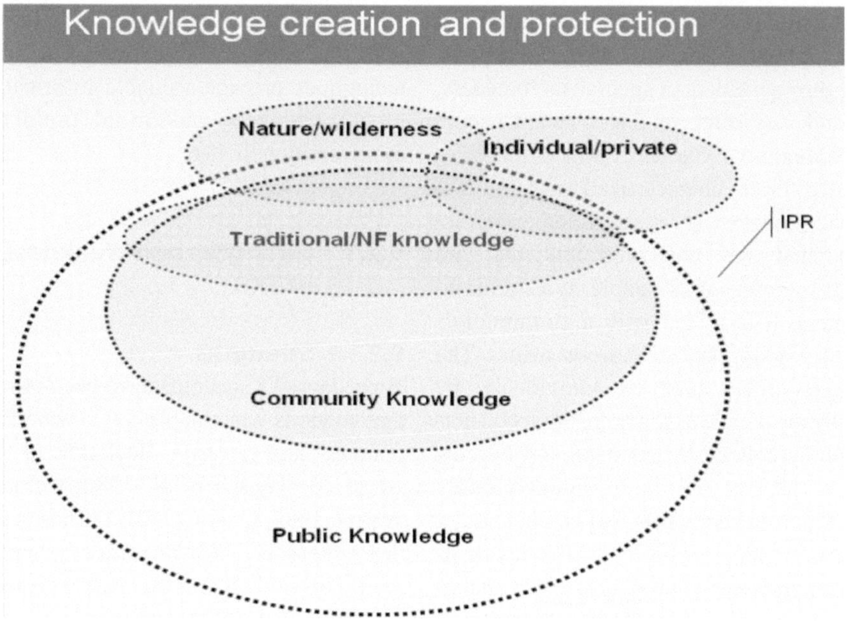

Fig. 9.1 Domain of traditional knowledge – creation and protection (Source: Oli 2009)

outside the particular region, country or environment.

In this context, traditional societies can be referred to as indigenous, as explained by Nikolas (2000). So, TK refers to the knowledge, innovations and practices of indigenous and local communities around the world. It has been developed from experience gained over the centuries and adapted to the local culture and environment that has been transmitted orally from generation to generation. TK is mainly of a practical nature in fields like agriculture, fisheries, health, horticulture and forestry. In general, it tends to be collectively owned that takes in the form of stories, songs, proverbs, cultural values, beliefs, rituals, community laws, local language and agricultural practices, including the development of plant species and animal breeds (Lawas and Luning 1997; Gupta 1992). Today, there is a growing appreciation of the value of TK, not only to those who depend on it in their daily lives but to modern industry as well as agriculture. Many commonly used products, such as plant-based medicines and cosmetics, are derived from TK, including agricultural as well as handicraft products. It has been noticed that indigenous and local communities are mostly situated in areas where the majority of the world's plant genetic resources are available (Meredith 1997). Many of these communities have cultivated and used the existing biological diversity in a sustainable way for years together. However, the contribution of these communities in conservation and sustainable use of biological diversity has gone far beyond their role as natural resource managers. Their skilled and technical hands provide very valuable and useful information to the world. TK has made a significant contribution towards sustainable developments as indigenous and local communities with extensive knowledge of local environments have been directly involved with conservation and their sustainable use. Since TK is collective in nature, it is often considered as the property of the entire community and not belonging to any single individual within the community. This is notwithstanding the fact that certain individuals within a community may be the ones endowed with some special skills like herbalists, iron mongers and blacksmiths. The available knowledge is being used to sustain the community and its culture and to maintain the resources necessary for the community in its continued survival. TK is shared through specific cultural and traditional information-exchange mechanisms, e.g. orally through elders or specialists (breeders, healers,) and very often to few selected people within a community (Nikolas 2000).

Generally, TK is characterized with the elements having knowledge, innovations and practices generated and preserved and that are transmitted by indigenous people in a tradition-based context. It is linked with a community's culture and is expressed in various forms. The knowledge is either held by individuals, by groups or by the people as a whole, but its ownership is usually collective and regulated by customary law. Mostly, TK is transmitted orally from generation to generation and is not static; it further evolves and generates new information as a result of improvements or adaptation to changing circumstances. The formal characteristics of TK focus on its context, method and qualities instead of its actual contents, such as medicine, food and agriculture, environmental management, biodiversity conservation, nutrition and cultural objects. In the framework of the international regime, TK is interpreted within the context of Article 8(j) and Article 15, as knowledge, innovations and practices associated with genetic resources that in question is held by indigenous and local communities.

In recent days, there is a great appreciation for the values of TK, which is valuable not only to those who depend on it for their livelihood but also to modern industry and agriculture as well. A good number of widely used products like plant-based medicines, health products and cosmetics are derived from TK, besides other valuable products that include agricultural and non-wood forest products including handicrafts. It reflects the significant contribution of TK towards sustainable development (Mathias 1994). Most indigenous and local communities are situated in areas where the vast majority of the world's genetic resources are found and many of them have cultivated and used such biological diversity in a sustainable way for years together. However, the contribution of indigenous and local communities towards the conservation and sustainable use of biological diversity has gone far beyond their role as natural resource managers (Maundu 1996). Though, their skills and techniques provide valuable information to the global community and towards useful models for biodiversity policies.

9.2.1 Strengths and Weaknesses of TK

9.2.1.1 Strengths

Nowadays, TK systems have been proven as pivotal towards sustainable socio-economic developments and poverty alleviation in developing world (Brokensha et al. 1980; Compton 1989; Warren 1990; Gupta 1992). TK has been increasingly regarded as a precious resource in both Northern and Southern parts (Warren 1991; Buttimer et al. 1991), which clearly focuses on heralds of it, a long overdue move. It represents a way forward from the preoccupation of the centralized, technically oriented solutions of the past, which reflects its failure to improve the prospects of most of the world's peasants and small farmers (Agrawal 1996).

The literature on traditional or indigenous knowledge, agricultural development and environmental management provides abundant evidences of human activities which utilize complex but implicit scientific principles (Atte 1992). It has also been demonstrated that the exclusion of such knowledge from development activities has had disastrous consequences in every region of the world where outsider knowledge has been imposed without regard to TK (Cashman 1989).

9.2.1.2 Weaknesses

A good quantum of literature have indicated that the TK has its own limitations (Leach and Mearns 1988; Bebbington 1993; Howard and Widdowson 1996) and it is not itself capable of addressing various issues related to sustainable socio-economic developments and poverty alleviation (Murdoch and Clark 1994). Some workers emphasized that TK needed to be formalized, since it is fragmentary and provisional in nature (Arce and Long 1992). It is still in its formalization phase as problems with respect to the application of TK are most likely to arise. Such knowledge is still not as well known as the coded and circulated objective language and the printed products of scientific discourse.

9.2.1.3 Who Are Indigenous People?

It has been estimated that about 370 million indigenous people that live across 70 countries worldwide are the owners of some unique cultures, languages and knowledge systems. These individuals have inhabited in their territories with historical continuities and have developed some unique relationship to their lands and resources, which forms the basis of their cultures (Bijoy 2007a). Mostly indigenous people have been subjected to experiences of subjugation, marginalization, dispossession, exclusion or discrimination by the mainstream societies that have become dominant through conquest, occupation or settlement. However, such people have managed to retain their distinct social, cultural and political characteristics, in particular, which they have maintained as wealth of TK. It has been generated and transmitted in a tradition-based context, and that forms an intrinsic part of peoples' self-identity. It has been accumulated over generations and has been constantly adapting to changing environments (Brown 1985).

For protection of TK and its equitable benefit sharing, it is worthwhile to consider all the ways in which indigenous people have been recognized as the knowledge owners and custodians of it. It is impressed upon here that indigenous people must have the right fundamental to their self-identity and self-determination to define them (Bijoy 2007b). However, there have been some attempts in various international standard setting developments to derive a universal 'definition' of indigenous people. In a 1986 report, the Special Rapporteur of the United Nations Economic and Social Council Sub-Commission on the Prevention of Discrimination of Minorities defined indigenous people in different following ways:

Indigenous communities, people and nations are those who are having a historical continuity with preinvasion and precolonial societies that have developed on their territories and consider themselves distinct from other sectors of the societies now prevailing in those territories or parts of them (Fingleton 1998). At present they have formed non-dominant sectors of society and are committed to preserve, develop and transmit it to future generations their ancestral territories in accordance with their own cultural patterns, social institutions and legal systems.

Further, ownership and property are the western legal concepts that do not easily transpose onto traditional and indigenous systems. It is due to the fact that traditional and indigenous people identify themselves within communities who are associated to the land, specifically based on some spiritual connections, customary values and responsibilities (Warren et al. 1993). Moreover, there is often no single identifiable individual that could be considered as the property owner of lands and biological resources. These features of indigenous societies and cultures form the basis upon which knowledge systems of indigenous peoples must be understood as being intrinsically different to western systems of property and knowledge.

Since, TK is transmitted through specific cultural and traditional information-exchange mechanisms; it includes mental inventories of local biological resources, animal breeds, local plants, and crop and tree species. It also includes practices and technologies, such as seed treatment and storage methods and tools used for planting and harvesting. TK also encompasses belief systems that play a fundamental role in people's livelihood, maintaining their health and protecting and replenishing the environment (Ronnie et al. 2007). TK is dynamic in nature and may include experimentation in the integration of new plant or tree species into existing farming systems. The rich and complex systems of TK have been protected and transmitted through customary law regimes and are vulnerable to be undermined by dominant legal systems, especially Intellectual Property Rights (IPRs) regimes. Efforts of indigenous people to assert and protect their own approaches for managing their TK have often given rise to discussions about the appropriateness, use and abuse of intellectual and cultural property rights (Vivas-Eugui and

Ruiz-Muller 2001). In this context, arguments about the need to develop a unique (sui generis) system of legal protection recognize the variety of situations in which TK is held, transmitted and evolved. The emphasis of sui generis systems is that TK should not be analogous to the 'public domain' as this concept is understood within the western legal framework of intellectual property laws (Krishna 2007). Instead, protection should be provided to it through the recognition of customary laws that govern TK. Such type of system would result in legal pluralism and the operation of different legal systems in parallel with each other. These systems of knowledge management might be referred to as an indigenous domain, or customary economy, enabling indigenous communities to maintain viable, dynamic and sustainable livelihoods attuned to their environments and which incorporated complex rules and codes regulating exchange, trade and appropriate use of knowledge and resources (Simpson 1997). It is important to appreciate that these systems continue to shape indigenous peoples' relationships to their country and its natural resources.

9.3 Access and Benefit Sharing (ABS)

Genetic resources include living organisms, plants, animals and microbes that carry genetic material and are potentially useful to humans. These resources could be wild, domesticated or cultivated ones and are sourced from environments in which they occur naturally (in situ), or from human-managed collections such as botanical gardens, gene banks, seed banks and microbial culture collections (ex situ). These resources provide a crucial source of information to better understand the natural world and can be used to develop a wide range of products and services for the benefit of human beings. However, they are not evenly distributed like many key resources in the world. Niches in which plants, animals and microbes are found often have make up complex and delicately balanced ecosystems which can be threatened or endangered. The way in which genetic resources are accessed and how the benefits of their use are shared should create incentives for their conservation and sustainable use which can contribute to the creation of a fairer and more equitable economy to support sustainable development. So, it is essential that the value of TK is understood and valued appropriately by those who use it and the rights of indigenous and local communities (ILCs) be considered during negotiations over access and use of genetic resources. Its failure can put the knowledge, the resources and the communities at risk.

Access and Benefit Sharing (ABS) refers to the way in which genetic resources may be accessed and how the benefits that result from their use are shared between the people or countries using the resources (users) and the people or countries that provide them (providers). The access and benefit-sharing provisions of the CBD were designed to ensure that the physical access to genetic resources was facilitated and that the benefits obtained from their use were shared equitably with the providers. The benefits to be shared can be monetary, such as sharing royalties when the resources are used to create a commercial product, or nonmonetary, such as the development of research skills and knowledge. It is vital that both users and providers understand and respect institutional frameworks such as those outlined by the CBD and in the Bonn guidelines. These help governments to establish their own national frameworks which ensure that access and benefit sharing happens in a fair and equitable way.

Prior to the commencement of the CBD in 1992, access to genetic resources and associated traditional knowledge was free for all mankind. The research and commercialization of genetic resources and associated TK has existed in many forms for hundreds of years. In the eighteenth century, European colonial explorers travelled across different parts of the world seeking exotic plants. They brought back decorative flowers, medicinal herbs and new types of food. These expeditions were a one-way transfer of knowledge, with biological explorers taking knowledge from local communities. However, there was little or no exchange of knowledge and no offer of compensation to such communities (ICIMOD 2009).

During the later part of the 1900s, a few countries developed legal provisions for ABS. However, benefits were usually defined as tangible benefits (such as royalties), and benefit sharing was largely carried out at the government level. Benefits did not reach the traditional owners of genetic resources and associated TK even; local communities and countries of origin were often not informed about the use of their genetic resources and associated TK, limiting their bargaining power and preventing them from sharing in the benefits of their own resources.

9.3.1 How Does It Work?

ABS is based on Prior Informed Consent (PIC) being granted by a provider to a user and negotiations between both parties to develop mutually agreed terms (MAT) to ensure the fair and equitable sharing of genetic resources and associated benefits. PIC is the permission given by the competent national authority of a provider country to a user prior to accessing genetic resources, in line with an appropriate national legal and institutional framework. MAT is an agreement reached between the providers of genetic resources and users on the conditions of access and use of the resources and the benefits to be shared between both parties. These conditions are required under Article 15 of the CBD, which was adopted in 1992 and provides a global set of principles for access to genetic resources, as well as the fair and equitable distribution of the benefits that result from their use.

9.3.2 Who Is Involved?

All countries have sovereign rights over natural resources under their jurisdiction. They are obligated to put in place conditions that facilitate access to these resources for environmentally sound uses. Providers agree terms, which include PIC and MAT, for granting access and sharing benefits equitably. The participation of ILCs is necessary in cases where TK associated with genetic resources is being accessed. Users are responsible for sharing the benefits derived from genetic resources with the providers. They seek access to genetic resources for a wide range of purposes, from basic research to the development of new products.

The term ABS also refers to the questions arising in the context of (a) the access to genetic resources and (b) the fair and equitable sharing of benefits arising out of their utilization as well as the equitable sharing of benefits arising from the utilization of TK. Though both issues are generally discussed under the umbrella term 'access and benefit sharing', they can be illustrated individually (Reed 1990). Benefit sharing as one of the three objectives of the CBD embraces the fair and equitable sharing of the results of research and development and of the benefits arising from the commercial and other utilization of genetic resources, the equitable sharing of the benefits arising from the utilization of TK, technology transfer and intellectual property issues arising thereof and priority access to the results and benefits arising from biotechnologies based on genetic resources (New Scientist 2007). With regard to TK, it seems worth noting that the above-mentioned access provisions do not apply to it. The wording of CBD Article 8 (j) regarding the promotion of wider application of TK however is often interpreted as a tool for facilitating access to TK. Some important timeline events related to ABS are presented below (Table 9.1).

9.4 ABS: A Regional Context

The Hindu Kush-Himalayan (HKH) region, more specifically the Eastern Himalayas, is 1 of the 34 global biodiversity hotspots and a treasure house of genetic resources and TK. It also has great linguistic and biocultural diversity. There are many common genetic resources and associated traditional knowledge across the countries in this region of the world. Sharing benefits from such common knowledge at the time of bioprospecting was a major challenge.

The genetic resources provide the basis for the livelihood security of mountain communities. These communities have used local plants and

Table 9.1 Important timeline events related to ABS

S. no	Year	Event
1	1992	The text of the CBD is opened for signature at the Rio Earth Summit
2	1993	The CBD is ratified and comes into force 29th of December
3	1998	A panel of experts is established to clarify principles and concepts related to access and benefit sharing
4	2000	The Conference of the Parties establishes the ad hoc open-ended working group on ABS with the mandate to develop guidelines to assist with the implementation of the ABS provisions of the CBD
5	2002	The Conference of the Parties adopts the Bonn guidelines on access to genetic resources and fair and equitable sharing of the benefits arising out of their utilization
		At the World Summit on Sustainable Development, governments call for action to negotiate an international regime to promote the fair and equitable sharing of benefits arising from the use of genetic resources
6	2004	The Working Group on ABS is given the mandate to negotiate an international regime on access and benefit sharing, in accordance with decision VII/19 D of the Conference of the Parties
7	2008	The Conference of the Parties establishes a clear process for the finalization of the international regime on access and benefit sharing
8	2010	The Nagoya Protocol on Access to Genetic Resources and the Fair and Equitable Sharing of Benefits Arising from their Utilization was adopted by the Conference of the Parties at its tenth meeting in Nagoya, Japan in October 2010

wildlife since time immemorial (collecting, selecting, growing and raising varieties of food crops, livestock and medicinal plants) for their livelihood and well-being. In recent years, awareness has grown about the value of these genetic resources and associated TK. The challenge now is to convert these resources into meaningful economic wealth in an ecologically sustainable and socially equitable way and to channel benefits to the communities that are the conservers and creators of these genetic resources and TK.

9.4.1 ICIMOD's Access and Benefit Sharing Programme

To facilitate the implementation of the CBD in letter and spirit in all the major biodiversity hotspots for its conservation and usage in the HKH region, International Centre for Integrated Mountain Development (ICIMOD) has launched a regional programme on ABS from Genetic Resources and Associated Traditional Knowledge in Eastern Himalayas, with support from the German Agency for Technical Cooperation (GTZ) since 2004. The overall objective of the programme was to facilitate the development and implementation of the CBD's ABS regime in these regions.

The main goals of the programme were:

1. To enhance the capacity of governments in the region at the national, provincial and local levels to design and implement practical strategies, legislation and guidelines on ABS in favour of those most in need; to improve the relationship between local, provincial and central levels of governments and to strengthen reciprocal trust and social legitimacy between countries in the region
2. To create conditions necessary for local marginalized communities to contribute towards the realization of government policies on ABS; to participate in the development and implementation of national and/or regional ABS regimes; to define their own solutions; to protect the rights of the most marginalized groups within the community; and finally to develop partnerships and other mechanisms to strengthen existing human, biological, cultural, financial and territorial resources

The thrust of ICIMODs has been to raise awareness on ABS with reference to the various marginalized segments of society in the Eastern Himalayan region. In this context, ICIMOD is working with 13 partners active in the field of biodiversity management and the promotion of equitable access rights for various marginalized groups in 7 project sites in 4 Eastern Himalayan countries (Nepal, India, Bangladesh and Bhutan). The focus in these project sites has been to reach out to hitherto marginalized groups like mountain women, dalits, indigenous and local communities and tribal communities. At the same time, support has been provided to governments to develop policies and legislation on ABS. The programme has been working to ensure that the awareness of policymakers, civil society groups and marginalized local communities is raised following which they can begin the journey towards equitable ABS agreements reflecting mutual concerns with the different organs of the state and with bioprospectors.

Recently in collaboration with the concerned government, national and local level NGOs are working towards documenting TK.

These policy-based research documents will influence the long-term implementation of ABS regimes in ICIMOD's regional member countries through the creation of an empirical foundation for the rights of marginalized communities to genetic resources and associated TK.

9.5 TK Access and Benefit Sharing

Though TK has not received an international definition yet, it is generally understood to cover a wide range of fields including environmental knowledge, agricultural practices, medical treatments, literary and artistic expressions and cultural practices. The CBD refers to a specific part of TK which means the knowledge, innovations and practices of indigenous people and local communities embodying traditional lifestyles relevant for the conservation and sustainable use of biodiversity. In other words, it is the knowledge built through generations by a group living in an intimate connection with nature. It is a living form of knowledge that is part and parcel of the identity, cultural and/or spiritual practices and natural resource management of indigenous people and local communities. Sometimes, TK has been seen as a unique and precious source of information to develop new pharmaceutical, cosmetic or other products derived from natural ingredients. In addition, TK is increasingly seen from a broader perspective as a form of knowledge that should be considered on the same level as modern/western science and that can provide unique approaches to global environmental challenges, such as nature conservation, sustainable use and adaptation to climate change. The ratification of CBD in 1993 brought forwards the agenda of ABS from the use of TK. As this concept is entwined with the concept of fair and equitable sharing of benefits from the use of TK and biological resources, it has become crucial matter of debate in the conservation and development of biological resources. Earlier, intellectual traditions of free exchange of biological/genetic resources and TK by the indigenous and local communities to the outside world was viewed differently, but after the Convention, perceptions have changed. The reason being the convention's Article 8(J), 10 (c) 15, which focuses on respecting traditional knowledge, customary rights of the indigenous local communities and fair and equitable sharing of benefits with PIC and MAT of the holders of resources by the users (Secretariat of Convention on Biological Diversity (UNEP 2001). The specific articles concerning the fair and equitable sharing of benefits from the use of biological resources have been analyzed and debated for the last 20 years, in which only 10 % of the Parties to the Convention (COP) have adopted regulatory measures or practices (Oli 2009; Young 2008).

Since, TK is inextricably linked to indigenous people's culture and own governance systems. As a result, it has been threatened and eroded in the past by colonization and mandatory assimilation and, recently, by relocation policies and globalization forces that marginalize indigenous people and local communities and deprive them in various ways of tenure over lands, territories and natural resources that they have traditionally owned, occupied and/or used (Ronnie et al. 2007). In addition, traditional lifestyles are rapidly disappearing as the younger members of the communities are reluctant to continue with traditional practices. Furthermore, use of the IPR system has resulted in a series of famous biopiracy cases involving misappropriation of TK.

At the fifth COP meeting, it was realized that a functional ABS system was essential that has been lacking. The COP sixth meeting adopted the Bonn guidelines to facilitate the process of developing legal procedures among the parties (Convention on Biological Diversity 2002). Many complex issues on the ownership over resources and benefit sharing are yet to be made clear. Contracting parties need to understand the mechanics and structure of development and implementation of ABS policy and legislation. The development and implementation of ABS policy and legislation in the Himalayan region have been slow. Only two countries, India and Bhutan, in the region have developed ABS legislation, while others are in the process of developing it. The delay in the development of ABS policy and law, however, has not prevented in raising awareness among the government civil society groups and to the local community. The challenges during intellectual deliberations are to convince communities to understand ABS and link policymakers into methodologies of biological and genetic resources trade (Milind and Kothari 2007). For a common man, ABS seems to be a simple agreement, but the CBD requires its Parties to respect, preserve and maintain TK. Finally, it needs its Parties to encourage benefit sharing from the use of TK with indigenous people and local communities. It could be thus inferred that a bidirectional relationship needs to be sought which can be expressed in benefit-sharing terms: applying TK with the approval of its holders and thus sharing its benefits with society as whole, by the government, private sector or research institutions, with the original holders of this knowledge (Sahai 2003). There may eventually be financial profits resulting from products or processes that are based on TK. Though these 'profits' may be directly related to the access and use of TK, in reality they are often far away from indigenous people and have more to do with the structures and processes of corporations in a globalized economy. Another mechanism potentially available to indigenous people as a means to derive monetary returns from or to protect their TK is the existing IPR regime. However, this is a limited option in context of TK, as discussed below:

The CBD presents the potential for opportunities in which the Secretariat of the CBD has developed an important body of work to advance the implementation of provisions concerning indigenous people. The CBD working groups on Article 8(j) and related provisions and on access and benefit sharing are the key developments in this regard. The introduction of the Bonn guidelines has been an important step to evolve more appropriate regimes for indigenous people (Secretariat of the Convention on Biological Diversity 2002).

There have also been some worthwhile developments at regional and national levels. Examples include the Andean Pact Decision 391 - Common Regime on Access to Genetic Resources 1996, national laws in a few developing, biodiversity rich countries such as Costa Rica that have attempted to recognise the distinct nature and values of TK. Other useful regional developments are the Organisation of African Unity (OAU) Model Law and the Pacific Model Law.

The OAU model legislation provides a relatively comprehensive scheme for ensuring conservation and sustainable use of biological resources. Its objectives include:

1. Support for the inalienable rights of local communities over their biological resources, knowledge and technologies
2. Access to biological resources subject to prior informed consent of local communities
3. Fair and equitable sharing of benefits
4. Effective participation of local communities with a particular focus on women
5. Appropriate institutional mechanisms for the effective implementation and enforcement of the rights of local communities

The OAU Model Law sets out comprehensive provisions provided for the community rights including requirements for prior informed consent; the right to refuse consent and access; the right to traditional access, use and exchange; the right to benefit; and the recognition of community intellectual rights. The Pacific Region Model Law for the Protection of TK, innovations and practices is an innovative development that offers useful guidance for introducing measures for protecting intangible knowledge. This Model defines TK as:

Generation sold knowledge whether embodied in tangible forms or not, gained over generations of living in close association with nature regarding living things, their constituent parts, their life cycles, behaviour and functions, their effects on and interactions with other living things (including humans) and with their physical environment; the physical environment including water, soils, corals, weather, solar and lunar effects, processes and cycles and obtaining and utilizing of living or nonliving things for the purpose of maintaining, facilitating or improving human life.

This Model Law contains provisions for ownership of TK, innovations and practices by a group or an individual and that reaffirm ownership rights in TK as inalienable and non-transferable. The Model Law also proposes for prior informed consent to be sought if the wider use or commercialization of TK is proposed.

Various legal mechanisms provide opportunities not only for protection but importantly also for equitable benefit sharing and the prospect of providing economic returns to indigenous people, where TK and associated practices are utilized by the wider societies. In CBD context, the protection of TK and associated practices contributes towards sustainable use, protection and management of biological diversity that leads to protection of TK. The viable options for developing effective benefit-sharing arrangements based on wider utilization of TK will depend on indigenous people's decisions as to how, when and to what extent they will allow this wider use. Benefit-sharing provisions may be very diverse and often will relate to micro economies of small-scale sustainable livelihoods based on the use and management of biological resources and participation in research and monitoring.

9.6 Challenges of ABS Regime

Though legal arrangements for protecting community rights over biological resources and associated TK are emerging in the developing countries and some are being enforced in practice, however, the enforcement of these laws has not yet been clear about how these communities will be benefited from bioprospecting. This is because genetic resources cannot be owned and the knowledge holders on the use of these resources are difficult to identify/trace and mostly knowledge are under collective ownership. Preliminary findings to ascertain the status of TK has suggested that major part of the knowledge was under public domain and collectively extending to more than one country. What the medicinal practices have been used by the traditional communities has been documented either in Ayurvedh, Unani or Chinese Pharmacopeias.

In addition to the problem in determining the ownership over genetic resources and knowledge, sources of knowledge and conflict of interest among the same ethnic groups in benefit sharing have also been reported. Benefit sharing is particularly important in the contexts in genetics: access to non-human genetic resources and associated TK under ABS regime has not been granted as yet. Biological resources are still taken out of the country for bioprospecting under the standard material transfer agreement through the government and research centres. Benefit sharing with the indigenous and local communities under such agreements is nonexistent. For example, in India, the Tropical Botanical Garden Research Institute (TBGRI) and Kani Tribe model of access to knowledge and benefit sharing have been much touted as one of the best known community benefit-sharing mechanism from the use of biological resources and associated TK in the region. The knowledge informers for the formulation of Jeevani drug who were members of Kani tribe in Kerala south India themselves violated customary rights of the Kanis by not obtaining the free and prior informed consent either of knowledge holders of their traditional chiefs or of the community. Kani tribes in state of Tamil Nadu also claimed the right over the knowledge (Bijoy 2007b).

Another difficult issue is that the knowledge and genetic resource providers have very less knowledge on the market value of biological resources which directly impacts the benefit-sharing mechanism. The users have knowledge on market value of the resources, but are not transparent to the providers. Once the biological resource is accessed, it is uncertain that it will not be sold to the third party. The providers will have very little information on the benefits that their resources will bring especially if the agreement has been undertaken with the community. This is mainly because of the laws that have been drafted and implemented so far, considering biological resources as physical substances, and treated as patentable information when the user obtains a patent for his work with them (Nyasha and Young 2003).

Determining the knowledge and common biological resources and establishing the ownership over knowledge, the biological resources are a major challenge in the Himalayan countries. The development of ABS policy and law their enforcement is urgently required for the conservation of biological diversity and getting benefits from it in the region. It has not received priority as many countries are indulged into their internal political processes, conflicts and nation building processes. Their priority agenda is poverty reduction and infrastructure development; the ABS agenda has been in less priority.

In countries, where ABS laws are in place, reaching into agreement and negotiating with the government and indigenous community within the stipulated time and financial resources is also a major challenge. There are many mountain areas with biological resources that are devoid of communication and with illiterate indigenous communities who do not know about ABS. The community members are spread over vast transnational areas; therefore, getting PIC and making MAT with the indigenous community poses challenges.

9.7 Future Directions

The Access and Benefit Sharing Clearing-House (ABS-CH) provides a platform for exchanging information on ABS established as part of the Clearing-House of the convention established under Article 18, paragraph 3 of the Convention. It is being a tool that facilitates the implementation of protocol by enhancing the legal certainty and transparency on procedures for ABS as well as monitors the utilization of genetic resources along with the value chain and through the internationally recognized certificate of compliance. By hosting relevant information regarding ABS, the ABS-CH will offer opportunities for connecting users and providers of genetic resources and associated TK.

A programme of work addressing the commitments embodied in Article 8 (j) and other provisions of the Convention dealing with TK, governments and contracting parties have undertaken:

1. To establish mechanisms to ensure the effective participation of indigenous and local communities in decision-making and policy planning
2. To respect, preserve and maintain TK relevant to the conservation and sustainable use of biological diversity
3. To promote its wider application with the approval and involvement of the indigenous and local communities concerned
4. To encourage the equitable sharing of the benefits arising from the utilization of such TK

Though these elements are equally important, the last one has taken on a special significance for indigenous and local communities, because TK has often been used in recent years by modern industry to develop new products and techniques without the involvement and consent of the holders of such knowledge, who have also received none of the resulting benefits. Governments and contracting parties have established a working group under the Convention with a mandate to make concrete proposals on how to translate all of these commitments into reality. Within the current framework, the ad hoc open-ended working group on Article 8(j) and related provisions' main tasks include:

1. The development of elements of sui generis systems
2. Developing indicators for the retention of TK and methods and measures to address the

underlying causes of the loss of such knowledge
3. The development of an ethical code of conduct to ensure respect for the cultural and intellectual heritage of indigenous and local communities relevant to the conservation and sustainable use of biological diversity
4. Contribute to the negotiation of an international regime on ABS, research on the impact of climate change into highly vulnerable indigenous and local communities, among others

Other elements to be considered at a later stage include the development of guidelines:

1. To ensure that indigenous and local communities obtain a fair and equitable share of the benefits arising from the use and application of their TK
2. To ensure that private and public institutions interested in using such knowledge obtain the prior informed approval of indigenous and local communities
3. To regulate how impact assessments are carried out regarding any proposed development on sacred sites or on land and waters occupied or used by indigenous and local communities
4. To assist governments in the development of legislation or other mechanisms to ensure that traditional knowledge, and its wider applications, is respected, preserved and maintained

9.8 Engaging Indigenous and Local Communities

Given the commitment by Parties to the Convention to respect, preserve, maintain and promote the wider use of TK with the approval and involvement of users of such knowledge, indigenous and local communities have a direct interest in the work of the convention. Consequently, their representatives have been invited to participate fully in the working group on TK, including in the group's decision-making. Indigenous and local community representatives also participate in other meetings of the Convention of relevance to them, and recently, in decision VIII/5/D/I, selection criteria of the voluntary fund to facilitate the participation of indigenous and local communities in meetings held under the convention was adopted.

Due to Convention's adoption and the work being conducted under its auspices, governments have already undertaken to facilitate the participation of indigenous and local communities in developing policies for the conservation and sustainable use of resources, access to genetic resources, sharing of benefits and the designation and management of protected areas. A growing respect for TK has led modern science to adapt its procedures for assessing the impact of development projects on biological diversity; for monitoring of ecosystems and species, particularly genetic resources and species at risk; for controlling alien species; and for promoting the in situ conservation and sustainable management of biological diversity. Governments are also seeking to involve indigenous and local communities more actively and to apply their knowledge and technologies in the conservation and sustainable use of forests, agricultural biodiversity, inland waters, coastal and marine ecosystems, rangelands and ecotourism.

Recently, mandatory international obligations have been established under the CBD in 2010 with specific regard to TK associated with genetic resources (i.e. TK that may spark research and development interest in certain living organisms to develop pharmaceutical, cosmetic or other products). The Nagoya Protocol on ABS to the CBD has made it mandatory for governments to obtain the prior informed consent (or approval and involvement) of indigenous people and local communities and to ensure benefit sharing from the use of their TK.

9.8.1 Need for International Guidelines

Albeit different legal developments, CBD Parties have recently come to the conclusion that more international guidance is needed on how to ensure benefit sharing from the use of TK. In October 2013, the CBD working group on TK noted the lack of a 'centralized mechanism for indigenous and local communities to report unauthorized access to their TK' and recommended that work should begin on developing international guidelines for the development of 'mechanisms, legislation or other appropriate initiatives' to ensure that private and public institutions interested in using TK obtain the PIC of indigenous and local communities; these communities obtain a fair and equitable share of benefits arising from the use and application of their knowledge; and unlawful appropriation of TK is reported and prevented. The international guidelines on benefit sharing will focus not only on TK associated with genetic resources (which is addressed in the Nagoya Protocol) but also on other TKs associated with ecosystems and biological resources.

Various documents outlined the difficulties for protecting TK through the existing system of IPRs. If indigenous people and local communities choose to commercialize traditional cultural expressions, such as textile designs, tradition-based crafts, sound recordings and traditional food products, the use of certain intellectual property tools including trademarks, copyright and geographical indications may prove to be useful. Trademarks and geographical indications may be especially suitable to protect such expressions of TK due to the possibility for collective ownership and indefinite extension of protection. However, their use can only prevent the unauthorized use of the protected mark or indication; it does not protect the knowledge as such. In addition, most biodiversity-related TK is not only intangible but also dynamic and adapted to a specific environment. Therefore, it does not lend itself to nor satisfy the protection requirements of conventional intellectual property systems, such as those governing copyrights, patents, trademarks and designs. The documents explain that:

1. TK is collectively held by communities and in many cases widely shared, thereby making it difficult to identify exclusive owners.
2. TK is often not 'owned' in the conventional sense but collectively held, developed and shared in accordance with customary norms and laws; in many instances, it is holistic and develops organically, thereby making it difficult to distinguish between 'new' and 'old' knowledge.
3. TK is integrally connected to a way of life – its development is not motivated by the possibility of personal reward but on the contrary develops in response to the needs of the community; the sharing and exchange of TK builds and binds the community, and the rules that govern its use are not based on 'ownership rights' but on 'stewardship duties or obligations'.
4. Indigenous and local communities regard their rights to their knowledge as inalienable and held in perpetuity for future generations; and TK is often transferred between generations to recipients, who earn the right to acquire the knowledge obligations.

Some national laws have provided for direct payments to indigenous and local communities or payments to trust funds kept on behalf of indigenous and local communities. On the other hand, the only source identified as providing 'elements of good process' was a 1999 report by the Swedish Scientific Council, which suggested that the definition of 'fair and equitable benefit sharing' is non-exhaustive and inclusive and should encompass certain minimum conditions as follow:

1. Should contribute to strengthen the situation of the less powerful party/parties at all levels in the sharing relation, by enabling equal access to information, effective participation by all relevant stakeholders, capacity building and privileged access to new technology and products
2. Should contribute towards, or as a minimum not counteract, the two other objectives of the CBD: conservation of biological diversity and the sustainable use of its components
3. Must not interfere with existing forms of fair and equitable benefit sharing, including customary benefit-sharing mechanisms

4. Must respect basic human rights
5. Must respect value and legal systems across cultural borders, including customary law and indigenous intellectual property systems
6. Must allow democratic and meaningful participation in policy decisions and contract negotiation by all stakeholders, including stakeholders at the local level
7. Must be transparent enough that all parties understand the process equally well, especially local and indigenous communities, and have time and opportunity to make informed decisions
8. Must not unnecessarily restrict access to nonrival goods and resources
9. Must, if contractual relations are involved, include provisions for independent third party review to ensure that all transactions are on MAT and proceeded by effective PIC
10. Must, if contractual relations are involved, provide for identification of the origin of genetic resources and related knowledge
11. Must, if contractual relations are involved, make information about agreed terms publicly available

9.9 General Principles for Implementation of Article 8(j) and Related Provisions

The work on Article 8(j) and related provisions is the main instrument that Parties to the Convention on Biological Diversity have given themselves to achieve the commitments in Article 8(j) to respect, preserve and maintain the knowledge, innovations and practices of indigenous and local communities embodying traditional lifestyles relevant for the conservation and sustainable use of biological diversity, to promote their wider application with the approval and involvement of the holders of such knowledge and to encourage the equitable sharing of the benefits arising from the utilization of such knowledge. The objective of this work is to promote within the framework of the Convention a just implementation of Article 8(j) and related provisions, at local, national, regional and international levels, and to ensure the full and effective participation of indigenous and local communities at all stages and levels of its implementation.

1. Full and effective participation of indigenous and local communities in all stages of the identification and implementation of the elements of the programme. Full and effective participation of women of indigenous and local communities in all activities of the programme of work.
2. TK should be valued, given the same respect and considered as useful and necessary as other forms of knowledge.
3. A holistic approach consistent with the spiritual and cultural values and customary practices of the indigenous and local communities and their rights to have control over their traditional knowledge, innovations and practices.
4. The ecosystem approach is a strategy for the integrated management of land, water and living resources that promotes conservation and sustainable use of biological diversity in an equitable way.
5. Access to traditional knowledge, innovations and practices of indigenous and local communities should be subject to prior informed consent or prior informed approval from the holders of such knowledge, innovations and practices.

9.10 Case Studies

To illustrate some of the concepts of how TK can be incorporated into projects and project planning, several case studies are presented below that illustrates a different aspect of how perspective is gained by including TK in the project. These case studies are drawn primarily from examples provided by the International Labour Organization (ILO), the World Bank (WB), Indian and Northern Affairs Canada (INAC) and the United Nations Education, Scientific and Cultural Organization (UNESCO).

9.10.1 Access and Benefit Sharing of Traditional Knowledge of the Hoodia Plant

Hoodia is a succulent plant indigenous to southern Africa that has been used for centuries by indigenous San people to save off hunger and thirst. In 1996, the South African-based Council for Scientific and Industrial Research (CSIR) patented active compounds of Hoodia for potential commercialization of an appetite suppressant. This led to a licensing agreement between CSIR and some large pharmaceutical companies to develop and commercialize a Hoodia-based product. Initial actions were taken without the consent of the San people which led to an outcry by NGOs and media attention. As a result, a benefit-sharing agreement was signed with the San people. The agreement was made which include the monetary benefits as well as the non-monetary benefits.

The agreement included:

Monetary benefits:

- Milestone payments during product development
- Royalty payment in case of commercialization

Nonmonetary benefits:

- Funds for development, education and training of the San community
- Funds to support projects and institutions working to improve research and protection of the San traditional knowledge and heritage

9.10.2 Promotion of Local Communities' Strategies for the Conservation of Medicinal-Plant Genetic Resources in Africa

In Africa, more than 80 % of the continent's population relies on plant- and animal-based medicine to meet their health-care requirements. For most part, the plants and animals used in traditional medicine are collected from the wild, and in many cases, demand exceeds supply. As African's population grows, demand for traditional medicines increased, and pressure on natural resources became greater than ever. Africa has a history of conserving biodiversity in medicinal plants for at least two reasons: traditional practices surrounding their use reflect local knowledge and wisdom, and the plants are readily available and relatively cheap – being either easy to gather in the wild or simple to cultivate.

Many plants continue to provide most of the rural population of Africa with ingredients for traditional medicines. They are used widely to prevent and treat common ailments, but their conservation also means that the indigenous knowledge associated with their unique properties and correct application will be preserved.

Through a combination of participatory research and development action involving local communities, project workers first learn about the local communities' own solutions for conserving medicinal plants and for putting them to safe and effective use for traditional health care. Appropriate incentives then provide further encouragement of community efforts to safeguard biodiversity at the village level. Economic incentives include seed funds, the promotion of income-generating activities and help with marketing. Social incentives include technical assistance and training, information and consciousness raising related to conservation, the provision of equipment and technical and scientific advice and assistance. Recognition for the value of traditional medicine and medicinal plants will foster sustainable methods of propagation and cultivation. TK and practices pertaining to medicinal plants will be preserved as herbal medicines are increasingly used to complement other forms of community health care.

9.10.3 Enhancing Pastoralist Self-Reliance through Sustainable Economic Development in Kenya

An integrated development programme for pastoralists in Kenya brought together traditional (indigenous) knowledge and modern technical knowledge in training; handbooks for treatment of cattle diseases also aimed at bringing together indigenous knowledge from different ethnic groups, sharing indigenous knowledge and practices and promoting pastoralism as a valid mode of production and way of life. This project was based on disseminating indigenous knowledge. In all project activities, the Kenya Economic Pastoralist Development Association (KEPDA) brings together traditional and modern technical knowledge, through publications and networking, to promote understanding and awareness on key issues. Such an approach offers considerable potential for improving dry land productivity in a sustainable manner.

9.10.4 Indian Craftsmen and Artists of Quebec: Customer Service Is the Key to Success

Indian Craftsmen and Artists of Quebec (ICAQC) is a corporation that was established more than 20 years ago to meet the needs of Aboriginal arts and crafts producers, located on the Hurons Wendat reserve, approximately 8 km outside of Quebec City; the corporation has remained successful in challenging economic times by expanding its customer base and by marketing the finished products of Aboriginal crafts producers. ICAQC's customer service practices include calling customers on the same day to confirm orders and shipments and training employees in French and English so that they can serve customers anywhere in Canada. As a result, ICAQC has become one of Canada's largest suppliers and distributors of basic handicraft materials for Aboriginal communities across the country and an important marketer of Aboriginal arts and crafts. ICAQC helps Aboriginal crafts producers and artists continue to produce arts and crafts and profit from their talents.

9.11 Conclusion

A right is a collective term that requires the States to respect, protect and promote the governmental and property interests of indigenous people (as collectivities) in their natural resources. The right to permanent sovereignty of indigenous people over natural resources is implicit in international law particularly in the right of ownership of the lands they historically or traditionally use and occupy and the rights to self-determination and autonomy. The authority of indigenous people to manage, conserve and develop their resources according to their own institutions and laws must be adhered to. Where locals for valid legal reasons do not own or control the natural resources pertaining to all or a part of their lands or territories, the indigenous concern must nevertheless share in the benefits from the development or use of these resources without any discrimination and must be fairly compensated for any damage that may result from development or use of the resources. Self-determination under certain conditions, collective ownership of lands and territories, exercise of customary law according to social and cultural practices, legal and political representation through their own institutions and control over their own indigenous knowledge are rights claimed by indigenous peoples that are not claimed normally by other sections. This calls for securing legal recognition and live experiences of the rights of governance of communities (1) over their biological resources; (2) to collectively decide over development project/programmes by recognizing the free, prior and informed consent of the IPs through the use of indigenous customary laws; (3) to collectively benefit from the use of their biological resources; (4) to collectively benefit from their innovations, practices, knowledge and technologies acquired through generations; (5) to collectively benefit from the utilization of their innovations, practices, knowledge and technologies; (6) to use their innovations, practices, knowledge and technologies in the conservation and sustainable use of biological diversity; and (7) to the exercise of collective rights as legitimate custodians and users of their biological resources. Increasing, international interest in genetic resources and associated TK raises the challenge of protecting and respecting aboriginal interests (in genetic resources and TK). An empirical analysis of such agreements, combined with discussions with the parties about the extent to which their interests and expectations are reflected over the long term, will add to our understanding of best practice. Such analysis will enable the drafting of research agreements that promote fair and equitable access to and benefit sharing of genetic resources and associated TK.

References

Agrawal A (1996) Indigenous and scientific knowledge: some critical comments. Indigenous Knowl Dev Monit 3(3):33–41

Arce A, Long N (1992) The dynamics of knowledge: interfaces between bureaucrats and peasants. In: Long N, Long A (eds) Battlefields of knowledge: the interlocking of theory and practice in social research and development. Routledge, London, pp 211–247

Atte DO (1992) Indigenous local knowledge as a key to local level development: possibilities, constraints and planning issues, vol 20, Studies in Technology and Social Change. Iowa State University, Technology and Social Change Program, Ames

Bebbington A (1993) Modernization from below: an alternative indigenous development. Econ Geogr 69(3):274–292

Bijoy CR (2007a) Access and benefit sharing from the indigenous people perspective: the TBGRI- KANI model. Law Environ Dev J 3(1):1–18

Bijoy CR (2007b) Access and benefit sharing from the indigenous people perspective: the TBGRI- KANI model. Law Environ Dev J 3(1):1–18

Brokensha D, Warren D, Werner O (1980) Indigenous knowledge systems and development. University Press of America, Lanham

Brown LD (1985) People-centered development and participatory research. Harv Educ Rev 55(1):69–75

Buttimer A, Van Buren J, Hudson-Rodd N (1991) Land, life, lumber, leisure: local and global concern in the human use woodland. University of Ottawa, Ottawa

Cashman K (1989) Agricultural research centers and indigenous knowledge systems in a world perspective: where do we go from here? In: Warren DM, Slikkerveer LJ, Titilola SO (eds) Indigenous knowledge systems: implications for agriculture and international development, vol 11, Studies in Technology and Social Change. Iowa State University, Technology and Social Change Program, Ames

CBD (1992) The convention on biological diversity united nations. https://www.cbd.int/doc/legal/cbd-en.pd

Compton J (1989) The integration of research and extension. In: Compton J (ed) The transformation of international agricultural research and development. Lynne Reinner, Environmental Assessment Research Council, Boulder

Correa CM (2001) Traditional knowledge and intellectual property – issues and options surrounding the protection of traditional knowledge a discussion paper. Quaker United Nations Office, Geneva

Fingleton JS (1998) Legal recognition of indigenous groups. FAO legal papers. http://www.fao.org, pp 113–136

Gupta A (1992) Building upon peoples's ecological knowledge: framework for studying culturally embedded CPR institutions. Indian Institute of Management, Center for Management in Agriculture, Ahmedabad

Howard A, Widdowson F (1996) Traditional knowledge threatens environmental assessment. Policy Options 17(9):34–36

Howes M, Chambers R (1979) Indigenous technical knowledge: analysis, implications and issues. IDS Bull 10(2):5–11

ICIMOD (2009) International Centre for Integrated Mountain Development. http://www.icimod.org

Krishna RS (2007) Intellectual property rights and traditional knowledge: the case of yoga. Econ Polit Wkly 47(27–28):2866–2871

Labatut GM, Akhtar S (1992) Traditional environmental knowledge: a resource to manage and share. Philos Soc Sci 7:277–287

Lawas CM, Luning HA (1997) Farmer's knowledge and GIS. Indigenous Knowl Dev Monit 4(1):12–19

Leach G, Mearns R (1988) Trees for rural people. In: Leach G, Mearns R (eds) Beyond the wood fuel crisis: people, land and trees in Africa. Earthscan Publications Ltd, London, pp 27–29

Mathias E (1994) Indigenous knowledge and sustainable development. International Institute of Rural Reconstruction, Silang

Maundu P (1996) Methodology for collecting and sharing indigenous knowledge: a case study. Indigenous Knowl Dev Mon 3(2):14–17

Meredith TC (1997) Making knowledge powerful. Altern J 23(4):28–35

Milind W, Kothari A (2007) Conservation and peoples's livelihood rights in India. Final report of a research project conducted under the UNESCO small grants programme. Kalpavriksh, Pune

Murdoch J, Clark J (1994) Sustainable knowledge. Geoforum 25(2):115–132

New Scientist (2007) CBD article 2. Use of terms. Article 15 and Article 8 J of CBD 22–29 Dec, p 46

Nikolas G (2000) Indigenous knowledge for development opportunities and challenges. www.unctad.org/trade_env/docs/gorjestani.doc

Nyasha C, Young TR (2003) Access to biological resources and sharing the benefits of their use: international and sub regional issues. Workshop paper for the Southern Africa biodiversity programme. IUCN, ROSA

Oli KP (2009) Access and benefit sharing from biological resources and associated traditional knowledge in the HKH region – protecting community interests. Int J Biodivers Conserv 1(5):105–118

Reed MG (1990) Environmental assessment and aboriginal claims. Canadian, Ottawa

Roach T (1994) Ancient ways guide modern methods. IDRC Rep 22(2):9–10

Ronnie V, Yiching S, Li J (2007) Local agricultural innovation in China. Ensuring a fair share and rights and benefits for farming communities. Tech Monit (Spl issue):27–33

Sahai S (2003) Indigenous knowledge and its protection in India. In: Bellmann C, Dutfield G, Melendez – Ortiz R (eds) Trading in knowledge. Development perspectives on TRIPS, trade and sustainability. Earthscan Publication Ltd, London/Sterling, pp 166–183

Secretariat of Convention on Biological Diversity UNEP (2001) Hand book of the convention on biological diversity. Part 1. Earthscan Publication Ltd, London

Secretariat of the Convention on Biological Diversity (2002) Bonn guidelines on access to genetic resources and fair and equitable sharing of the benefits arising out of their utilization. Secretariat of the Convention on Biological Diversity, Montreal: Secretariat of the Convention on Biological Diversity (ISBN: 92-807-2255-7)

Showers KB (1996) Popularizing science education in developing countries through indigenous knowledge. Indigenous Knowl Dev Monit 4(1):33–40

Simpson T (1997) Indigenous heritage and self-determination: the cultural and intellectual property rights of indigenous people, international work group for indigenous affairs. Copenh Indigenous Knowl Dev Monit 3(2):14–17

Tandon R (1981) Participatory research in the empowerment of people. Convergence 14(3):20–27

Vivas-Eugui D, Ruiz-Muller M (2001) Handbook on mechanisms to protect the traditional knowledge of the Andean Region Indigenous Communities. Prepared for the UNCTAD BIOTRADE initiative. UNCTAD, Geneva

Warren DM (1990) Indigenous knowledge systems and development. Background paper for seminar series on sociology and natural resource management. The World Bank, Washington, DC

Warren DM (1991) Using indigenous knowledge in agricultural development, vol 127, World Bank Discussion Paper. The World Bank, Washington, DC

Warren DM, von Liebenstein GW, Slikkerveer L (1993) Networking for indigenous knowledge. Indigenous Knowl Dev Monit 1(1):2–4

Young TR (2008) The challenges of new Regime: the quest for certainty in access to genetic resources and benefit sharing. Asian Biotechnol Dev Rev 10(3):113–136

10 Plant Genetic Resources and Traditional Knowledge in Light of Recent Policy Developments

Javaid Akhter Bhat and Steve Taiwo

Abstract

Plant genetic resource and associated traditional knowledge are the invaluable assets determining the global food security especially with expanding global population and climate change. They provide important genes/alleles governing resistance to biotic and abiotic stress that are usually available in wild species and landraces. A number of activities such as increasing population growth, urbanization, clearing of land, overgrazing, the cutting and smoldering of forests, indiscriminate use of fertilizers and pesticides, loss of habitat, climate change, war, and civil strife have impacted negatively that destroyed natural habitats and threatened the genetic diversity of crop species as well as associated traditional knowledge. The plant genetic diversity has become highly vulnerable to "genetic erosion," and traditional knowledge too faces serious levels of erosion. In addition, the plant genetic resources and associated traditional knowledge are subjected to misuse and misappropriation known as "biopiracy." Therefore, to address these complex issues of plant genetic resources and traditional knowledge, a number of international instruments/policies have been developed over the years in the form of treaties, conventions, agreements, etc., to promote conservation and access of plant genetic resources and traditional knowledge. These policies also promote benefit sharing arising out of the utilization of these resources as well as prevent the misuse of plant genetic resources and traditional knowledge through intellectual property protection.

J.A. Bhat (✉)
Division of Plant Breeding and Genetics, Sher-e-Kashmir University of Agricultural Sciences and Technology of Jammu,
Chatha, Jammu, Jammu and Kashmir, India
e-mail: javid.akhter69@gmail.com

S. Taiwo, Ph.D.
Federal Ministry of Agriculture and Rural Development, Garki, Abuja, Nigeria
e-mail: staiwo9@gmail.com

R.K. Salgotra, B.B. Gupta (eds.), *Plant Genetic Resources and Traditional Knowledge for Food Security*, DOI 10.1007/978-981-10-0060-7_10

10.1 Introduction

Plant Genetic Resources (PGR) are the backbone upon which world food security and agriculture depends, especially with expanding global population and climate change. Plant Genetic Resources for Food and Agriculture (PGRFA) comprise of the genetic diversity available in landraces, traditional varieties and improved cultivars, and in the wild relatives of crops as well as other wild plant species that can be used for crop improvement now or in the future. It has been perceived that the hereditary adjustment and the rate of evolutionary reaction of a species to particular selective forces such as abiotic (drought, heat, frost, and so forth) and biotic stresses (new diseases and pests) as well as new climatic conditions rely on inherent levels of genetic diversity present within species (Hammer and Teklu 2008).

The vitality of PGR will be reflected clinched alongside each feature from claiming human attempt as it provides the gene pool from which resistant and improved varieties can be engineered. The diffusion of improved modern varieties leading to increasing crop productivity as well as economy has been extensively documented, particularly in the context of industrialized agriculture (Alston et al. 2000; Evenson and Gollin 2003). For developing and underdeveloped countries, PGR assumed a significant role in guaranteeing income security in view that a greater part of livelihood in these countries depends upon these natural resources. The loss of genetic diversity in the past was largely due to natural processes (climate changes), but it is now accelerated by the activities of earth's dominant species. These activities which include increasing population growth, urbanization, clearing of land, overgrazing, the cutting and smoldering of forests, indiscriminate use of fertilizers and pesticides, loss of habitat, climate change, war, and civil strife have all impacted negatively that destroyed natural habitats and threatened the genetic diversity of crop species (Hvalkoff 2001; Sodhi and Erhlich 2010; Ogwu et al. 2014).

Plant genetics and breeding are the major driving force for science-based productivity enhancements in major food, feed, and industrial crops. However, these disciplines of crop improvement are to be empowered in view of ever-increasing threats and vagaries of global climate challenges (Jackson et al. 2011). The science-based crop improvement efforts will attempt to enhance yields for the increasing population, and the impacts of climate change such as deterioration of biophysical resource base, drought, changed pest spectrum, etc., will dent our efforts. The population of the world is also expected to rise by 50 % by 2050, and therefore, feeding this population will require an astonishing increase in food production. For instance, cereal grain yields must increase by 70 % before 2050 (Furbank and Tester 2011). Therefore, to meet these challenges, the new and diverse genes are required to maintain as well as improve the productivity of crop species which are usually available in unutilized breeding stock and sometimes in wild relatives and landraces (Duvick 1986). The new varieties are resistant to pests and diseases for an average period of 5 years, but it usually takes 8–11 years for a plant breeder to develop new varieties, and hence in order to maintain the resistance of crop species against particular biotic stress, new varieties are continuously developed (NPM 1990). It has been highly recognized that sustainable agriculture production as well as adaptation of crop species to climate changes rely on the level, use, and access of genetic diversity. The plant genetic diversity is highly vulnerable to "genetic erosion," according to the Food and Agriculture Organization (FAO); replacement of local varieties by modern varieties results in genetic erosion and is further intensified due to emergence of new pests, weeds, and diseases, climate change, environment disturbances, urbanization, and clearing of land.

Indigenous traditional knowledge (ITK) also referred to as traditional knowledge (TK), Indigenous Knowledge (IK), or Local Knowledge (LK) generally refers to developed long-standing information, wisdom, customs, and tones of certain indigenous *people* or local communities. Generally, traditional knowledge is passed orally to generations from the persnickety will of man. TK also exists in the structure of stories, legends, folklore, rituals, songs, art, and even laws, and also other forms of traditional knowledge are communicated through diverse means (Khandelwal and Mishra 2015). Recently, the role of TK has been perceived in a range of sectors including mixed cropping techniques, pest and disease control, species diversity, seed and plant varieties in agriculture, plant/animal breeding techniques in biology, traditional medicine for human health, soil conservation, irrigation, water conservation in natural resource management, and oral traditions and local languages in education and has led to an increasing enthusiasm in it by academicians and policymakers. Moreover, the traditional knowledge of indigenous and farming communities is considered to be a key factor in the conservation and development of genetic resources. The indigenous knowledge is a precious groundwork to combat environmental as well as other forms of change because it helps for the development of adaptation and natural resource management strategies against them. This was reaffirmed in that "indigenous or traditional knowledge" might demonstrate service for seeing those possibilities about certain adaptation strategies that would be cost effective, participatory, and sustainable (Nakashima et al. 2012).

TK has also been accounted for to face not-kidding levels of erosion, as the *people* and communities possessing TK face an extent from claiming dangers from out-and-out destruction will "assimilate" under "mainstream" society and also the knowledge they possess slips away. This is further intensified by the single model of Western education, and by means of mass communication that is disseminated across the world, newer generations of customary people groups would essentially not guzzle TK in a way that their guardians or progenitors finished. It has been estimated by many researchers/scholars that half of about 6,000 languages spoken in the present-day world will become extinct in 2050 or 2100. A language (oral or written) will not be mainly an implication of correspondence between individuals or alternately the community; it additionally processes inside it the pith of significant information and knowledge and wisdom of the people or community. Hence, the loss of language means the loss of TK. The risk has been felt highest in the case of TK that is passed over to generations orally because the next generation usually receives only a part of the knowledge from their parents which leads to its disappearance in every generation.

Hence, if the PGR and TK are not managed appropriately at the national and international levels, they may result in the misappropriation and biopiracy of biological and genetic resources and TK. Therefore, to promote the conservation and access of PGR and to prevent their misuse and misappropriation, a number of international policies have been developed over the recent years in the form of treaties, conventions, agreements, etc., to address these complex issues which are discussed below.

10.2 PGR and TK: Complementary Phenomena

PGR and TK are an essential analytic quality of present-day developing countries especially in the field of agriculture and horticulture. TK transforms biodiversity under bio-resources. Indigenous people had developed plant varieties of several agricultural and horticultural crops from their corresponding wild species called landraces/indigenous varieties over generations which are the basic foundations of modern plant breeding and world food security. A series of genes have also been identified and managed by the local farming communities through selection and crossbreeding, possessing suitable traits for biotic and abiotic stresses. Development of crop varieties that are tolerant to rising global temperature and other environmental changes often forces the international scientists to visit tropical areas for identification of sources for drought tolerance/resistance, for which they largely depend on TK and local farmers. Using them in hybridization with local varieties and subsequent selection results in the development of a high-yielding crop variety having a combination of other desirable traits such as tolerance to drought, resistance to pests, etc. Hence, the TK of local people plays an important role in the identification of different strains of crop species possessing a unique set of traits that lead to increased efficiency of plant genetic resource conservation and its utilization. The loss of biodiversity in turn leads to the loss of TK, which negatively impacts the global food security and fight against hunger.

The location of indigenous and local communities is usually in areas possessing most of the world's genetic resources and that too in fragile and vulnerable environments. They have cultivated and used biological diversity since agriculture began, and their practices have been recognized to improve and enhance biodiversity and are a major contributor of the global diversity in domesticated plants and animals (Blackburn and Anderson 1993; Harlan 1995; Nabhan 1997). These communities contributed considerable measure for the conservation and sustainable use of biological diversity that goes far beyond their role as natural resource managers, and they process extensive knowledge of their local environments which are directly involved with conservation of biodiversity and PGR. Therefore, maintenance and transmission of their TK and practices relating to these ecosystems are especially vital (Craig and Davis 2005).

The complementary nature of traditional knowledge and biodiversity is fundamental to mankind. Recently, a huge advance has been aggravated to understand the complementarity of cultural diversity (reflected as traditional knowledge of natural resource management including that of plants and animals) and biodiversity. The Global Biodiversity Strategy is included as one of its ten principles for conserving biodiversity and humanity's aggregate knowledge in regard to biodiversity, and its use and management rest in cultural diversity (World Resources Institute et al. 1992). Of the two methods of PGR conservation, the ex situ conservation of PGR has assumed great significance and was through well-established institutions such as the International Board for Plant Genetic Resources (IBPGR), the Plant Introduction Stations (PIS) in the United States, and the Consultative Group on International Agricultural Research (CGIAR), whereas institutions that parallels the ex situ conservation used for documenting indigenous knowledge about the environment are now being established (traditional knowledge resource centers) and might assume an important role for in situ conservation. Therefore, for the effective establishment of national programs for in situ conservation of PGR, the comprehensive understanding of indigenous management strategies of local people and farmers is necessary that has encouraged the diversity in plant and animal species, and it will complement the ex situ programs' official presence (Altieri and Merrick 1988; Juma 1989; Wilkes 1991). To get thoroughly built for learning PGR conservation, the genetic establishment must acknowledge a mandate where concern is not only for the germplasm but also the knowledge frameworks that produce it (Brush 1989).

10.3 PGR and TK: Key to Agriculture Sustainability

Sustainability refers to the use of natural resources that meets the needs of the present without compromising the needs of future generations. It is a prerequisite to have an effective system for the conservation of PGR and associated TK for maintaining sustainability in agriculture. The role of climate change in relation to crop production and world food security is discussed currently all over the world (Reynolds 2010). IPCC (2007) has revealed that the average global surface temperature has increased by 0.74 °C since the late nineteenth century and is expected to increase by 1.4 °C to 5.8 °C by 2100 AD with significant regional variations. The climate change include higher temperatures, changes in precipitation (droughts and floods), and higher atmospheric CO_2 concentrations which may negatively affect the yield (both quality and quantity), growth rates, photosynthesis, and transpiration rates, and also disease and pest epidemics become frequent (Mahato 2014). Therefore, to fight against the climate change and to maintain sustainable agriculture, we need genotypes which could withstand better abiotic and biotic pressures. Genes for such traits need aid frequently in all the accessible wild species and landraces and thus have to be exploited fully to achieve desired objectives. Recently, biotechnological approaches have enhanced the scope of locating and transferring useful genes from wild species (Malik and Singh 2006). A tremendous reduction of genetic diversity has been recognized in all the agricultural crops mostly due to modern agriculture, and in this connection, significant deliberations have been made around the world to collect, characterize, and conserve crop genetic diversity, although for minor crops this conservation has been sensitively lower (Keneni et al. 2012). The ex situ conservation, in situ conservation, and on-farm conservation are different methods for the conservation of PGR and TK; the on-farm and in situ conservation methods are mostly through local farmers and communities. This on-farm type of conservation has expanded considerably at the international level, since it allows the dynamic process of crop evolution to proceed and also safeguards the TK associated to biodiversity (CUBIC 2000), which is fundamental for celebrating and appreciating crop diversity today and in the future. An individual genotype with an apparently futile set of characters today may suddenly become crucial tomorrow due to changing climatic conditions or outbreaks of disease. Therefore, it has long been acknowledged that we "conserve" all the diversity we have. The modern intensive agriculture calls for uniformity and consequently has a narrow genetic base. In contrast, traditional agriculture and the associated TK had a large number of diverse landraces. Hence, it is evident that the PGR and associated TK provide the useful genes to meet the challenges of the present as well as the future to ensure food security and fight against hunger.

10.4 PGR and TK: Recent International Policies

International instruments/policies that have been developed over the recent years in the form of treaties, conventions, agreements, etc., to address the complex issues of erosion, benefit sharing, and biopiracy of genetic resource and associated TK are discussed hereunder. These instruments make the world aware by emphasizing about the conservation and benefit sharing arising out of the utilization of PGR and associated TK and preventing the misuse and exploitation of PGR and associated TK as well as protecting the PGR and associated TK through intellectual property rights (Table 10.1 and Fig. 10.1).

10.4.1 Convention on Biological Diversity (CBD)

The Convention on Biological Diversity (CBD) was initially signed by 150 governments in 1992 in Rio de Janeiro, and to date the number of parties that became members of this convention has reached to 196 (168 signatures). Being a worldwide legitimately binding treaty, it has three fundamental objectives: protection of biodiversity, sustainable utilization of biodiversity, and the sharing of benefits in a fair and equitable way arising from the use of genetic resources. It also has an overall objective of sustainable future. The CBD in regard to biodiversity focuses on all three levels such as ecosystems, species, and genetic resources, and also biotechnology is covered through the Cartagena Protocol on Biosafety. In fact, all possible issues related to biodiversity as well as its role in development, ranging from science, politics, and education to agriculture, business, culture, etc., are covered under it. The CBD provides a framework of general standards for regulating the exchange of genetic resources at a global level, known as "mutually agreed terms" (MAT) and "prior informed consent" (PIC), and also gives national sovereignty to every nation over genetic resources that originate within their jurisdiction (Eaton et al. 2004). The Bonn Guidelines on access to genetic resources and the fair and equitable sharing of the benefits arising out of their utilization were adopted in 2002 by the countries that signed the CBD, and these provide the necessary directions for the development of national legislation and on contracts for benefit sharing. These guidelines are voluntary and do not strictly involve lawful commitments on parties involved in negotiating access and benefit sharing.

Table 10.1 Name of the policy, adopted year, objectives, and number of member states of each policy

S. no.	Policy	Adopted year	Objectives	No. of member states
1	Convention on Biological Diversity (CBD)	1992	Conservation and sustainable use of biodiversity, benefit sharing arising from the use of genetic resources and associated TK, protection of TK	196
2	Global Plan of Action (GPA)	1996	Conservation and sustainable use of plant genetic resources, benefit sharing arising from the use of plant genetic resources	155
3	The International Treaty on Plant Genetic Resources for Food and Agriculture (ITPGRFA)	2001	Conservation and sustainable use of PGRFA, benefit sharing arising from the use of PGRFA and associated TK, protection of traditional knowledge relevant to PGRFA	135
4	Trade-Related Aspects of Intellectual Property Rights (TRIPS)	1995	Protection of plant varieties through a *sui generis system*	158
5	UPOV Convention	1961	Protection of plant varieties through PBR and acceleration of plant breeding research	72
6	Intergovernmental Committee on Intellectual Property and Genetic Resources, Traditional Knowledge and Folklore (ICGTK)	2000	Access to genetic resources and benefit sharing as well as the protection of traditional knowledge and traditional cultural expressions	188
7	Global Crop Diversity Trust (GCDT)	2002	Ex situ conservation of crop germplasm relevant to food and agriculture	–
8	International Undertaking on Plant Genetic Resources (IUPGR)	1983	Conservation and sustainable use of PGRFA, protection of traditional knowledge, equitable sharing of benefits, and the right to participate in decision-making of local farmers and communities	113

(continued)

Table 10.1 (continued)

S. no.	Policy	Adopted year	Objectives	No. of member states
9	UN Declaration on Rights of Indigenous People	2007	To protect traditional knowledge and provide rights of indigenous people to maintain, control, protect, and develop their cultural heritage, traditional knowledge, and traditional cultural expressions	148

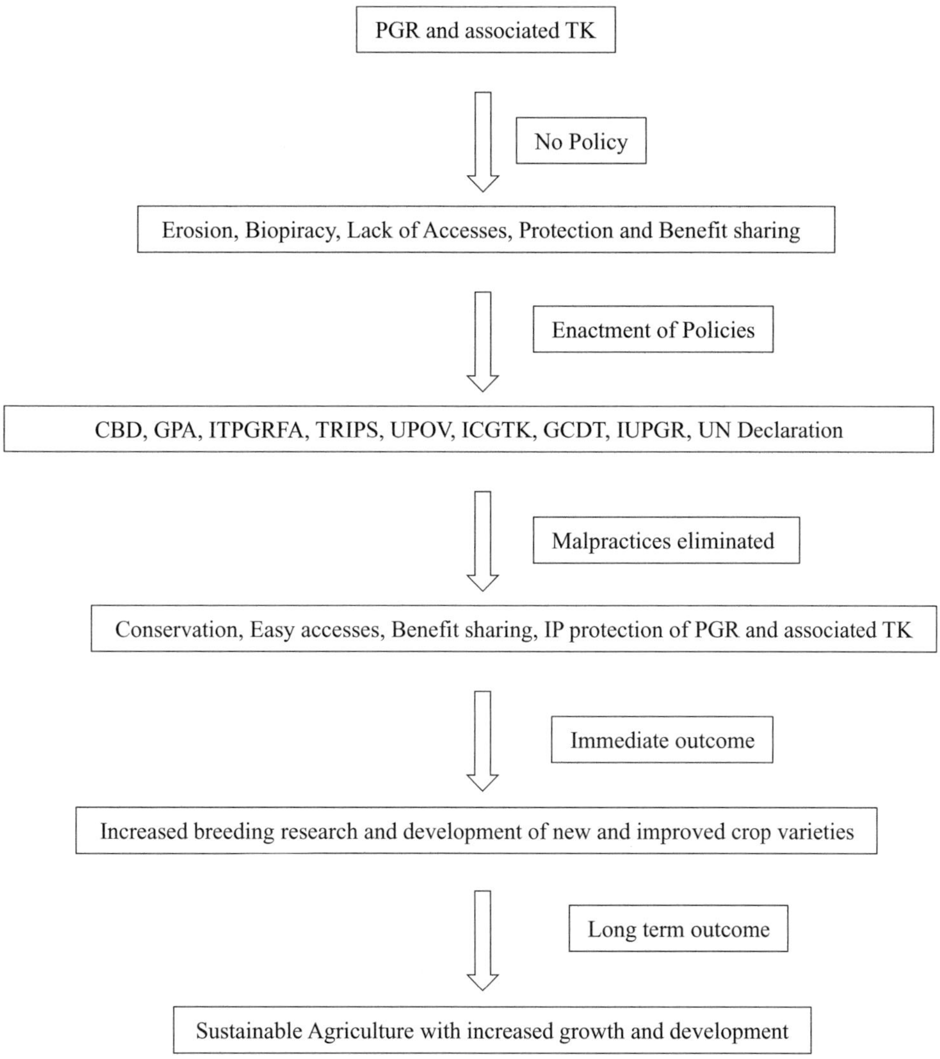

Fig. 10.1 Showing impact of international policies/instruments on global PGR and associated TK as well as on agriculture

The Conference of the Parties (COP) is the ultimate administering figure of the CBD and has ratified the treaty meets every 2 years to know progress and set priorities and also purposes of the work plan. In 2010, the Strategic Plan for Biodiversity 2011–2020 was adopted by the parties to the CBD which is a 10-year framework emphasizing all countries and stakeholders for taking action to safeguard biodiversity and the benefits it provides to people. This plan has a vision that biodiversity is conserved, restored, valued, and effectively used till 2050, as well as maintaining ecosystem services and sustaining a healthy planet and delivering benefits essential for all people, and it also has a mission of taking effective and immediate action to prevent the loss of biodiversity in order that by 2020 ecosystems are resilient and continue to provide essential services, thereby securing the planet's variety of life and contributing to human well-being and poverty eradication. The CBD Secretariat is located in Montreal, Canada, that assists the governments in the implementation of the CBD and its programs at work, organizes meetings, drafts documents, coordinates with other global organizations, and collects and spreads information.

The Nagoya Protocol on Access to Genetic Resources and the Fair and Equitable Sharing of Benefits Arising from Their Utilization was adopted in the Tenth Conference of the Parties, held in Nagoya, Japan, in 2010. The aim of this protocol is to offer the benefits arising from the utilization of genetic resources in a fair and equitable way, hence contributing to sustainable use and conservation of biodiversity. The protocol strengthens the access and benefit sharing provisions of the CBD by creating more stupendous lawful assurance and transparency for both providers and users of genetic resources, because it establishes more predictable conditions for access to genetic resources helping to ensure benefit sharing when genetic resources leave the contracting party providing the genetic resources.

The CBD unequivocally recognizes the vitality of TK and the rights of indigenous and local *people* in that knowledge. It creates a structure for guaranteeing that local people share benefits arising from appropriation and use of their knowledge and preamble. The preamble of the CBD recognizes "the close and traditional dependence of indigenous and local communities on biological resources and the desirability of sharing in the benefits derived from the use of TK, innovations and practices." National commitments toward local and indigenous communities occur in Article 8 (in situ conservation), Article 10 (sustainable use of components of biodiversity), Article 17 (exchange of information), and Article 18 (technical and scientific cooperation) (UNEP 1992; Mauro and Hardison 2000). The most authoritative provision dealing with TK is perhaps Article 8, and it provides that each member party shall, as much as possible and as appropriate, "Subject to its national legislation, respect, preserve and maintain knowledge, innovations and practices of indigenous and local communities embodying traditional lifestyles relevant for the conservation and sustainable use of biological diversity and promote their wider application with the approval and involvement of the holders of such knowledge, innovations and practices and encourage the equitable sharing of the benefits arising from the utilization of such knowledge, innovations and practices."

10.4.2 Global Plan of Action (GPA)

The Global Plan of Action (GPA) for the conservation and sustainable use of plant genetic resources for food and agriculture was formally adopted by 150 countries at the Fourth International Technical Conference held in Leipzig in 1996. In the same conference, the "Leipzig Declaration" was also adopted, which keeps tabs, considering the importance of PGR for global food security and emphasizing the member countries to implement the plan. The GPA provides an integrated global framework for action toward the conservation and sustainable utilization of PGRFA. This plan also provides a thorough framework for the activities related to ex situ and in situ conservation and sustainable utilization of PGR as well as institution and capacity building. GPA is the breakthrough, as it leads to global awareness of plant genetic resources for food and agriculture, and is the primary reference report for national, regional, and global deliberations to protect and use plant genetic resources for food and agriculture sustainably and to share benefits equitably and fairly arising from their utilization. GPA, being part of the FAO global framework for the conservation and sustainable use of plant genetic resources for food and agriculture, has been used by the FAO Commission on Genetic Resources for Food and Agriculture as an essential element to satisfy its mandatory admiration to plant genetic resources. It has assisted nations in the detailing of extensive policies and strategies on PGRFA.

Twenty priority activities of the original GPA described in four areas, viz., in situ conservation and development, ex situ conservation, utilization of plant genetic resources, and institution and capacity building, were reduced to 18 in the second GPA adopted by the FAO council. The former priority activities 5 and 8 have been merged to new priority activity 6, sustaining and expanding ex situ conservation of germplasm; similarly the earlier priority activities 12 and 14 were combined to new priority activity 11, promoting development and commercialization of all varieties, primarily farmers' varieties/landraces and underutilized species. In addition, the aim of many other priority activities has been adjusted so as to accommodate newly defined priorities. The second GPA gives greater emphasis and visibility to plant breeding, as reflected in priority activity 9, supporting plant breeding, genetic enhancement, and base-broadening efforts. An effort has also been made, based on guidance from the regional consultations, to simplify and clarify the document.

In the second GPA, new challenges and opportunities have been tended through 18 priority activities. A series of regional consultation meetings, the Second Report on the State of the World's PGRFA, as well as inputs from experts throughout the world have provided necessary inputs to make the second GPA current, forward looking, and relevant to global, regional, and national perspectives and priorities. The role of GPA as a supporting component of the International Treaty on Plant Genetic Resources for Food and Agriculture (ITPGRFA) gets stronger due to updating of the global plan of action.

10.4.3 The International Treaty on Plant Genetic Resources for Food and Agriculture (ITPGRFA)

The International Treaty on Plant Genetic Resources for Food and Agriculture (ITPGRFA) was adopted in 2001 by the members of the Food and Agriculture Organization (FAO) but came into force in 2004. It is considered a powerful tool to tackle simultaneous challenges of food security, climate change, and loss of agricultural biodiversity. Till now there are 135 contracting parties of this treaty and has the following three main objectives:

1. Conservation of Plant Genetic Resources for Food and Agriculture (PGRFA)
2. Sustainable use of PGRFA
3. Sharing the benefits arising from the use of PGRFA in a fair and equitable way

The ITPGRFA is in harmony with the CBD. ITPGRFA implementation is upheld and monitored in the individual signatory states by a governing body made up of delegates of all the contracting states. It was the first international treaty that recognizes the contribution the farmers and local communities have made and will continue to make in the conservation and developments of plant genetic resources. In a world where most countries depend strongly upon crops originating elsewhere, the treaty facilitates the exchange and conservation of crop genetic resources among member nations, as well as the sharing of benefits in a fair and equitable way arising from their utilization. The treaty has made rapid advancement in its implementation between 2007 and 2015 improving the utilization of plant genetic resources along the seed quality chain at a global level. The treaty has created a global gene pool for food security under the immediate control of all contracting parties with 1.6 million specimens of genetic material that facilitate research for major agriculture crops including maize, rice, wheat, and cassava, among others. ITPGRFA has two core elements of multilateral system and farmer's rights.

The multilateral system (Article 10–13) has been created by contracting parties of the international treaty in the schema of which they allow increased access among the member parties for their most important 35 food crops and 29 forages covered under it, which are recorded in Annex I of the treaty and are under the superintendence and dispense of the contracting parties and in the public space. Alternate contributors for PGRFA of the multilateral system up to date are the CGIAR centers and other international institutions which have established applicable agreements with the governing body of the ITPGRFA. It has been estimated that more than 80 % of human calories from plants are provided by the crops of Annex I globally (Moller et al. 2014). The PGRFA can be acquired from the multilateral system through a standard contract called the Standard Material Transfer Agreement (SMTA) embraced by the governing body to regulate transfers of material as well as prevent the abuse of the material that is exchanged and ensure that the benefits shall be shared in a fair and equitable way arising from commercial utilization of material from the multilateral system among the member parties (Moore and Tymowski 2005).

This international treaty through Article 9 recognized for the first time the efforts and enormous contribution the farmers including local and indigenous communities have made and continues to make the conservation and development of crop diversity in the world form the basis of farmer's rights. The international treaty advises contracting parties on detracting measures to protect and promote farmer's rights in accordance with national laws and provides farmers a basis to advocate their rights. These include the protection of traditional knowledge relevant to PGRFA, the right to participate in the sharing of benefits arising from the use of PGRFA, and the right to participate in decision-making related to PGRFA. Also, the rights of the farmers to save, use, exchange, and sell farm-saved seed are well addressed.

10.4.4 Agreement on Trade-Related Aspects of Intellectual Property Rights (TRIPS)

This international agreement concerns Intellectual Property Rights (IPRs) and came into force in 1995 under the umbrageous agreement established by the World Trade Organization (WTO). Thus, the TRIPS agreement applies to every member country of WTO. TRIPS provide a minimum intellectual property protection standard for living matter such as plant varieties, microorganisms, and microbiological processes. TRIPS accentuate member countries to furnish patent security for all the inventions but not for plants and animals other than microorganisms. For plant varieties, Article 27(3) of TRIPS allows the member countries to make an effective *sui generis system* for their protection. In spite of the fact that TRIPS do not permit member countries to provide patents for plants but only allow inventions pertaining to processes such as nonbiological and microbiological to be used for the production of plants and animals, various biotechnological applications are developed, including genetic modification.

10.4.5 Plant Breeder's Rights and the UPOV Convention

The International Union for the Protection of New Varieties of Plants (UPOV) was established by the International Convention for the Protection of New Varieties of Plants and adopted in Paris in 1961 and revised in 1972, 1978, and 1991. UPOV is an intergovernmental organization with headquarters in Geneva, Switzerland, and has a mission to provide and promote an effective system of plant variety protection, with the aim of encouraging the development of new varieties of plants, for the benefit of society. Most countries and intergovernmental organizations which are acquainted with a Plant Variety Protection (PVP) system decided that they will build their system based on the UPOV Convention, so as to provide an acceptable internationally recognized system. Till June 2015, there were 72 members of UPOV; 17 states and one intergovernmental organization have initiated the procedure for acceding to the UPOV Convention, and 23 states and one intergovernmental organization have been in contact with the office of the union for assistance in the development of laws based on the UPOV Convention. The UPOV Convention facilitates plant breeding in the UPOV members by providing intellectual property protection to new plant varieties developed by breeders called breeder's right. The breeder has to file individual applications for getting protection from the authorities of UPOV members endowed with those assignments of granting breeder's rights. Other agencies needed the consent of the breeder when the variety is used to propagate for commercial purpose. The convention defines the acts that oblige the authorization of breeders in the case of a propagating material of a protected variety and, under certain conditions, in case of the harvested material. Under the UPOV Convention, only those varieties which fulfill the criteria of being new, distinct, uniform, and stable and have a suitable denomination are subjected to breeder's rights. However, the convention also provides "breeder's exemption" in case the variety is needed for breeding other varieties and for experimental, nonprofit, and private purposes. The rights are granted for at least 20 years in the case of agricultural crops and for at least 25 years in the case of varieties of trees and vines (Das 2011).

The UPOV has a relationship with the CBD and ITPGRFA in that all the three global instruments are meant to promote plant breeding activities as well as to accelerate the development of improved varieties in crop plants. The CBD and ITPGRFA achieve this by providing a framework of increased access to PGR, while the UPOV Convention achieves this by creating a system for plant variety protection. When implemented by members of UPOV, the pertinent legislations managing these matters should be compatible and mutually supportive.

According to the 1991 revision of UPOV, the rights of plant breeders have been further extended, curtailing the rights of the farmers and the community over the plant genetic resources. Before 1991, the protected varieties could be used by farmers among others as an initial source of variation for the creation of new varieties, and farmers could plant, share, and exchange the varieties they produced. The protected variety may still be used as an initial source of variation for the creation of new varieties, but such new varieties cannot be marketed or sold without the Plant Breeder's Right (PBR) holder allowing it. It is understood that the PBR holder will want to retain the share and control over the market to maximize his sales and profit. It is unlikely that he will ever authorize the new variety to compete in the market. In the 1991 revision, the breeder's rights now cover not only the production for sale but also reproduction, multiplication, conditioning for the purpose of propagation, and exporting/importing and stocking for these purposes.

The introduction of the concept of "farmer's rights" in the undertaking on PGR has opened up a new horizon. In the March 1987 meeting of the FAO commission on PGR, delegates argued that if plant breeders had rights of ownership, control, and compensation by virtue of laboring for a decade to develop a new variety, then farmers also had rights since they had domesticated agricultural crops and observed, developed, and safeguarded the enormous biodiversity that the breeders and seed industry use as "raw material." It has been accepted that farmers have right to the Germplasm, Information, Funds, Technologies, and Farming/Marketing Systems (GIFTS).

Despite limitation of the status of farmer's rights at FAO, it is recognition of the fact that farmers and rural communities have contributed greatly to the creation, conservation, exchange, and knowledge of genetic resource since the origin of agriculture. It is therefore logical and a matter of principle that the farming communities are recognized for their ingenuity, invention, and scientific discovery. It is also an attempt to correct the injustice and imbalance caused by the one-sided recognition of the plant breeder's right depriving the farmer and the farming community from the benefit of new invention. It implicitly challenges the notion that the genetic materials that are used as "raw materials" by the breeders are "naturally" available as "wild species" or "landraces." The fact is they are conserved and maintained by the community's conscious and creative system of management and care. These systems evolved through thousands of years of experience and knowledge. Although it was not politically easy to establish the principle of farmer's rights, the acceptance of the notion in the policy discourse of the international community has also opened up the broader question of the rights of the community. In particular, it has brought to the forefront the margin between individual and collective rights and the rights of the private seed companies to privatize and monopolize the seed business in contrast to the rights of the community to hold the knowledge and innovation as common property to be shared and used for common good. In fact the rich biodiversity of the countries with a large agrarian community is absolutely linked to this rich and collective concept of property rights and community needs which are totally opposite to the egoistic societies based on private rights and unending greed.

Farmer's rights extend beyond the issue of compensation for farmers and farming communities: it includes rights to land and secure tenure, implying the need for land reforms. In the context of plant genetic resources, it is absolutely crucial to recognize that the farmers have the fundamental right to save seeds and exchange germplasm, as well as the right to refuse to disclose their knowledge and make their germplasm available.

The critique of the fund-centered notion of farmer's rights as construed in the FAO commission has also demonstrated the lack of institutions or space for farmers to negotiate their rights and privileges in the local, national, and global management of biodiversity. Most importantly, they have no institutional space to determine the utilization of the biological materials and genetic resources of which they are the main custodians. Nor can they decide the nature of transactions of knowledge, intellect, and skilled development over thousands of years. The "knowledge" itself has now been rendered as "raw material" to be robbed freely to make new Intellectual Property (IP) as a commodity in the world market. Government and nongovernment agencies involved in collection receive the compensation for information and resource transfer while local communities stay excluded.

To reconstruct the broader meaning of farmer's rights and indigenous knowledge within the framework of legally binding international instruments, the CBD makes a strategic document available to the communities in order to claim their sovereign right over the biological resources, knowledge, and skills. Moreover, the IPR regimes through GATT/TRIPs have instituted internationally binding trade agreements in 1994. Interestingly, it is true despite the fact that no article in the convention explicitly mentions farmer's rights or explicitly addresses mechanisms for the compensation or reward for the indigenous knowledge.

10.4.6 WIPO Intergovernmental Committee on Intellectual Property and Genetic Resources, Traditional Knowledge and Folklore (ICGTK)

Intergovernmental Committee on Intellectual Property and Genetic Resources, Traditional Knowledge and Folklore (ICGTK), established in September 2000, serves as a gathering place where World Intellectual Property Organization (WIPO) member states can examine the intellectual property issues that emerge in connection to right to access to genetic resources and benefit sharing as well as the protection of TK and traditional cultural expressions, including disclosure requirements in patent applications. TK will be lawfully secured and affiliated with the biological resources. In 2009, WIPO members have concluded that the intergovernmental committee (IGC) should start formal negotiations with the goal of arriving agreement on one or more international lawful instruments that would guarantee the successful protection of genetic resources, traditional knowledge, and traditional cultural expressions. Such an instrument or instruments might reach from a suggestion to WIPO members to a formal treaty that would bind countries choosing to endorse it.

10.4.7 Global Crop Diversity Trust (GCDT)

The Global Crop Diversity Trust (GCDT) is an international organization attempting for the conservation and accessibility of crop diversity for food security around the world. It was established in 2002, with an endeavor made by the FAO of the United Nations and the World Bank to create a trust fund for global ex situ collections of germplasm of pertinence for food and agriculture. The crop trust is raising an enrichment to fund the global system for the conservation of crop diversity, in perpetuity, through existing institutions. The funds are used in particular to support the preservation of globally valuable collections of plant genetic material in gene banks or living collections as well as to train staff and other stakeholders in this field. In addition, it also supports data frameworks for the concerns of agricultural biodiversity, including databases, documentation of collections, and the exchange of information through networks. Hence, it serves to secure our food and agriculture.

The Crop Trust and governing body of ITPGRFA entered into agreement in 2006, which recognizes the Crop Trust as an "essential element" of the treaty's funding strategy concerning the ex situ conservation and availability of plant genetic resources for food and agriculture.

10.4.8 International Undertaking on Plant Genetic Resources (IUPGR)

The international undertaking was adopted by the FAO Conference in 1983 and is an important instrument to support global harmony over matters in regard to access on PGRFA (Rose 2004). The landraces and modern cultivars as well as wild forms and relatives of crop species and also varieties developed by researchers in laboratories are covered under this undertaking. The IUPGR recognizes the rights of both breeders and farmers and provides the sovereign rights to countries over their plant genetic resources. The draft article with respect to farmer's rights focuses on the protection of TK, the fair sharing of benefits emerging from the use of biological resources, and the right to take part in decision-making. The undertaking is amicable with those of CBD and serves as an instrument for the conservation and sustainable use of plant genetic resources and the fair and equitable sharing of the benefits arising by the utilization of these resources.

10.4.9 UN Declaration on Rights of Indigenous People

The UN Declaration on Rights of Indigenous People was adopted by the UN General Assembly on 13 September 2007 during its 61st session, with a majority of 143 states in favor, 4 votes against, and 11 abstentions. The UN General Assembly declaration will not only be a legitimately binding instrument but provides a universal framework of norms for the living, dignity, prosperity, and rights of the world's indigenous people. It represents a vital tool in abolishing violations of human rights for 370 million indigenous people harboring earth by assisting them in combating discrimination and disparaging. For all those matters that are concerned to these people, this instrument supports their full and effective participation. To appreciate and promote the inherent rights of indigenous people, their culture, institutions, traditions, identity, dialect, and other issues identified with their social structures are well recognized by the UN General Assembly. Article 24 of the declaration recognizes that indigenous people have the right to their customary medicines and to administer their health practices, including the conservation of their crucial medicinal plants, animals, and minerals.

Article 31 of the declaration attempts to protect TK by emphasizing the member states to recognize the rights of indigenous people to preserve, control, secure, and develop their social heritage, indigenous knowledge, and traditional cultural expressions as well as the expressions of their innovations, sciences, and cultures, including human and genetic resources, medicines, knowledge of the properties of fauna and flora, seeds, oral traditions, literatures, designs, traditional games and sports, and visual and performing arts. These people also have rights to preserve, control, secure, and develop their intellectual property over such social heritage, indigenous knowledge, and traditional cultural expressions.

10.5 PGR and TK: Indian Initiatives

By knowing the importance of plant genetic resource and associated traditional knowledge in food and agriculture, the Indian government has enacted the Biological Diversity Act of 2002 (BDA) and Protection of Plant Varieties and Farmers' Right Act of 2001(PPV&FR Act), which are in harmony with several international treaties and are briefly discussed below.

10.5.1 Biological Diversity Act (BDA)

The Biological Diversity Act of 2002 is enacted by the government of India with the target of preservation of biological diversity and benefit sharing arising out of the utilization of biological diversity and associated traditional knowledge as well as prevention of biopiracy and protection, respect, and conservation of TK related to biological diversity. This act is in harmony with the United Nations CBD, to which India is a member state. The National Biodiversity Authority (NBA) is the nodal agency of this act established in 2003 with headquarter in Chennai and has the obligation to execute the provisions under the act.

The act in its Section 3(1), 3(2), and 4 prohibits any person not possessing a citizenship of India or any corporate association/organization not registered in India or registered in India but has non-Indian citizen participation in equity or management from obtaining any biological resource and associated knowledge existing within India for commercial utilization, research or bio-survey and bio-utilization without prior approval of the national authority.

Any person who violates the administrative provisions of this act will be punished with imprisonment for a period of 5 years or has to deposit a fine that extends to 1 million rupees or more depending upon the damage caused.

10.5.2 Protection of Plant Varieties and Farmers' Right Act (PPV&FR Act)

The Protection of Plant Variety and Farmers' Right Act of 2001 (PPV&FR Act) was enacted by the government of India to provide for the establishment of an effective system for protection of plant varieties and the rights of farmers and plant breeders and to encourage the development of new varieties of plants. To satisfy the obligations under the TRIPS agreement which states that member countries shall provide either patents or an effective *sui generis system* for the protection of plant varieties, India has implemented the first *sui generis system* which is "the Protection of Plant Varieties and Farmers' Rights Act of 2001." Although the act is in harmony with the UPOV Convention, it involves some provisions that are not included in the convention (Dhar 2002). The act recognizes and secures the rights of farmers in regard to their contribution in conserving, improving, and making available plant genetic resources for the development of new plant varieties. To make available new and improved plant varieties timely, the act also recognizes plant breeder's rights (PBRs) that promotes research and accelerated agriculture growth. The act provides PBR to those plant varieties that fulfill the criteria of being novel, distinct, uniform, and stable. This act also has a provision of any person claim over particular plant varieties and hence creates a platform through which the plant varieties of indigenous communities can be protected. Though the PBR holder has sole rights to multiply, market, sell, or disseminate seeds of his variety, the act has no provision that precludes the use of the protected variety for research or breeding of other new varieties by the breeders or researchers. By taking into consideration the contribution made by the farming and tribal communities to genetic diversity used by the breeders, the act recognizes the introduction of benefit sharing between the two groups.

10.6 Conclusion

Before the introduction of above international agreements/policies related to plant genetic resources and associated traditional knowledge, the global community cleared little consideration for their conservation. The PGR and associated TK were subjected to a number of malpractices including genetic erosion, lack of benefit sharing and accesses of crop diversity, lack of recognition, lack of legal protection, and biopiracy. But the implementation of these instruments has facilitated globally the conservation, accesses, and benefit sharing arising out of the utilization of PGR and associated TK, which in turn increased the agriculture growth and development. In addition, the rights of indigenous and local communities over their resources and other things drawn from their social structure are protected by these instruments. Although these policies have increased the PGR conservation especially through the ex situ method, the in situ conservation is still in its immature stage. Therefore, further exertions are necessary around the world to ensure in situ conservation that allows the dynamic process of crop evolution to proceed and also safeguard the TK associated to biodiversity. In addition, it is pertinent that all the countries in the world should implement these policies to promote a global network of preventing the above malpractices identified with PGR and associated TK.

References

Alston JM, Chan-Kang MC, Marra PP, Wyatt TJ (2000) A meta-analysis of rates of return to agricultural R &D: expede herculem. Research report 113. Intl Food Policy Res Institute, Washington, DC, p 118

Altieri MA, Merrick LC (1988) Agro-ecology and in situ conservation of native crop diversity in the Third World. In: Wilson EO, Peter FM (eds) Biodiversity. National Academy Press, Washington, DC, pp 15–23

Blackburn TC, Anderson K (1993) Before the wilderness: environmental management by native Californians. Ballena Press, Menlo Park

Brush SB (1989) Rethinking crop genetic resource conservation. Conserv Biol 3(1):19–29

Craig D, Davis M (2005) Ethical relationships for biodiversity research and benefit-sharing with indigenous people. MqJICEL 2:31

CUBIC (2000) Links between cultures and biodiversity. Proceedings of the Cultures and Biodiversity Congress, Cultures and Biodiversity Congress, Yunnan

Das S (2011) Seeds and trees: an insight into agriculture, biotechnology and intellectual property rights. Int J Bus Soc Sci 2:205–211

Dhar B (2002) *Sui generis* systems for plant variety protection. Quaker United Nations Office, Geneva

Duvick DN (1986) Plant breeding: past achievements and expectations for the future. Econ Bot 40(3):289–297

Eaton DJ, Kalaugher E, Bijman J (2004) International agreements relating to plant genetic resources for food and agriculture and implications for Dutch policy. Agricultural Economics Research Institute, No. 29132

Evenson RE, Gollin D (2003) Assessing the impact of the Green Revolution, 1960 to 2000. Science 5620:758–762

Furbank RT, Tester M (2011) Phenomics-technologies to relieve the phenotyping bottleneck. Trends Plant Sci 16:635–644

Hammer K, Teklu T (2008) Plant genetic resource: selected issues from genetic erosion to genetic engineering. J Agric Rural Dev Trop Subtrop 109:15–50

Harlan J (1995) The living fields: our agricultural heritage. Cambridge University Press, New York

Hvalkoff S (2001) Outrage in rubber and oil. In: Zerner C (ed) People, plants and justice. Columbia University Press, New York, p 56

IPCC (2007) Intergovernmental Panel on Climate Change: Climate change impacts, adaptation, and vulnerability. Summary of Policymakers, Geneva 2, Switzerland

Jackson SA, Iwata A, Lee SH, Schmutz J, Shoemaker R (2011) Sequencing crop genomes: approaches and applications. New Phytol 191(4):915–925

Juma C (1989) Biological diversity and innovation: conserving and utilizing genetic resources in Kenya. African Centre for Technology Studies, Nairobi

Keneni G, Bekele E, Imtiaz M, Dagne K (2012) Genetic vulnerability of modern crop cultivars: causes, mechanism and remedies. Int J Plant Res 2(3):69–79

Khandelwal A, Mishra Y (2015) Traditional knowledge: significance and its protection. Int J Res Anal 2(6):1–11

Mahato A (2014) Climate change and its impact on agriculture. Int J Sci Res Publ 4(4):1–6

Malik SS, Singh SP (2006) Role of plant genetic resources in sustainable agriculture. Indian J Crop Sci 1(1–2):21–28

Mauro F, Hardison PD (2000) Traditional knowledge of indigenous and local communities: international debate and policy initiatives. Ecol Appl 10(5):1263–1269

Moller K, Isbruch F, Flinspach T (2014) An investigation of the preferences and behaviour of users of the SMTA when making decisions to use the alternative payment options of Articles 6.7 and 6.11 of the SMTA. Foodand Agriculture Organisation (FAO), Rome, Italy

Moore G, Tymowski W (2005) Explanatory guide to the International Treaty on Plant Genetic Resources for Food and Agriculture, IUCN Environmental Policy and Law 57

Nabhan GP (1997) Cultures of habitat: on nature, culture, and story. Counterpoint, Washington, DC

Nakashima DJ, Galloway MK, Thulstrup HD, Ramos CA, Rubis JT (2012) Weathering uncertainty: traditional knowledge for climate change assessment and adaptation. UNESCO/UNU, Paris/Darwin, p 100

NPM (1990) National Planning Manual. SCS Draft USDA, USA

Ogwu MC, Osawaru ME, Ahana CM (2014) Challenges in conserving and utilizing plant genetic resources (PGR). Int J Genet Mol Biol 6(2):16–22

Reynolds MP (2010) Climate change and crop production. CABI International, Oxfordshire

Rose GL (2004) The international undertaking on plant genetic resources for food and agriculture: will the paper be worth the trees? Faculty of Law-Papers 32

Sodhi NS, Erlich PR (2010) Introduction. In: Sodhi NS, Erlich PR (eds) Conservation biology for all. Oxford University Press, Oxford, p 1

UNEP (1992) Convention on biological diversity. United Nations Environment Programme, Nairobi, Kenya, VNEP No. 92–8314

Wilkes G (1991) In situ conservation of agricultural systems. In: Oldfield ML, Alcorn JB (eds) Biodiversity: culture, conservations, and eco-development. Westview Press, Boulder, pp 86–101

World Resources Institute, The World Conservation Union, The United Nations Environment Programme (1992) Global biodiversity strategy: policy-makers' guide. WRI Publications, Baltimore

Plant Genetic Resources and Indigenous Traditional Knowledge Conservation Toward Resilience to Climate Change

11

Manmohan Sharma and Rajinder Parshad Kaushik

Abstract

Climate change is a stark reality and its effects are discernible even today. Changing climatic parameters will have impact on the global agriculture conditions with variable severity from region to region. On the one hand, unpredicted changes in climatic factors are quite likely to threaten the production and productivity of economically important plant species particularly that of food crops; at the same time, agriculture productivity requires significant increase to meet the expected growth in demand for food by the ever-increasing world population. Crop diversity, an important component of total plant biodiversity, is considered essential for sustainability of agriculture systems and for achieving food security under changing climate scenario. Plant genetic resources (PGRs) serve as the key ingredients for breeding crop varieties with resilience to different types of biotic and abiotic stresses resulting due to changes in climatic parameters. It makes the conservation and multiplication of plant genetic resources an essential strategy for development of crop genotypes with resilience to climate change and for achieving the overall goal of food security. Indigenous traditional knowledge, handed down from generation to generation over centuries, has been recognized as a treasure of information which can be utilized for developing resilience to climate change and for framing strategies for natural resource management in response to environmental changes. Indigenous traditional knowledge and judicious use of PGRs underline indigenous and local communities' understanding of phenomenon of climate change and strategies to develop resilience in response to such changes for global sustainable development. Documentation, recognition, and scientific validation of traditional practices of local communities for conservation of biodiversity and sustainability of agriculture are essential for resilience to changing climatic conditions.

M. Sharma, Ph.D. (✉)
School of Biotechnology, Sher-e-Kashmir University of Agricultural Sciences and Technology of Jammu, Chatha, Jammu 180009, Jammu and Kashmir, India
e-mail: man_2007@yahoo.co.in

R.P. Kaushik, Ph.D.
Rice Research Station, CSK Himachal Pradesh Krishi Vishwavidayala, Palampur 176062, Himachal Pradesh, India
e-mail: kaushikplprks@rediffmail.com

R.K. Salgotra, B.B. Gupta (eds.), *Plant Genetic Resources and Traditional Knowledge for Food Security*, DOI 10.1007/978-981-10-0060-7_11

11.1 Introduction

Climate change refers to a change in the state of the climate that can be identified (e.g., by using statistical tests) by changes in the mean and/or the variability of its properties, and that persists for an extended period, typically decades or longer (IPCC 2014). Climate change may be due to natural internal processes or external forces such as modulations of the solar cycles, volcanic eruptions, and persistent anthropogenic changes in the composition of the atmosphere or in land use. Framework Convention on Climate Change (UNFCCC), in its Article 1, defines climate change as: "a change of climate which is attributed directly or indirectly to human activity that alters the composition of the global atmosphere and which is in addition to natural climate variability observed over comparable time periods." According to latest observations of IPCC WG-II on climate change (IPCC 2014), in recent decades, changes in climate have caused impacts on natural and human systems on all continents and across the oceans. Global and regional weather conditions are expected to become more variable, with more frequent extreme events, expansion of areas with high climate variability, and significant changes in precipitation patterns and rise in temperatures. In many regions, changing precipitation or melting snow and ice is altering hydrological systems, affecting water resources in terms of quantity and quality. Warming of the climate system is unequivocal, and many of the observed changes are unprecedented over decades to millennia. The atmosphere and oceans have warmed and amounts of snow and ice have diminished, and sea level has risen. The Intergovernmental Panel on Climate Change (IPCC), in its Fourth Report, predicts average rise in the global temperature between 1.8 and 4 °C by the end of the current century compared to an increase of about 0.75 °C over the last century. According to Fifth Assessment Report of IPCC (2014), Earth surface temperature is projected to rise over the twenty-first century, and as per the Summary Report of IPCC (2014), it is very likely that heat waves will occur more often and last longer and that extreme precipitation events will become more intense and frequent in many regions. The ocean will continue to warm and acidify and global mean sea level to rise. Climate change will amplify existing risks and create new risks for natural and human systems. Risks are unevenly distributed and are generally greater for disadvantaged people and communities in countries at all levels of development. The Fifth Assessment Report of the Intergovernmental Panel on Climate Change (IPCC) 2014 has given focused attention to climate change adaptation and mitigation.

11.2 Causes of Climate Change

The climate change is a movement in the climate system because of internal changes within the climate system or in the interaction of its components or because of changes in external forcing either by natural factors or anthropogenic activities (IPCC 1992). There are factors which cause the natural variation of climate systems such as:

1. Changes in solar output and total solar irradiance over a long period of time
2. Variation in Earth's orbit around the Sun as well as the position of rotational axis with respect to the plane of the orbit
3. Water vapors and cloud variation influencing Earth atmosphere radiation balance
4. Volcanic aerosols and injection of sulfur-containing gases which induces a cooling effect

These changes have longer timescales above 10,000 years, but the other causes due to human activities are more serious for climate change activities such as:

1. Changes in atmospheric aerosols leading to their interaction with solar and terrestrial radiation in absorption and scattering
2. Hydrosphere – variation in liquid water on Earth's surface
3. Cryosphere – changes in global snow and ice cover as a source of water for sea level rise

4. Land surface changes – large-scale afforestation or deforestation influencing the albedo, aerodynamic roughness, and evapotranspiration in the region
5. Internal dynamics of climate systems – changes in climate system components such as ENSO phenomena over shorter periods influencing regional climate system, abnormal climatic extreme events of tropical cyclones, storms, etc.
6. North Atlantic Oscillation – changes in sea surface temperatures are associated with sudden change in oceanic circulation.
7. Anthropogenic causes of climate variations including combustion of fossil fuels, emissions of greenhouse gases and deforestation by burning of forests and agricultural waste

However, if the present situation continues where emission of anthropogenic gases is double than removal rate, the situation goes on cumulatively build up from bad to worse to attain alarming rate of global warming. Similar features are supported with the results of Sterman and Sweeney (2007).

11.3 Impact and Implication of Climate Change on Agriculture and Food Security

Changing climatic parameters are expected to have strong impact on agriculture worldwide including increased incidence of different types of abiotic and biotic stresses such as increased intensity and dynamics of weed flora, pests, and diseases in cultivated areas. Under higher temperatures in temperate latitudes, particularly in Europe and North America, there is likelihood of increase in growing season and the total area suitable for cultivation. However, anticipated gains in yield are to be viewed in the background of losses due to the spread of weeds, pests, and diseases. Contrary to this, regions in lower latitudes will be most severely affected by a decline in land suitable for cultivation, especially sub-Saharan Africa and the Caribbean. In tropical and subtropical regions of Africa, Asia, and South America, extreme seasonal heat is expected to severely lower agricultural outputs. Food security in many countries is under threat from unpredictable changes in rainfall pattern and more frequent extreme weather conditions. Farming communities in poorer countries with erratic climatic conditions will most likely be affected. Many studies conducted across different regions on diverse crops have predominantly indicated negative impacts of climate change on crop yields as compared to the positive impacts, and very few studies indicate positive impacts especially those focusing on high-latitude regions though it is not yet clear whether the balance of impacts has been negative or positive in these regions (IPCC 2014). Studies also indicate negative impact of climate change on wheat and maize yields for many regions and at global level. However, effects on rice and soybean yield have been smaller in major production regions both at regional level and globally. Observed impacts relate mainly to production aspects of food security rather than access or other components of food security. For the major crops (wheat, rice, and maize) in tropical and temperate regions, climate change without adaptation is projected to negatively impact production, although individual locations may benefit. Climate change is a fast-happening phenomenon, and studies indicate that by the year 2050, many countries, making up about 35 % of the global land area, will experience novel climates, they never experienced before. Many of the crops that have evolved and sustained well in a particular geographical region over centuries may have to be substituted by other crops leading to changes in food habits of the people. For example, 90 % of global rice is produced and consumed in Asia at present. The changing rainfall patterns and other climatic factors may no longer support rice cultivation in this region by 2050.

With world population expected to touch nine billion mark by 2050, the global agriculture is required to support production of 70 % more food than is produced today (Godfray et al. 2010). On one hand, unpredicted changes in climatic parameters are likely to threaten the production and productivity of economically important plant species particularly that of food crops; at the same time there is challenging task of significantly increasing agriculture productivity to meet the expected increase in demand for food by the ever-increasing world population in the next few decades. The major constraint in achieving targeted levels of food production is posed by the unpredictable changes happening in climatic parameters which have bearing on agricultural productivity. This necessitates the need to develop crop genotypes having resilience to climate change for food security in the future. Climate resilience refers to the capacity for a socio-ecological system to absorb stresses and maintain function in the face of external stresses imposed upon it by climate change and to adapt, reorganize, and evolve into more desirable configurations that improve the sustainability of the system, leaving it better prepared for future climate change impacts (Nelson et al. 2007).

Agriculture and food systems will suffer the most due to climatic changes over the next few decades (Brown and Funk 2008), and it will adversely affect food security (Schmidhuber and Tubiello 2007). Rising temperatures, unpredictable rainfall patterns, frequently occurring intense weather conditions, and increased incidence of pests and diseases would have impact on food production (Parry et al. 2007; Kotschi 2007; Morton 2007; Brown and Funk 2008; Lobell et al. 2008). The predominant community suffering due to global climate change would be smallholder farmers, particularly in developing and underdeveloped countries (Morton 2007). The latest IPCC report predicts the probability of prolonged and more severe heat waves, increased precipitation, increase in Earth's surface temperature, droughts, and other extreme weather throughout the twenty-first century (IPCC 2014). It has been shown that by 2080, the 40 poorest countries, located predominantly in tropical Africa and Latin America, could lose 10–20 % of their basic grain-growing capacity due to drought (Kotschi 2007). The biggest problem for food security will be the predicted increase in extreme weather, which will damage crops at particular developmental stages and make the timing of farming more difficult, reducing farmers' incentives to cultivate (Morton 2007). The natural habitats of the plants are at the highest risk of being disturbed due to climatic variations, and therefore, the wild relatives of cultivated plants face the threat of extinction. Up to 61 % of peanut species, 12 % of potato species, and 8 % of cowpea species are likely to be lost within the next 50 years, and16–22 % of wild relatives of species with direct value to agriculture may be in danger of extinction (Jarvis et al. 2008).

11.4 Plant Biodiversity in Climate Change Resilience

Plant diversity on Earth is represented by an estimated 300,000 species of higher plants. However, only about 7,000 species have been domesticated and cultivated by humans over the millennia for food, fodder, and feed. While the number of cultivated plant species is relatively small and seemingly insignificant, nature has evolved an extraordinary intraspecific genetic diversity in crop plants and their wild relatives. Diversity provides a natural insurance policy against major ecosystem changes, be it in the wild or in agriculture (McNaughton 1977; Chapin et al. 2000; Diaz et al. 2006). It is now predicted that genetic diversity will be most crucial in highly variable environments and those under rapid human-induced climate change (Reusch et al. 2005; Hajjar et al. 2008; Hughes et al. 2008). The larger the number of different species or varieties present in one field or in an ecosystem, the greater the probability that at least some of them can cope with changing conditions. Species diversity also reduces the probability of pests and diseases by diluting the availability of their hosts (Chapin et al. 2000). It is an age-old insurance policy of farming communities to hedge their risks and plant diverse crops or varieties. The strategy is not to maximize yield in an optimum year, but to maximize yield over years, good and bad, by decreasing the chance of crop failure in a bad year (Altieri 1990).

Crop biodiversity is essential for achieving food security under changing climate scenario. Farmers have always relied on crop genetic diversity to mitigate the effects of variability in climatic situations. There is abundant scientific evidence that crop biodiversity has an important role to play in the adaptation to our changing environment. The farming systems predominated by monocultures of genetically identical plants would not be able to cope with a changing climate; therefore, increasing the biodiversity of an agroecosystem can help maintain its long-term productivity and contribute significantly to food security. Genetic diversity within a field provides a buffer against losses caused by environmental change, pests, and diseases. Genetic diversity provides the resilience needed for a reliable and stable, long-term food production (Diaz et al. 2006). In addition to enhancing food security and climate resilience, diversity in the field also delivers important ecosystem services. Varietal mixtures that are tolerant to drought and flood not only increase productivity but also prevent soil erosion and desertification, increase soil organic matter, and help stabilize slopes (Hajjar et al. 2008). Benefits for farmers include reducing the need for costly pesticides, receiving price premiums for valued traditional varieties, and improving their dietary diversity and health (Hajjar et al. 2008). Narrowly adapted species and endemics are especially vulnerable to the direct effects of climate change. Indirect effects may also have important impacts, through changes in biotic interactions, including changes in pest and disease pressure (Newton et al. 2008), competition and successional dynamics, and changes in symbiotic compositional interactions.

Thus, to meet the challenges of climate change, exploring and conserving natural diversity in the existing plant genetic resource pool are needed for developing climate-resilient elite genotypes. The conservation and enhanced utilization of agricultural biodiversity will increase the adaptability, production, productivity, and resilience potential of agroecosystems in view of expected changes in climatic factors.

11.5 Plant Genetic Resources (PGRs)

International Undertaking on PGR (FAO 1983) defines PGR as the reproductive or vegetative propagating material of (1) cultivated varieties (cultivars) in current use and newly developed varieties; (2) obsolete cultivars; (3) primitive cultivars (landraces); (4) wild and weed species, near relatives of cultivated varieties; and (5) special genetic stocks (including elite and current breeder's lines and mutants). PGRs, according to the CBD (1992), are any living material of present and potential value for humans. Plant genetic resources include all our agricultural crops and some of their wild relatives because they possess valuable traits. CBD implies genetic resource to be any "genetic material of actual and potential value" and genetic material to be a "functional unit of heredity." Occasionally, genes, DNA fragments, and RNA are also included under the purview of genetic resources. Plant genetic resources (PGRs) thus comprise of wild and weedy relatives of crop plants, germplasm complexes, land races, genetic stocks, and improved varieties. Plant Genetic Resources for Food and Agriculture (PGRFA) are the raw material that farmers and plant breeders use to improve the quality and productivity of crops. They can be defined as any genetic material of plant origin of actual or potential value for food and agriculture, e.g., seeds, tubers, mature plants, etc. Crop Wild Relatives (CWR) are highly genetically diverse and locally adapted. PGRs are considered as the treasure of genes for agronomically important traits and serve as raw material for breeding the next generation of crops to mitigate the effect of biotic and abiotic stresses resulting due to climate change (Toledo and Manzella 2012; Maxted et al. 2012). Increase in yields of major food crops – or even maintaining them – in the face of climate change will depend on combining genetic traits found in materials of a wide range of origins (Petit 2001), including wild species.

The unpredictable changes in climate are expected to have strong impact on biodiversity including plant genetic resources (PGRs) which constitute the raw material for providing reliance to the changing climatic conditions. Unfortunately, wild species are especially vulnerable to climate change because they do not receive management interventions that help them adapt to changing conditions. Crop wild relatives have saved agriculture millions of dollars, both directly and indirectly, by improving crop resilience to biotic and abiotic stresses (Dwivedi et al. 2008). A number of crops, such as sugar cane, tomatoes and tobacco, could not be grown on substantial commercial scales were it not for the contribution to disease resistance made by wild relatives of those crops (FAO 1997). However, crop wild relatives (CWRs) themselves are now under threat of extinction due to climate change. Under elevated CO_2 levels, CWRs produce relatively less fruit and seed than domesticated crops (Jablonski et al. 2002), increasing their risk of extinction. Moreover, CWRs remain a relatively low priority in germplasm collection due to financial and political impediments. Increasing threats to natural habitats and farming systems make it imperative to collect, conserve, and characterize traditional varieties (landraces) and wild relatives in order to make available for use in mitigating the effects of biotic and abiotic stresses caused by climate change (Lane and Jarvis 2007). The extent to which climate affects the spatial distribution of wild species makes this an especially important theme when examining the impact of climate change on PGR.

Realizing the importance of plant genetic resources, efforts have been made to collect, conserve, and evaluate plant genetic resources (PGRs), to support the plant breeders with diverse genetic materials to create new crop varieties to combat the climate change (Bansal et al. 2014). It is essential to conduct comprehensive analysis of PGRs for resistance to biotic and abiotic stresses and their properties such as adaptability, plasticity, and resilience under fluctuating environmental conditions, so as to achieve higher production and productivity. They form the basis of all crop varieties that are bred to produce more, withstand stresses, and yield quality output. Advances in modern technologies including biotechnology are also contributing for developing climate-resilient cultivars. Molecular breeding and genetic engineering approaches utilizing the PGRs are gaining momentum (Varshney et al. 2012). Thus, accelerated research efforts are required to harness the genetic potential of PGRs utilizing prebreeding and modern genomics approaches to develop superior stress-tolerant cultivars. Genomics tools provide new options for accelerated domestication of new species to allow adaptation of agriculture to climate change (Shapter et al. 2013).

11.5.1 Conserving PGRs for Climate Change Resilience

Out of 300,000 species of plants estimated to grow on Earth, only three crop plants, viz., rice, wheat, and maize together contribute about 75 % of the total global grain production. This necessitates the need to conserve, multiply, and judiciously utilize the existing PGRs, comprising of cultivars, landraces, and wild relatives, to combat not only the food shortage but also to mitigate the crop losses occurring due to climate change. Wild and weedy relatives of crops possess genes for mitigating the impact of various biotic and abiotic stresses resulting due to climate change and developing improved cultivars with resilience to unpredictable changes in climatic parameters. Indirectly, they are the key factors in livelihood and food security in the future. Diversity in PGRs and farming are the most important factors to achieve food security in a changing climate. Crop wild relatives (CWRs) are a key resource for climate change adaptation, providing researchers with genes and traits for biotic and abiotic resistance (Maxted et al. 2008; Lane and Jarvis 2007). There is a need to conserve the PGRs because of the following reasons:

1. The modern intensive agriculture calls for uniformity and relies heavily on few crops and their high-yielding varieties and consequently has a narrow genetic base. In contrast, traditional agriculture had a large number of diverse landraces with a substantial genetic variation. If we do not counteract the increasing genetic impoverishment, it may have serious consequences, especially when facing a changed climate.
2. Plant genetic diversity is vulnerable to "genetic erosion," the loss of individual alleles/genes and of combinations of alleles/genes, such as those found in locally adapted landraces. According to FAO, replacement of local varieties by modern varieties resulting in

reduction of the sheer number of cultivars is the main cause of genetic erosion. This is intensified by the emergence of new pests, weeds and diseases, environmental degradation, urbanization, and land clearing.

3. PGR, the only source of plant genetic diversity, provides valuable traits needed for meeting the challenges of adapting crop varieties to climate change. An individual genotype with seemingly useless set of characters today may suddenly become essential tomorrow due to changing climatic conditions or outbreaks of disease. Therefore, it has been long realized that we "conserve" all the diversity we have.
4. To meet the challenges of climate change, exploring and conserving natural diversity in the existing plant genetic resource pool are needed for developing climate-resilient elite genotypes.
5. Ever-increasing area under existing abiotic stresses and emerging abiotic stress factors and their combinations have further added to the problems of the current crop improvement programs. Feeding the ever-increasing world population in the era of climate change demands the development of stress-tolerant crop cultivars. Development of such tolerant cultivars through conventional breeding methods depends heavily on the availability of natural genetic variations in a given crop species.
6. Various abiotic stresses such as high and low temperature, excess and deficient water stress, salinity, heavy metal toxicity, high radiations, high and low nutrient content in the soil, etc. are consequences of the rapidly changing climate and are responsible for loss in crop production and productivity. Thus, there is an urgent need to accelerate research efforts to conserve the PGRs and harness their genetic potential.

11.6 Strategies for Conservation of PGR

While climate change is one of the drivers of crop diversity loss, it is also an important reason to conserve PGRs, exchange them, and use them in a sustainable way. The broader the genetic base our civilization can rely on, the better equipped it will be to adapt to changing climate conditions and to guarantee global food security. There are a number of potential approaches to achieve systematic global or sub-global (regional, national, and local) PGR conservation. For conservation of PGRs, a balanced combination of in situ and ex situ conservation strategies will have to be adapted for securing biodiversity threatened by changing climate and altered production practices. Regardless of the approach, the systematic conservation of PGR diversity involves the complementary application of in situ and ex situ strategies. The appropriateness of each is determined by the conservation context, the nature of the diversity to be conserved, and the nature of the conserver. The use of both strategies provides the optimal condition of conservation.

The in situ approach allows for the conservation of diversity at all levels – ecosystem, species, and the genetic diversity within species – at specific locations, with continuation of the evolutionary processes (gene flow and natural selection, farmer selection, and exchange), and accessibility for use, in particular by local communities. Ex situ methods enable the conservation of alleles, genotypes, and populations, at any location, with easy access for breeders and other users, and safeguard them from loss or change due to continual evolutionary pressures. Each conservation method has advantages and disadvantages with respect to the different conservation factors, and it is necessary to balance these by integrating two or more methods. Whenever possible, in situ conservation should be part of the strategy in order to enable change under continual evolutionary processes. In situ conservation is concerned with preserving the adaptive capacity of the plant population. Its purpose is to ensure the continuous adaptation of local genetic material to environmental micro-niches and the changing socioeconomic conditions of agricultural production. However, the emphasis on in situ and ex situ approaches will depend on the conservation context – the object, aims, and location of conservation. For each situation, an understanding of genetic diversity is fundamental.

The application of complementary conservation strategies and their design criteria requires research which should include studies on the scientific and institutional linkage between in situ and ex situ conservation. Critical in this regard is the issue of accessibility to the diversity conserved in situ or ex situ on the part of users, be they local or institutional. Conservation in situ with local communities and farmers ensures that resources remain directly in the hands of the primary users. Provision of new resources to communities through ex situ conservation programs may be beneficial and could have implications for diversity maintenance. Mechanisms by which diversity can move from in situ to ex situ conservation or which can enable continual access to in situ-conserved diversity by institutional users (e.g., formal-sector breeders) will help overcome the limitations of the "snapshot" character of ex situ collections. Such access may best be facilitated by the development of more structured ex situ collections which provide an entry point to in situ populations – a modification of the core collection concept.

11.7 Role of Participatory Approaches in the Era of Climate Change

Community participation in the conservation and use of plant genetic resources is an integral part of the management of natural resources by many agrarian communities, and it becomes even more relevant in view of changes in global climate vis-a-vis their negative impact on agriculture. In participatory plant breeding activities, scientific and farming communities work together for generation, conservation, and utilization of plant genetic resources suiting local climatic situations. By including decentralized breeding, farmers and crop biologists can become partners in local crop improvement efforts for marginal agroclimatic zones. Both from a scientific and developmental perspective, there are convincing arguments suggesting the potential benefits to gene banks of their engaging in participatory approaches and establishing a dialogue with farmers and their communities. Local collectors often have extensive knowledge of ecogeographical, climatic, biological, and cultural issues and are present year round and can collect throughout the fruiting season and are better placed to include important microenvironments within complex agricultural systems which may otherwise be overlooked. Such programs which provide farmers greater access to appropriate genetic resources and training could assist farmers in improving various characteristics of their planting materials such as disease or pest resistance and drought tolerance, hence making them more adapted to the unpredicted changes in weather factors, thereby increasing food production (Friss-Hansen and Sthapit 2000).

11.8 Biotechnological Interventions in PGR Conservation and Improvement

Since the beginning of agriculture 8–10,000 years ago, farmers have been altering the genetic makeup of the crops they grow by selecting the features such as faster growth, higher yields, pest and disease resistance, larger seeds, or sweeter fruits, and that has dramatically changed domesticated plant species compared to their wild relatives. When the science of conventional plant breeding was further developed in the twentieth century, plant breeders understood better how to select superior plants and breed them to create new and improved varieties of different crops. This has dramatically increased the productivity and quality of the plants we grow for food, feed, and fiber. Conventional plant breeding has been the method used to develop new varieties of crops for hundreds of years. However, conventional plant breeding can no longer sustain the global demand with the increasing population, decline in agricultural resources such as land and water, and the apparent plateauing of the yield curve of the staple crops.

The recent integration of advances in biotechnology, the molecular marker applications, genetic engineering, and genomic research in combination with conventional plant breeding practices demonstrably hold great promise for crop improvement which is revolutionizing twenty-first-century crop improvement. Biotechnology is a viable option for developing genotypes that can perform better under harsh environmental conditions. In addition, advances in genomics coupled with bioinformatics and stress biology can provide useful genes or alleles for conferring stress tolerance. Superior genes or alleles where they have been identified in the same species can be transferred into elite genotypes through molecular breeding. Moreover, genetic engineering approaches dilute the genetic barriers to transfer useful genes or alleles across different species from the animal or plant kingdoms. Molecular breeding approaches have the potential to enhance crop production under different stress conditions. On the one hand, abiotic stresses are complex in nature; on the other hand, there are several challenges that have restricted the realization of the full potential of using biotechnology approaches in crop breeding.

11.8.1 Genomic Resources for Enhancing Resilience to Climate Change

In the era of precision agriculture, crop improvement objectives are also becoming more and more trait oriented. To meet these objectives, it is not only necessary to conserve available genetic variability but also important to utilize it. Genome variation found in wild crop relatives may be directly relevant to the breeding of environmentally adapted and climate-resilient crops (Henry 2014). Analysis of the genomes of plant genetic resources will become a key tool to enable their utilization in crop improvement for climate adaptation, and utilization of PGRs can be enhanced significantly by generating genomic resources. Recent advancements in "omics" techniques have enabled generation of genomic resources much more efficient (Mondal and Sutoh 2013). Ultimately, these genomic resources are utilized for developing better cultivars resilient to climate change either through marker-assisted breeding or by genetic transformation for transgenic crop development. Next-generation gene banking based on advanced DNA sequencing technologies is considered as an integral component of generation and conservation of plant genetic resources (Bansal et al. 2014; Treuren and Hintum 2014). Large-scale sequencing of accessions in plant germplasm collections will provide a platform to enable these approaches (Henry 2013).

11.9 International Policy Frameworks for Promotion of PGR Conservation in View of Climate Change

11.9.1 International Treaty on Plant Genetic Resources for Food and Agriculture (ITPGRFA 2001)

The International Treaty on Plant Genetic Resources for Food and Agriculture (ITPGRFA) is a global leader in dealing with the issues of crop diversity preservation, food security, and climate change adaptation. It is in fact the only operational international agreement of a legally binding nature with the overall goal of achieving global food security through the management of crop diversity. The Treaty provides an effective policy response to global challenges through:

1. Its comprehensive provisions providing guidance to countries regarding the measures and activities to be undertaken at the national level for the conservation and the sustainable use of crop diversity
2. Its provisions on farmers' rights which aim at supporting farmers and local and indigenous peoples in conserving crop diversity on their farms
3. The multilateral system that facilitates access to a global gene pool of crop genetic resources for agricultural research and breeding of new crop varieties that may achieve higher yields and nutritional values and that are adapted to new climate conditions

4. The field initiative and its benefit-sharing fund that supports initiatives for the conservation and the sustainable use of crop diversity in developing countries, with a focus on helping ensure sustainable food security by assisting farmers adapt to climate change

The International Treaty is an instrument that is uniquely designed to promote and facilitate essential international germplasm exchanges needed to enhance farmers' resilience to climate change. In addition, the Treaty implementation is supported by an extensive network of institutions in more than 127 countries, as well as with significant prior partnerships with FAO, CGIAR, and the Global Crop Diversity Trust in support of Treaty implementation.

11.9.2 Multilateral System for Exchange of PGRFA

Global interdependence with regard to PGRFA will increase as a result of climate change, and the wide international exchange of agricultural genetic materials will become ever more necessary. Many agroecological zones or geographical regions will be unable to ensure food security with the crops and varieties they currently grow and will be forced to rely on resources from elsewhere including exchanges and imports from other countries. In turn, the resources these areas currently hold are likely to be of great importance for farmers in yet other areas. The fully operational multilateral system has established an ever-expanding global gene pool of more than 1.2 million samples. The gene pool of the multilateral system consists of samples of genetic material from a set of crops, which are listed in Annex 1 of the Treaty document. Samples are included in the gene pool by the contracting parties to the Treaty (i.e., the state governments) and the institutions that they control. Samples also come into the gene pool from international institutions, such as the international gene banks of the Consultative Group on International Agricultural Research (CGIAR), as well as other stakeholders. These samples are pooled and administered under a common set of rules in the Treaty and further specified in a contractual instrument, namely, the Standard Material Transfer Agreement (SMTA). The rules apply to individual transfers of these samples (e.g., from a gene bank to a breeder) for certain purposes, namely, utilization and conservation for research, breeding and training for food, and agriculture. The rules regulate not only how to obtain access to the plant genetic material but also how to share the results of research and breeding on that material. In practice, the multilateral system works as a common pooling, distributing, and benefit-sharing system for the genetic material that it covers. Access to such resources is facilitated in the sense that those who want to access the genetic material in the system do not need to negotiate access agreements on a case-by-case basis with national competent authorities. Instead, the resources are available to anyone who wants them under a standard contract, the SMTA. The use of the SMTA cuts out all of the costs involved in the bilateral process, including the negotiations, legal contracts, and other transactions, for the benefit of farmers and gene bank managers who typically provide the genetic material and for the plant breeders and researchers who typically seek access to this material to develop new climate-ready plant varieties. As much as access is multilateral, so is benefit sharing. Since genetic material in the multilateral system is treated as pooled goods, there is no individual owner with whom individual contracts for access and benefit sharing must be negotiated. As such, benefits resulting from their use do not go back to the provider but, rather, must be shared in multilateral ways.

11.9.3 Benefit-Sharing Fund: Leading the Field in Climate Change Adaptation

The International Treaty leads the "Leading the Field" initiative by creating a multilateral fund referred to as the Benefit-sharing Fund (BSF), supported by member governments, the private sector, and international foundations. This multilateral fund is currently supporting high-impact projects aimed at keeping farmers ahead of the climate change challenge and food secure. The operation of the initiative is facilitated by the Treaty that issues a call for proposals supported by the Treaty's extensive network of National Focal Points in more than 127 countries. The Governing Body, at its second meeting in 2007, identified three priority areas for support from the BSF, building on the Global Plan of Action for the Conservation and Sustainable Utilization of Plant Genetic Resources for Food and Agriculture. These include:

1. Information exchange, technology transfer, and capacity building
2. Managing and conserving plant genetic resources on farm
3. The sustainable use of plant genetic resources

11.10 Traditional Knowledge (TK)

Another subject of global importance in view of climate change is traditional knowledge (TK). Understanding the relevance and importance of TK in context of climate change adaptation is an emerging area of collaborative investigation, involving indigenous people, local communities, and scientists. UNFCCC (2011) has recognized the conservation of TK as co-benefits of ecosystem-based approaches to adaptation. The observations of local communities regarding impacts of climate change vis-a-vis customary practices and mechanisms may provide a strong basis for devising strategies to develop climate resilience agriculture. Besides being diverse, holistic, cultural, and intergenerational in character, TK is a practice-based intellectual in nature and bears close relationship with the natural environment. It makes TK more relevant in decision-making on climate change adaptation. The tough task before nations, associated research, and social organizations and institutions is to frame policies and take decisions and actions that encourage local communities to enhance and refine their climate change assessment knowledge and adaptation through the mobilization of their TK. It requires more investment to enable effective contributions of TK toward sustainable development as well as to encourage effective participation of traditional knowledge holders, i.e., indigenous peoples and local communities.

TK practices are increasingly recognized and used as a valuable resource for planning climate change adaptation (IPCC 2010; UNFCC 2013). Indigenous and traditional knowledge refers to the knowledge, innovations, and practices of indigenous and local communities around the world. Developed from experience gained over the centuries and adapted to the local culture and environment, TK is transmitted orally from generation to generation. It tends to be collectively owned and takes the form of stories, songs, folklore, proverbs, cultural values, beliefs, rituals, community laws, local language, and agricultural practices, including the development of plant species and animal breeds. Indigenous traditional knowledge has been defined by different international organizations. Some of the standard definitions have been described here. According to CBD, TK is the knowledge, innovations, and practices of indigenous and local communities around the world. Developed from experience gained over the centuries and adapted to the local culture and environment, TK is transmitted orally from generation to generation. Sometimes, it is referred to as an oral tradition for it is practiced, sung, danced, painted, carved, chanted, and performed down through millennia. TK is mainly of a practical nature, particularly in such fields as agriculture, fisheries, health, horticulture, forestry, and environmental management in general.

According to UNESCO, local and indigenous knowledge systems refer to the understandings, skills, and philosophies developed by societies with long histories of interaction with their natural surroundings. For rural and indigenous peoples, local knowledge informs decision-making about fundamental aspects of day-to-day life. This knowledge is integral to a cultural complex that also encompasses language, systems of classification, resource use practices, social interactions, ritual, and spirituality. These unique ways of knowing are important facets of the world's cultural diversity and provide a foundation for locally appropriate sustainable development. WIPO defines traditional knowledge as the knowledge, know-how, skills, innovations, or practices that are passed between generations in a traditional context and that form part of the traditional lifestyle of indigenous and local communities who act as their guardian or custodian. According to IPBES, indigenous and local knowledge refers to the multifaceted arrays of knowledge, know-how, practices, and representations that guide societies in their innumerable interactions with their natural surroundings. This interplay between people and place has given rise to a diversity of knowledge systems that are at once empirical and symbolic, pragmatic and intellectual, and traditional and adaptive. The IPBES working definition goes on to further propose using Berkes' definition of indigenous and local knowledge, which is: A cumulative body of knowledge, practice and belief, evolving by adaptive processes and handed down through generations by cultural transmission, about the relationship of living beings (including humans) with one another and with their environment.

In addition to highlighting the diverse, holistic, cultural, and intergenerational character of TK, its practice-based and intellectual nature, and the relationship of TK with the natural environment, there are several other concepts related to TK that contribute to understanding its relevance to decision-making on climate change adaptation. Recognizing that TK is accumulated, reviewed, and adapted over generations, it is erroneous to view it as immutable and fixed in time and/or space. Instead, it emerges through a history of involvement in an environment, and just as ecosystems shift and alter, so does the resultant knowledge gained for consistent interaction with that system (Ingold and Kurttila 2000). Berkes (2009) makes the explicit point that indigenous elders cannot transmit an actual knowledge of climate change; what they can do is to teach what to look for and how to look for what is important. That caution illustrates the dangers in assuming that TK is static content, when it is in fact a dynamic process that involves observing, discussing, and assessing new information. Indeed, the knowledge itself, viewed as entrenched sets of information, may be less important in the context of change than the values and traits embedded in and transmitted through knowledge systems. In China and the Philippines, for example, the United Nations Institute for Sustainability and Peace (2013) found that the effective use of water through water storage and infiltration facilities enhances groundwater recharge to reduce the drying up and subsequent collapse of rice terraces. Such traditional practices may enhance resilience in the face of a changing climate. Similarly, in the Andes, communities traditionally develop qochas, rectangular excavations with two side channels that allow them to collect and store runoff water during the rainy season. That water can then be used for livestock, to improve pastures, or to irrigate small gardens. Qochas as a practice are embedded in ayllu, a traditional community structure that provides for local-level action and consensual decision-making (Conservation International 2013). In Egypt, traditional storage methods, such as burying grain in the sand or constructing granary rooms in houses, may assist farmers in the face of increasing climate unpredictability and resulting food insecurity (Parrish 1994).

Documentation and research is also uncovering initial efforts by communities to respond to climate change impacts. In Northeast India, the Asian Indigenous Peoples Pact (2012) noted that the Tangkhul Naga villages' attentive observation of a subtle shift in seasons has led to the revival of sustainable rotational agriculture, governed by traditional local leadership structures. In addition, research in the Sahel by Nyong et al. (2007) found that pastoralists integrate mobility and active management of their multispecies composition of herds to survive drought.

Finally, response to disasters, hazards, and other extreme events may also provide input relevant to indigenous knowledge and adaptation. Farmers in Haiti traditionally apply a number of actions, before and after natural hazards, in order to reduce their impacts, particularly on high-market value crops like bananas. Such actions include the removal of leaves and the immediate sale or storage of marketable fruit. In Tuvalu, families often have their food stores (kaufata) stocked with enough preserved foods to help them through a cyclone or drought event (Nakashima et al. 2012). A recent study in Nepal documents a spectrum of knowledge, practices, and insights relevant to the four thematic sectors identified through research across 18 socially and economically diverse districts in Nepal. These practices were found to be practical, flexible, and cost-effective since they involved the use of local materials, local technical know-how, and local institutions in their development and promotion. These practices were not only community initiated, owned, and managed but were also situation specific and adaptive to changes in human environmental systems (MoSTE 2015).

The indigenous or traditional knowledge has contributed tremendously over centuries for prediction and management of climatic variables for benefit of mankind, although it seems to be new to climate science. TK has long been recognized as a key source of information and insight in areas such as traditional medicine and medicure, biodiversity conservation, customary resource management, impact assessment, and natural disaster preparedness and response. A lot of natural biodiversity is preserved due to knowledge of local communities regarding importance of plant genetic resources for food and livelihood security. Indigenous observations and interpretations of meteorological phenomena are most of the times very precise and accurate. Their observations contribute significantly to advancing climate science, by ensuring that assessments of climate change impacts and policies for climate change adaptation are meaningful and practically applicable. Indigenous peoples in most of the societies the world over have known that maintaining a diversity of plant varieties and animal races provides a low-risk buffer in uncertain weather environments. Traditional systems of governance and social networks had efficient strategies to manage diversity and share resources at community levels, while still conserving the resources for sustainable food and livelihood security.

11.11 Relevance of TK in View of Climate Change Resilience

Farmers, out of their centuries of experience, are devising strategies and experimenting with the climatic variability. Their knowledge and experience can provide basic inputs in developing technologies to mitigate the impact of risks posed by climatic variations. For instance, farming communities belonging to the different regions of the world follow different indigenous practices for absorption and conservation of soil moisture, rainwater harvesting and conservation, nutrient and weed management, insect pest and disease management, etc. Their local belief systems and observations, particularly with respect to weather parameters, have effectively contributed toward weather forecasting and risk management in crop husbandry. Relevance of TK for climate change resilience can be substantiated through the following points:

1. TK contributes fine-grained, long-term observations about the changing environment. It is essential for understanding not just local-level impacts of climate change but also the range of options that a community might consider for adaptation that are appropriate to its ecological and sociocultural environments and in conformity with its priorities, values, and worldviews (UNFCC 2013).
2. Indigenous people, particularly those in small islands, high altitudes, desert, and the Arctic, are already experiencing the impacts of climate change. Despite high exposure to the impacts of climate change, the knowledge of indigenous people offers valuable insights to observations of climate change which when coupled with scientific data and forecasts will provide enhanced understanding of climate change phenomenon and its predicted impacts on mankind.
3. Indigenous knowledge provides basis for understanding the adaptation priorities of a local community as well as the range of appropriate adaptation options to enhance resilience and food security of indigenous people.

4. The IPCC noted that indigenous knowledge is "an invaluable basis for developing adaptation and natural resource management strategies in response to environmental and other forms of change." This was reaffirmed at the 32nd Session of the IPCC in 2010 according to which indigenous or TK may prove useful for understanding the potential of certain adaptation strategies that are cost-effective, participatory, and sustainable.
5. Recognition of the significance of TK for climate change has only begun to emerge at the international level in the last few years. For the most part, however, that recognition is based on a documentation of actions that local communities are taking on their own initiative. The challenge for governments, organizations, and institutions is to identify policies, decisions, and actions that might support those communities and enhance their efforts to mobilize TK for climate change assessment and adaptation.

It can thus be concluded that there is now a growing awareness that TK has significant contributions to make within the climate change adaptation process, from observation and assessment to planning and implementation.

11.12 Documentation of TK

Due to social, economic, political, and natural factors, a large number of local communities have migrated to other places as well it has led to gradual uprooting of many such untapped resources from their native habitats resulting in loss and erosion of very rich indigenous knowledge. Rapid pace of urbanization and modernization has contributed to changes in lifestyles of indigenous communities which has resulted into loss of their peculiar culture and heritage (Girach 2007). The traditional knowledge has been transferred from generation to generation through verbal discourses only. With the passing away of an experienced old person, a library of knowledge is lost if not documented or recorded properly. Systematic documentation of practices and beliefs in different spheres of human life including observations regarding natural phenomena, environmental processes, religious ceremonies, human health, food, animal health, agriculture, etc. and their interactions have great practical utility in evolution of modern-day scientific innovations, development, and planning. It provides useful clue for planning projects for conservation of biological diversity, sustainable uses of natural resources, indigenous health practices, etc. Therefore, systematic documentation, cataloging, and recording of TK of different indigenous communities and local societies is essential for developing projects and technologies for resilience to climatic changes and resource conservation.

11.13 Conclusion

There is growing recognition that TK and customary sustainable use of PGRs reflect the indigenous peoples' and local communities' resilience to change including climate change, as well as contribute directly to biological and cultural diversity and global sustainable development. Therefore, developing policies at national and international level to promote conservation of PGRs and TK toward climate resilience is an important factor for sustainable development. PGR and TK are a strategic resource at the heart of sustainable crop production and food security in the face of adverse environments, rapidly fluctuating climate conditions and burgeoning population growth. Climate change is one of the major threats to food security. Climate/weather is an important component of crop production causing high variability in yield potentials even in highly favorable and high-technology agricultural ecosystems. In addition to an increasing world population, there are numerous reasons for serious concern about sufficient future global production of food from crop plants. Since the supply of land is fixed, the continuous expansion of agricultural areas by bringing more areas under cultivation is not possible in the long run; further growth will have to rely on productivity increases. Biodiverse farming is a proven, effective strategy to adapt to climate change. Through it, we can create farms that are able to maintain and increase food production in the face of increasingly unpredictable conditions. Meeting the food requirement of immensely growing population without any damage to the ecological aspects now is the topmost priority of agriculturists, environmentalists, or development planners to focus on sustainable food production strategies for livelihood security especially in the context of climate variability and change. There is a need to develop and apply a standard methodology across the board for various studies related to climate change and PGR.

References

Altieri MA (1990) Agroecology. In: Carrol CR, Vandermeer JH, Rosset PM (eds) Agroecology. McGraw Hill, New York, pp 551–564

Asia Indigenous Peoples Pact (2012) Indigenous peoples and climate change adaptation in Asia. Asia Indigenous Peoples Pact, Chiang Mai. http://www.adb.org/sites/default/files/climate-change-assessmentcoo.pdf

Bansal KC, Lenka S, Mondal TK (2014) Genomic resources for breeding crops with enhanced abiotic stress tolerance. Plant Breed 133:1–11

Berkes F (2009) Indigenous ways of knowing and the study of environmental change. J R Soc N Z 39(4):151–156

Brown ME, Funk CC (2008) Food security under climate change. Science 319:580–581

CBD (1992) United Nations Convention on Biological Diversity. Preamble. https://www.cbd.int/doc/legal/cbd-en.pdf

Chapin FS, Zavaleta ES, Eviner VT, Naylor RL, Vitousek PM, Reynolds HL et al (2000) Consequences of changing biodiversity. Nature 405:234–242

Conservation International (2013) Annual report. July 2012–June 2013, Arlington

Diaz S, Fargione J, Chapin FS, Tilman D (2006) Biodiversity loss threatens human well-being. PLoS Biol 4:1300–1306

Dwivedi SL, Stalker HT, Blair MW, Bertioli DJ, Upadhyaya H, Nielen S, Ortiz R (2008) Enhancing crop gene pools with beneficial traits using wild relatives. Plant Breed Rev 30:179–230

FAO (1983) International undertaking on plant genetic resources for food and agriculture. Food and Agriculture Organisation. Commission on Genetic Resources for Food and Agriculture. Rome, http://www.fao.org/Ag/cgrfa/iu.htm

FAO (1997) The State of the World's Plant Genetic Resources for Food and Agriculture. Rome, p 510

Friss-Hansen E, Sthapit B (2000) Participatory approaches to the conservation and use of plant genetic resources. International Plant Genetic Resources Institute, Rome

Girach RD (2007) Methods of documenting indigenous knowledge. Tradition 4:24–30

Godfray HCJ, Beddington JR, Crute IR, Haddad L, Lawrence D, Muir JF et al (2010) Food security: the challenge of feeding billion people. Science 327:812–818

Hajjar R, Jarvis DI, Gemmill-Herren B (2008) The utility of crop genetic diversity in maintaining ecosystem services. Agric Ecosyst Environ 123:261–270

Henry RJ (2013) Sequencing crop wild relatives to support the conservation and utilization of plant genetic resources. Plant Genet Resour. doi:10.1017/S1479262113000439

Henry JH (2014) Genomics strategies for germplasm characterization and the development of climate resilient crops. Front Plant Sci 5(68):1–4

Hughes AR, Inouye BD, Johnson MTJ, Underwood N, Vellend M (2008) Ecological consequences of genetic diversity. Ecol Lett 11:609–623

Ingold T, Kurttila T (2000) Perceiving the environment in Finnish Lapland. Body Soc 6:183–196

IPCC (1992) Climate change, the supplementary report of IPCC Scientific Assessment. Houghton J, Callendar BA, Varnay SK (eds). Cambridge University Press, Cambridge, p 198

IPCC (2010) Best practices and available tools for the use of indigenous and traditional knowledge and practices for adaptation, and the application of gender-sensitive approaches and tools for understanding and assessing impacts, vulnerability and adaptation to climate change. Cambridge University Press, Cambridge, United Kingdom, Geneva

IPCC (2014) Summary for policymakers. In: Field CB, Barros VR, Dokken DJ, Mach KJ, Mastrandrea MD, Bilir TE, Chatterjee M, Ebi KL, Estrada YO, Genova RC, Girma B, Kissel ES, Levy AN, MacCracken S, Mastrandrea PR, White LL (eds) Climate change 2014: impacts, adaptation, and vulnerability. Part A: Global and sectoral aspects. Contribution of Working Group II to the fifth assessment report of the Intergovernmental Panel on Climate Change. Cambridge University Press, Cambridge, UK, pp 1–32

ITPGRFA (2001) Secretariat of the international treaty on plant genetic resources for food and agriculture, food and agriculture organization of United Nations. Viale delle Terme di Caracalla, Rome. http://www.planttreaty.org

Jablonski LM, Wang X, Curtis PS (2002) Plant reproduction under elevated CO2 conditions: a meta-analysis of reports on 79 crop and wild species. New Phytol 156:9–26

Jarvis A, Lane A, Hijmans R (2008) The effect of climate change on crop wild relatives. Agr Ecosyst Environ 126:13–23

Kotschi J (2007) Agricultural biodiversity is essential for adapting to climate change. GAIA Ecol Perspect Sci Soc 16:98–101
Lane A, Jarvis A (2007) Changes in climate will modify the geography of crop suitability: agricultural biodiversity can help with adaptation. Paper presented at ICRISAT/CGIAR 35th Anniversary Symposium, "Climate-Proofing Innovation for Poverty Reduction and Food Security". ICRISAT, Patancheru, p 12
Lobell DB, Burke MB, Tebaldi C, Mastrandrea MD, Falcon WP, Naylor RL (2008) Prioritizing climate change adaptation needs for food security in 2030. Science 319:607–610
Maxted N, Ford-Lloyd BV, Kell SP, Iriondo JM, Dulloo ME, Turok J (2008) Crop wild relative conservation and use. CABI Publishing, Wallingford
Maxted N, Magos BJ, Kell S (2012) Conservation and sustainable use of plant genetic resources for food and agriculture: a toolkit for national strategy development. Food and Agriculture Organization of the UN, Rome
McNaughton SJ (1977) Diversity and stability of ecological communities: a comment on the role of empiricism in ecology. Am Nat 111:515–525
Mondal TK, Sutoh K (2013) Omics: application in biomedical, agriculture and environmental science. In: Bhar D, Zambare V, Azevedo V (eds) Application of next-generation sequencing for abiotic stress tolerance. CRC Press, Boca Raton, pp 347–365
Morton JF (2007) Climate change and food security special feature: the impact of climate change on smallholder and subsistence agriculture. Proc Natl Acad Sci 104:19680–19685
MoSTE (2015) Indigenous and local knowledge and practices for climate resilience in Nepal, mainstreaming climate change risk management in development. Ministry of Science, Technology and Environment (MoSTE), Kathmandu
Nakashima DJ, Galloway McLean K, Thulstrup HD, Ramos Castillo A, Rubis JT (2012) Weathering uncertainty: traditional knowledge for climate change assessment and adaptation. The United Nations Educational, Scientific and Cultural Organization, Paris
Nelson DR, Adger WN, Brown K (2007) Adaptation to environmental change: contributions of a resilience framework. Annu Rev Environ Resour 32:395–419
Newton AC, Johnson SN, Lyon GD, Hopkins DW, Gregory PJ (2008) Impacts of climate change on arable crops -adaptation challenges. Association for Crop Protection in Northern Britain. In: Proceedings of the crop protection in Northern Britain conference. Dundee
Nyong A, Adesina F, Osman E (2007) The value of indigenous knowledge in climate change mitigation and adaptation strategies in the African Sahel. Mitig Adapt Strat Glob Chang 12:787–797
Parrish AM (1994) Indigenous post-harvest knowledge in an Egyptian oasis. In: IK Monitor. http://maindb.unfccc.int/public/adaptation/adaptation_casestudy.pl?id
Parry ML, Canziani OF, Palutikof JP, Vander Linden PJ, Hanson CE (2007) Summary for policymakers. In: Climate change 2007: impacts, adaptation and vulnerability. Contribution of Working Group II to the fourth assessment report of the Intergovernmental Panel on Climate Change. Cambridge University Press, Cambridge, UK, pp 7–22
Petit M (2001) Why governments can't make policy: the case of plant genetic resources in the international arena. Commission on Intellectual Property, Lima
Reusch TBH, Ehlers A, Hammerli A, Worm B (2005) Ecosystem recovery after climatic extremes enhanced by genotypic diversity. Proc Natl Acad Sci 102:2826–2831
Schmidhuber J, Tubiello FN (2007) Global food security under climate change. Proc Natl Acad Sci 104:19703–19708
Shapter FM, Cross M, Ablett G, Malory S, Chivers IH, King GJ et al (2013) High-throughput sequencing and mutagenesis to accelerate the domestication of *Microlaena stipoides* as a new food crop. PLoS One 8:e82641. doi:10.1371/journal.pone.0082641
Sterman JD, Sweeney LB (2007) Understanding public complacency about climate change: adults' mental models of climate change violate conservation of matter. Clim Chang 82:225–251
Toledo A, Manzella D (2012) Building resilience for adaptation to climate change in the agriculture sector: the role of the International Treaty on Plant Genetic Resources for Food and Agriculture. Secretariat of the International Treaty on Plant Genetic Resources for Food and Agriculture, FAO, Rome
Treuren R, Hintum TJL (2014) Next-generation genebanking: plant genetic resources management and utilization in the sequencing era. Plant Genet Resour Charact Util 12(3):298–307
UNFCCC (2011) United Nations Framework Convention on Climate Change. Report of the conference of the parties on its seventeenth session. Durban
UNFCC (2013) UN climate change conference, Warsaw, Poland, 11–22 November 2013
UNUISP (2013) Eco-system based adaptation strategies for enhancing resilience of rice terrace farming systems against climate change. United Nations University Institute for Sustainability and Peace. http://unfccc.int/7769.php
Varshney RK, Chen W, Li Y, Bharti AK, Saxena RK, Schlueter JA et al (2012) Draft genome sequence of pigeonpea (*Cajanus cajan*), an orphan legume crop of resource-poor farmers. Nat Biotechnol 30:83–89

Rational of Participatory Approaches for the Management of Plant Genetic Resources and Traditional Knowledge

12

Bharat Bhushan Gupta
and Romesh Kumar Salgotra

Abstract

For timely and successful diffusion of any agro-technique to the farming communities, the participatory approach is one of the options. In participatory approach, the researchers can reach to number of farmers for diffusion of agricultural technology with in a period of time. Now participatory approaches are becoming the integral part of the researchers for diffusion of any technology. For conservation and sustainable utilisation of PGR and TK, participatory approaches play a significant role. The genetic diversity of crop species can be enhanced by conserving the PGR and TK through formal institutions and the farming communities. In developing countries, farmers are the main elements in the conservation of PGR and production of agricultural crops. The women also play a major role in the selection of plant species, their preservation and sustainable utilisation. Although women are not involved in policymaking and to take decisions, they play a main role identification of crop genetic resources for household uses and their conservation. Through participatory approaches, farmers are encouraged to take part in the selection of new varieties by using their own experiences. They can interact with the researchers freely while selecting the suitable crop genetic resources. Both participatory varietal selection and participatory plant-breeding approaches encourage the farming communities for active involvement in the selection of new genotypes from the developed and the segregating genetic materials. The farmers select the genetic material for conservation and use depending upon their need and to secure the food security.

B.B. Gupta, Ph.D. (✉)
Division of Plant Breeding and Genetics, Sher-e-Kashmir University of Agricultural Sciences and Technology of Jammu,
Chatha, Jammu 180009, Jammu and Kashmir, India
e-mail: dr_bbgupta@ymail.com

R.K. Salgotra, Ph.D.
School of Biotechnology, Sher-e-Kashmir University of Agricultural Sciences and Technology of Jammu,
Jammu 180009, Jammu and Kashmir, India
e-mail: rks_2959@rediffmail.com

R.K. Salgotra, B.B. Gupta (eds.), *Plant Genetic Resources and Traditional Knowledge for Food Security*, DOI 10.1007/978-981-10-0060-7_12

12.1 Introduction

Plant genetic resources (PGRs) and Traditional Knowledge (TK) are providing livelihood to farmers and food security to the world. PGR and TK are the basic input and important resources used by the farmers in crop production which are essential for sustainable agricultural production. The basic aims of the Convention on Biological Diversity (CBD) regarding food security are the conservation of PGR and TK, their sustainable utilisation and equal share of benefits. All the countries under Global Plan of Action (GPA) have rights over biological resources related to PGRFA for preservation and their sustainable utilisation.

In agriculture, participatory approach is more than to talk hundreds of different farmers. The participatory approach is a systematic talk between the researchers, policymakers or extension workers and indigenous farming communities to diffuse new agricultural production technologies effectively. In participatory approaches, researchers, policymakers or extension workers have to work very closely in association with the farming communities to know their behaviour and interest in the adoption of new technologies. However, this is not a very easy task to work with the farmers as their needs, knowledge and views are different.

In participatory approaches, the farmers are encouraged to participate to learn new techniques in association with their own traditional techniques. The indigenous farmers gain the experience through experimentations and then apply these techniques in their own fields and spread to other farmers. The researchers and extension workers are the facilitators and work very close with the farmers and try to engage the farming communities to adopt the new technologies related to preservation and sustainable utilisation of PGR and TK. Farmers are aware about the benefits of adopting new technologies and how these technologies help to secure their livelihood and food security. The researchers and extension workers follow up the farmers who have adopted the modern techniques related to conservation of PGR and TK. New techniques and concepts can be disseminated through participatory approaches.

Over the past 20 years, there has been a paradigm shift in participatory approaches to collect TK associated with PGR from extracting knowledge for a scientist to use and to a focus on how local communities can benefit from involvement with scientists. This reflects an increase in recognition of the fundamental right that local farmers continue to play in generating and maintaining the diversity of landraces. During this period, the PGR community has shifted from emerging recognition of the value of TK to general acceptance that a comprehensive approach, such as the methodology of a TK, is needed.

This methodology shifts the control of the documentation from the scientist to the community itself. Before collecting ethnobotanical data, scientists play a role in empowering farmers and communities to document their knowledge by themselves. Then, the scientists seek permission to jointly use the knowledge documented. Prior Informed Consent (PIC) and material transfer agreements (MTAs) are now an inherent part of ethnobotanical data collecting. The role of in situ and ex situ conservation has taken a new direction with an increased understanding of the links between local culture and the conservation of plants by the community.

The conservation of PGR and TK by the local and indigenous people is very important for sustainable utilisation and management of these resources. This is not new for the indigenous communities who are already conserving the PGR and TK, but a formal way of conservation of these resources is required to provide more recognitions. To save the agrodiversity, the underutilised and neglected crops should be conserved (CIAT 1997). There is a need to provide the compensation to the farming communities in the form of recognition, welfare and incentives to conserve more on-farm resources. This will give more moral support to the farming communities for the conservation of PGR through participatory approach. This approach will also help for the equal sharing of benefits to the farmers. For the last three decades, numbers of debates are going at the national and international levels for the successful preservation of PGR and TK. Later on, with the adoption of GPA, concrete steps were taken for conservation of genetic resources and their sustainable use. The previous recommendations of international forum are not showing any significant development, but after the implementation of GPA guidelines, farmers are aware about the conservation of PGR and TK and sharing of benefits. Moreover, earlier recommendations are not changing the socioeconomic conditions of indigenous communities (CIAT 1997).

To provide the sharing of the benefits to indigenous peoples out of utilisation of their PGR and TK helps in strengthening the in situ conservation and their sustainable utilisation. For the conservation of PGR and TK and plant genetic resources for food and agriculture (PGRFA), local and indigenous people play an imperative role. This will also increase the employment and food security of the rural people (FAO 2012).

The government institutions, policymakers and nongovernment organisations (NGOs) are emphasising the management and conservation of all the natural resources through participatory approaches (McGuire et al. 1999). This will not only help in the conservation and sustainable utilisation of PGR and TK, but it enhances the awareness about the importance of these resources among the indigenous people. Numerous projects were implemented to re-establish the equilibrium linkage between international organisations and the indigenous communities to conserve the biodiversity through participatory approach (CIAT 1997).

Most of NGOs, international communities and policymakers have emphasised to develop and entail participatory programmes for the development of project related to conservation of PGR and TK. These projects must be framed in such a way that the experiences of the local and indigenous communities should be included. Moreover, PGR and TK conservation-related participatory methods should be documented for further improvement and conservation of these resources (UPWARD 1996; Veldhuizen et al. 1997). The participatory approaches related to PGR and TK preservation and their sustainable utilisation have already been validated in developing countries (Eyzaguirre and Iwanaga 1996; Sperling and Loevinsohn 1996).

12.2 Farmers' Participatory Approaches to Sustain Genetic Diversity

The biodiversity created by the plant breeders and farmers is primarily used by the farmers. The farmers have the extensive knowledge of the biological diversity and production of agricultural crops. The farmers have different strategies and indigenous knowledge to conserve and maintain the biodiversity in their fields. There is no random occurrence of genetic diversity in the fields of the farmers. They play a significant and substantial role to understand the interactions of the agroecosystem which are guided by their traditional practices of crop cultivation and maintenance of genetic resources and related traditions (Teshome 1996; Teshome et al. 1997, 1999b). For example, farmers' indigenous knowledge on storage reduced the incidence of insect-pest attack (Teshome 1996; Teshome et al. 1999a).

The knowledge of indigenous peoples and their traditional practices significantly decrease the insect-pest infestation in storage of major food crop produced by the farmers. Their knowledge also helps in increasing the genetic diversity of major crop plants. The generation and maintenance of genetic diversity by the farmers by using indigenous knowledge appends the scientific communities' knowledge to understand how the farmers are generating and maintaining the biodiversity in their fields. Farmers have generated and maintained the genetic diversity to obtain the stable crop yields by using the accumulated experiences, knowledge and traditional practices. This is only due to wide cultivation of local and landrace varieties using their traditional experience and traditional practices of cultivation over the years. These practices are also linked with local festivals and cultural values of farmers. Moreover, the majority of the germplasm and other genetic resources are being maintained and conserved by the farmers. For their security, the farmers always save some seed material in stock to be used in emergency or failure of rains. The farmers have their own system to secure their livelihood and food security in different regions of the country. They have their own network and seed supply system. They are exchanging the saved seeds of landraces and knowledge of crop production to other farmer communities for their livelihood security. This practice of exchanging the seed material and related knowledge of production among the farmers facilitates to conserve more genetic diversity.

The farmers' varieties are the varieties which are selected by the farmers from local germplasm and are being cultivated over the years by them. The seed of these varieties needs to maintain and multiply by the farmers to enhance the genetic diversity and utility of these important materials. The seeds of these varieties should be multiplied at large scale and distributed among the farmers to meet out their demand of seeds for cultivation. This will meet the requirements of farmers who have yet not obtained the seeds adequately of high-yielding varieties (HYVs). Subsequently, the farmers can produce more crops per unit area and sell the surplus to market to improve their economy as well as livelihoods. In the seed supply system, numerous stakeholders are involved and actively participate in the preservation of crop genetic resources and their sustainable utilisation. Although farmers are conserving and maintaining these crop genetic resources in their fields, the introduction, monitoring and restoration are also supplemented by in situ and ex situ conservation by various government institutions. The diverse crop genetic resources have been generated, conserved and maintained by the farmers in their fields. They multiply these genetic resources on their field, store in community gene banks and distribute among the farmers through different farmers and social and local extension agencies. On the other hand, the germplasm has been stored in the national gene banks of a country. These materials have been scientifically selected using various breeding methods, demonstrations and multiplication and their distribution to farmers' community through government institutions, research institutions, NGOs and extension agencies. These systems ultimately take the genetic materials to farmers' fields, generating diversity in the germplasm and diversifying the crop production system at various levels. Compared to ex situ conservation method, in situ approach generates more diversity in crop genetic resources. However, ex situ conservation method acts as a backup of the important genetic material when on-farm crop fails due to biotic and abiotic stresses and other adverse climatic conditions. Therefore, we have to use both the approaches, i.e. in situ as well as ex situ methods of conservation, to meet any challenges and to secure the food security in the future on sustainable basis.

12.2.1 Ensuring Benefits to Farmers for Conserving Diversity

For conservation and sustainable use of PGR and TK, farmers' involvement is necessary to secure the food security. Farmers' participation in debate or framing policies and developmental programmes will help to bring the change in existing approaches of conservation and sustainable utilisation of resources. In olden days, germplasm was stored as ex situ gene banks by extracting the germplasm from the farmers only and then linked with agricultural production technology. With the use of modern technology, these techniques are transferred from research to farmers' field through extension programmes. Nowadays, all the techniques are used by the farmers in the agricultural crop production systems. All the research and innovations are carried out by the scientists; however, the farmers are still using the conventional approaches of crop production. At village and panchayat levels, farmers and their representatives are actively involved and participate in modern approaches of conserving the PGR and TK. The main aim is that the farmers and their representatives should take the advantages of both the approaches, i.e. modern and conventional for conserving PGR and TK (CIAT 1997). However, a number of obstacles are there for active participation of the farmers and their representatives. The ethical, social, cultural and institutional issues exist which discourage the active participation of the farmers in these approaches. The scientists and the policymakers have different approaches and methods of conservation and utilisation of genetic resources and related knowledge in which most of the cases the contribution of local farmers is ignored or discouraged. The traditional and local farmer-based participatory approach is more effective compared to the scientific-based approach which is not easily adopted by the indigenous communities. The participatory approaches are still immature and are at initial stage to implement for conservation and sustainable utilisation of PGR and TK (CIAT 1997).

For effective conservation and sustainable utilisation of PGR and TK and to ensure the benefits to farmers, the indigenous farmers should be given the responsibilities as conservers, protectors and enhancers of genetic diversity under certain rules and norms. This will help in better management of biological resources, improve farmers' seed supply system and rise in farmers' innovations coupled with capacity building of farmers. The capacity-building approaches of farmers increase the decision-making power of the farmers in developmental programmes to secure their livelihood and food security (Sthapit et al. 2008). The management of biodiversity by the local farmers and the stakeholder is linked with a number of other community-based programmes such as social network, institutional modalities, planning and monitoring and evaluation (Sthapit et al. 2006). To revive and conserve the PGR and TK, farmers' seed supply system of landrace varieties and other neglected crop varieties needs to strengthen through various programmes such as field demonstrations, seed minikits, seed fairs, agricultural fairs, community seed banks, community seed supply system, etc. Farmers' access and capacity building to crop genetic resources can be enhanced through Participatory Plant-Breeding (PPB) approach, varietal selection through participatory varietal selection method and exchange of seed material among the farmers (Sthapit et al. 2006). This will keep the farmers' community well versed with the advanced technologies, and exchange of crop genetic material further increases the plant genetic diversity in the ecosystem. For strengthening the diversity of PGR and TK, the role and responsibilities of the farmers should be well described. Moreover, some rewards and recognitions should be given to farmers for strengthening the seed supply systems and related traditional practices for conservation of crop genetic diversity.

12.3 Classification of Participation

In the last three decades, numerous national and international projects were initiated to enhance the participation of farming communities in various developmental programmes. In the initial stage, farmers are encouraged to participate in the common developmental programmes related to increase the income for their livelihood. After that, the farming communities can be encouraged to participate in social and biodiversity conservation and their sustainable utilisation programmes. The farmers' participation to conserve the PGR and TK is at different levels, i.e. local and national levels. The main objectives of the participatory approach are:

1. To give more responsibilities to farmers in conservation of PGR and TK
2. To involve the farmers in decision-making
3. To be aware about the accessing and sharing of benefits arising from the utilisation of these resources
4. To encourage farming communities to adopt the new technologies
5. To give importance of the conservation of genetic resources for their livelihood and food security in the future (Uphoff 1979)

In participatory approaches, the farmers are encouraged to adopt new technologies related to conserve their valuable PGR and TK. For successful implementation of project through participatory approach, the researchers, policymakers and extension workers should have knowledge about the socioeconomic conditions, rituals and cultures of farmers. While implementing any agricultural experiment, researchers and extension workers should have the knowledge of agronomical practices and control of diseases and insect pests. Moreover, they should have knowledge of farmers' socioeconomic conditions and other structures (Long and Long 1992). Today, participatory approaches are becoming the integral parts of any extension programmes.

There are different levels of farmers' participation which are generally classified into four categories. In the first category, the farmers rent their land and provide the labour to researchers and extension workers. In the second category, the farmers' knowledge and opinion are also considered while implementing the projects. In the third category, the farmers are directly involved in activities for successful implementation of the project. In the last category, the farmers are involved in designing the experiments and decide how to frame the objectives of the projects and actively participate to implement the project.

The main objective is not only to categorise the participation of the farmers at different levels, it is also important to know how their participation is going and at what level. Uphoff et al. (1979) classified the participation in different categories and gave due consideration to this deliberation (Table 12.1).

Table 12.1 Elements of indigenous farmers' improvement

Type of involvement and participation	Involvement in making the decision
	Involvement in implementing the programme
	Involvement in sharing of benefits
	Involvement in evaluation
Who are involved?	Persons/farmers from local areas
	Influential/leaders
	NGOs and government persons
	Persons from other countries
How is involvement there?	Basis of involvement
	Types of involvement
	Level of involvement
	Outcome of involvement

Source: Uphoff et al. (1979)

12.4 Participatory Approaches to Use and Conserve PGR and TK

For the conservation and sustainable utilisation of PGR and TK, there are two main players or actors. The first one is at the local level where indigenous farming communities are the main actors who are actively involved in the conservation of these resources, whereas, in the second one, government institutions and gene banks are the main actors. In the first one, the farmers are preserving the genetic resources by using the traditional practices; however, in the second method, gene banks and institutions with the help of plant breeders and other seed suppliers manage the PGR and TK. In order to streamline the system of conservation and utilisation of PGR and TK, support from the government system is essential. For effective implementation of the PGR, conservation standard methods or protocols are required. These should be modified as per the need and socioeconomic conditions of the farming communities. Both the systems, i.e. farming communities and the government institutions, should work together in a complementary way to achieve the maximum targets. Moreover, the government institutions can provide better support to the needs and grievances of local farmers for successful implementation of a project (Boef et al. 1997).

12.4.1 The Farmers' System

The farmers are the main elements who are engaged in the agricultural crop production system. The farmers are producing different crop varieties on their own fields, and the farmers obtained the seeds from relatives, exchange of seeds and purchase from the middlemen. In this system, farmers are also engaged in crop production, conservation of crop seeds and exchange or supply of seeds. All these components are well interlinked with each other and are carried out by the individual farmer or by the farming communities.

This system is still prevailing in the world, and farmers are engaged in production of seeds of different crop varieties and using the selection criteria for the selection of seeds. The farmers' selection criteria for seed help in shaping the gene flow from one population to another one. The farmers' system of agriculture is also maintaining the genetic diversity in crop species which will help the farmers to cope with the unpredictable weathers and to secure the food security in the future (Clawson 1985; Prain and Hagmann 2000).

12.4.2 The Institutional System

The institutional system was started when the system of breeding methods were used for the selection and development of a crop variety. Likewise, the farmers' system, institutional system of conserving and production of crop varieties, was established. In the institutional system, systematic crop characterisation was done, and the selected plants were used in crossing programmes to develop a new crop variety. The breeding methods such as introduction and selection, mass selection, pure line selection, pedigree methods and mutations resulted in crop improvement and development of new variety.

The plant breeders developed a large scale of genetic material, and gene banks were developed for the conservation of these genetic materials. Subsequently, explorations were made by a number of researchers and collected germplasm which was also conserved in gene banks. The gene banks act as repositories of genetic materials to avoid the loss of these valuable resources. These gene banks also maintain the genetic diversity in the germplasm. When the plant breeders develop a crop variety, the seed supply system begins to diffuse the variety in the farmers' fields.

12.4.2.1 Opportunities for the Institutional System

The institutional system for conservation of PGR and TK has to streamline for its approach to meet the demands of the farmers. To meet the different areas' requirements for crop genetic material or varieties, the plant breeders have to screen the genetic material at different locations. The supply of crop variety seeds and their distribution among the farmers are only effective when the farmers show different demands for different varieties. For effective and further improvement of seed supply system, the system should be decentralised instead of centralised one. This can be achieved only by bringing the farmers and the organisation close to each other complementarily.

To remove the major bottlenecks of farmers' systems and the institutional systems, it has been analysed that these bottlenecks can be removed only by a complementary way of collaborations and increased farmers' participations. The farmers should know all the agricultural production technologies and their mandate of crop production as the farmers are not always capable of producing the good quality seeds of various crops and distributing these seeds to other farmers of the indigenous communities using their traditional methods. It is analysed that the farmers' traditional practices can be improved to cope with the present need and new initiatives taken by the modern practices. The chances of introduction of new genetic material are very less in the farmers' system. The farmers system does not involve any plant-breeding practices such as crossing of different genotypes, genetic recombination and selection of best genotypes and identification of disease and insect-pest varieties. Whereas, in institutional system, new genes and genotypes can be introduced in crossing programmes to enhance the gene recombination, and new techniques can be introduced in systematic ways. The major drawback of institutional system is the farmers' choices and preferences and limited knowledge of organisers, researchers and extension workers about the conditions of indigenous farming communities.

The farmers have better opportunity to access the genetic diversity of PGR and TK under participatory approaches. The farmers can access and use this diverse genetic material for further improvement of traditional system and in certain adverse environmental conditions. This will help the plant breeders, policymakers and extension workers to explore more in detail about the preferences of the farmers, quality seed production and their distributions. The enhanced access to genetic diversity by the farmers has an advantageous outcome on the development and use of genetically diverse genotypes in the farmers' fields. This will increase the practice of in situ conservation of PGR in farmers' fields which helps in maintaining the genetic diversity on sustainable basis for future use.

12.5 Role of Gender in Participatory Management of PGR and TK

The women are playing a major role in conserving and exchanging the PGR and TK. The rural women have different roles and will change depending upon socioeconomic conditions and cultures. The potential roles of women and men in the conservation of PGR and TK depend upon the responsibilities allotted to them and relations between them (Andersson 1992). Both women and men have different responsibilities for preservation and sustainable utilisation of PGR and TK in a society.

Most of the minor and underutilised crops are conserved and maintained by the rural women. These minor crops are highly nutritious for human health. These crops are very imperative for human health. These crops are not generally targeted and used by the plant breeders in their breeding. The world's major landraces and wild species of crop plants have been domesticated by the women. During domestication, women have different vision and experience than men and select only those crop plants which have good characteristics. The women are generally concerned with selection of new plant species or landrace variety for domestication with the characteristics of cooking, processing and health.

The diverse crop varieties selected by women give the farming communities an opportunity to select and exchange new commodity to the farmers. The farmers can sell their produce in markets to meet the demands and preferences of the peoples. This is only possible where some specialised markets are available and farmers can produce the different crop species depending upon the demands of the consumers. These demands are different from personal preferences or home consumptions. However, the preferences and demands of landless communities totally depend on markets, i.e. what is being sold in the market and how they have to make their preferences on it.

The conservation and sustainable utilisation of PGR and TK through participatory approaches, depend upon how effectively the farmers and the stakeholders are participating in it. As participatory approaches depend upon the quality of persons involved, their qualification and how actively they are interacting with stakeholders during the process. There are some differences that do exist in family members and even among males and females of farming communities towards the conservation and management of PGR and TK. In such cases, there is a need to analyse and understand the variation and dissimilarities in their choices of management of these resources during the process of participatory approaches. The plant species identification and their management by the women are entirely different from the men. The women and men in farming communities have different responsibilities, jobs as well as knowledge of conservation of PGR and TK which helps to enhance the genetic diversity of crop species. The women use their knowledge to give different shapes to the selected and domesticated plant species compared to men's selection to create more genetic diversity among the crops (Eyzaguirre and Raymond 1995). Numerous studies have shown that women have played a pivotal role in the selection and domestication of diverse plant species and distribution of seed material to other women in a society (Tapia and De la Torre 1998; Iriarte et al. 1999; Shrestha 1998). Women in the household activities also conserve the unwanted and underutilised plants having medicinal and related qualities. These knowing or unknowing practices of farming women enhance the genetic diversity among the plant species (Wilde and Vainio-Mattila 1995). In the developing world, women are still not included in the formal and institutional participatory approaches. Women are not involved in decision-making processes and sharing of benefits arising out their conservation and sustainable use of resources. Sometimes, females are allowed to interact and talk with the males of the stakeholders in the participatory approaches. However, there are still constraints on the inclusion of women in the participatory research and development programmes.

For the conservation and better management of PGR and TK, the males and females should be given separate jobs, i.e. females (ladies and girls) have better experience to selection of plant species and their household uses compared to men. It has been seen that the field activities such as selection and farming of crop species of women are different from men. The distinct selection of plant species by women and men because of difference in their knowledge helps to enhance the biological diversity. The women and men have different on-farm and off-farm activities to conserve and the use of PGR and TK (Table 12.2).

Table 12.2 Different activities related to PGR and TK management

Activity	Gender	Location	Notations
Farming system			
Plough the land	Males	In the field	
Raising of nursery	Females	In the field or near the home	
Transplanting	Males and females	In the field	
Removing of weeds	Females	In the field	Some of the plant species can be used for household uses
Applying chemical fertiliser	Males	In the field	
Harvesting and threshing of crops	Males and females	In the field	
Postharvest management			
Collection of plants	Males and females	In the field and forest	Most of the plant species can be used for food purposes
Postharvest management of crop plants depending upon their values	Females	Around home	
Preparation of medicinal plants	Females and shamans	Home	Processing and local healers and home
Marketing			
Selling of plant species	Males and females	Markets	In the market, most buyers are females
Seed selection and conservation			
Selection in field	Females	Farm/home	
Postharvest dry	Females	Rooftops	This is done at different places depending upon the requirements and qualities of seeds
Storing the seed	Females	At home or seed banks	
Distribution of seeds among farmers	Females	In festivals and local functions	

Source: IDRC (1998)

12.6 Collection and Documentation of PGR and TK through Participatory Approaches

The process of documenting the TK of communities about PGR can be perceived as exploitation when communities do not benefit from it. The main issues include language, recognition and protection (Quek et al. 2009). From the community's perspective, the data collected and the papers produced are in the scientist's language, so the content is not available to the community. When scientists finish their work, there are usually no funds provided for any translation to benefit the community, and the community loses the possibility of complementing their indigenous knowledge with the scientific knowledge. For the scientists, the knowledge they have captured is in a translated or filtered form and makes reinterpretation difficult, and the original source of the knowledge might no longer be available. In addition, there is no recognition of the farmer as the source of the knowledge.

Before scientists can use the knowledge of communities that have oral traditions, they need to assist the community in documenting this knowledge in their own language for their own use. Then the community's knowledge may only be used with a citation. This process avoids the one-sided approach where TK is studied without any benefit to the community. The TK methodology was developed on the premise that farmers and community members have knowledge that is valid for scientific use and that empowering farmers to do their own documentation addresses the problem of knowledge erosion and enables scientists to have access to the information. This methodology was first proposed by Quek and Zhang (1997) and has since been developed further and used by several organisations in the Asian region. It begins when a paper was authorised by a farmer with the help of a researcher. The researcher then interprets the version of that paper or collects ethnobotanical data and cites the farmer's paper after obtaining permission to use it. The farmer's paper can be in any format for those who can write.

Having the farmers' knowledge documented in their own language enables communities to reuse their traditional knowledge in a manner that is similar to the way scientific communities reuse their knowledge. The farmer is recognised as the author of the paper because the scientist and others will cite it as the source of the information. This allows cultural and social information to become available to be validated by the community for reuse; other farmers can react as well. Knowledge captured in the farmer's paper is the original source and therefore can be reinterpreted by different scientists if the community allows. This makes reinterpretation of the knowledge possible as science progresses.

At the subsistence level, where most farmers are illiterate, their knowledge is usually passed on by word of mouth. Traditional groups of people have devised different methods of preserving their knowledge by embedding it in folklore, songs and sayings, as well as in their cultural arts and practices. In the process, one can expect the information to change along the way as some information will undoubtedly be omitted or added. Thus, any documentation has to take into account the possibility of such changes over time. Empowering communities to document their knowledge helps to promote in situ conservation because they then become more interested in maintaining their PGR. It also validates their knowledge and helps to promote the sharing of knowledge among the community, especially among the youth and children (Naming et al. 2010).

12.6.1 Maintaining Record of PGR and TK

The genetic resources of local and indigenous communities are conserved and maintained by the rural people and communities (Swaminathan 2000). On-farm PGR and TK conservation and maintenance of biological diversity are very important to secure the food security in the future. Moreover, genetic erosion has occurred in most of the underutilised and less important field, horticultural and medicinal crops (King et al. 2009).

For example, minor millets are highly nutritional crops which are being neglected by the farming communities since long and genetic erosion of these crops has occurred. These minor millets are the staple food of some African countries and provide the nutrition to the small farming communities (Bhag Mal et al. 2010). These minor millets also play a significant role to cope with the climate change. These crops are poor man's crops which provide food for their livelihood even under biotic and abiotic stresses and other unprecedented environments.

In India, Kolli Hills in Tamil Nadu, Koraput in Orissa and Wayanad in Kerala are three hot spots of biodiversity. The M.S. Swaminathan Research Foundation (MSSRF) has initiated the programmes for conservation of these minor millets. They have recorded all the factors related to the communities such as the socioeconomic conditions and cultural and rural festivals and traditions. They tried to correlate their cultures and traditions with the conservation and sustainable utilisation of minor millets and other biological resources. They use integrated approaches for the conservation and sustainable utilisation of these millets in these hot spots (Anil Kumar et al. 2010).

12.7 Participatory Approaches for In Situ Preservation

The modern plant-breeding and crop improvement programmes are based on the landraces and wild relative of crop species. The landraces are being cultivated by the farming communities as local varieties over the years. These local varieties are also termed as farmers' varieties and are developed over years without the involvement of researchers and plant breeders. The number of local varieties has been developed by the farming communities for different agro-ecological conditions depending upon the needs of the farmers. These landraces are the important components of farmers' livelihood and providing food security. These are providing the basic breeding material to a number of plant breeders. Moreover, landraces are the rich source of various diseases and insect pests and other abiotic stresses. These are the staple food of poor farmers in most of the developing and underdeveloped countries. For sustainable development of PGR and TK, these landraces have been identified as the main component of biodiversity. Therefore, the CBD has emphasised to on-farm conservation of these valuable resources through farmers' participatory approaches. The conservation of landraces through farmers' participatory approaches encourages the farmers to multiply and maintain these resources as such on their farm without much disturbing the agrodiversity (Jarvis et al. 1998). By doing this, the gene flow will occur between the wild relatives and the cultivated crop species. Moreover, both natural and artificial selection will practise for the selection of suitable genotypes.

The distribution of seed among the farmers is very fast, and sometimes farmers obtain the seeds even before the release of varieties. Farmers' indigenous method of seed supply is very reliable among the farmers and diffusion of seed is quick. In olden days, seeds of new varieties were exchanged between the farming communities through rural cultural events, marriages in different communities and employment of youth outside their locality. These traditional methods of seed exchange enhance the genetic diversity and gene flow from one village to another village and from one geographical area to other areas.

The conservation of biodiversity is also increased by seed exchange between the government institutions and local farmers. For example, public institutions most of the time do survey for the collection of local landraces and wild relatives on farmers' fields, and in exchange they provide the seeds of newly released varieties. This also enhances the gene flow from landraces to the newly released or exotic varieties. These landraces help in stabilising the yield potential of crop varieties because these landraces provide the sources of diseases and insect-pest resistance. The use of landraces also gives sustainable and stable crop production even under adverse climatic conditions.

12.7.1 Strengthening On-Farm Conservation

12.7.1.1 On-Farm Conservation through Participatory Variety Selection (PVS) Approach

In several countries, farmers' participatory approaches were launched to promote the adoption of modern varieties for marginal farmers. In these programmes, plant-breeding programmes are designed by identifying the needs of the farmers. A plant breeder designs programmes or develops varieties on the desire of the farmers, i.e. the characters they like. In the approach, farmers visit the experimental fields and select the desired genotypes or varieties depending upon their choice and need. The farmers select the varieties for their food and livelihood by using their own experiences. The participatory approach is used in this method for the selection of breeding lines by the farmers by their own experiences in an environment. Participatory varietal selection is a selection by farmers on their own fields of advanced breeding lines or varieties likely to be released. These may be the advanced breeding lines which are well-characterised material such as advanced nonsegregating lines in inbreeding crops or advanced populations in outbreeding crops. In the participatory variety selection approaches, the farmers are offered a number of crop varieties to select as per their needs.

The farmers are the real user of PGR, and PVS approach helps the farmers to select the varieties of their own choice and need. Before the inception of new technologies, farmers are not exposed to these technologies. They use traditional crop varieties for food and their livelihood (Horne and Stur 2003). The involvement of farmers and stakeholders in the process of selection of varieties enhances the likelihood. The farmers will adopt the technologies which would have greater impact in the farming communities. The participatory rice varietal improvement process has several stages that involve farmers and the community, such as the setting of breeding goals, evaluation of new rice lines, wide diffusion of seeds and assessment of benefits of PVS. There are many ways to ensure that socioeconomic and cultural aspects are taken into consideration in the conduct of PVS.

12.7.1.2 Participatory Plant Breeding (PPB)

In most of the cases, Participatory Plant-Breeding (PPB) is used when participatory varietal selection fails. The farmers select the suitable varieties for diverse agro-systems to secure the food security. In PPB approach, the researchers and the farmers select new varieties from the breeding material for certain targeted environments. The plant breeders cross the newly developed crop varieties with the local landrace varieties, and the farmers do maximum selection in the material at early stage under different environmental conditions (Witcombe et al. 1996). In this approach, the plant breeders use the landraces in breeding programmes to increase the genetic base of the breeding material for the release of variety. The farmers and the researchers/plant breeders jointly select the best breeding lines for cultivation suitable to a particular environment.

Nowadays, farming communities are responsible for conservation of PGR. The livelihood of the farmers is improved by the capacity building of farmers to select and conserve the diverse breeding material and other crop genetic resources. The strengthening of conservation and sustainable utilisation of crop genetic resources can be done through organisation of farmers' fairs, demonstrations and field-level trials (Fig. 12.1). There is a need of more collaboration between the public sector institutions and the private sectors for enhancing the genetic diversity among the crop genetic resources in the communities. Moreover, farmers have their own criteria for selection of new crop variety. The participation of farmers in PPB approach provides the farmers to access the early segregating material which has empowered the farmers to select the crop genetic resources depending upon their need

Fig. 12.1 Participatory Plant-Breeding (PPB) in rice field

in certain conditions. In some cases, farmers sell the seeds of newly developed crop varieties at higher rate compared to their local and landrace varieties (Sthapit et al. 1996). The farmers who have selected the material from early breeding material have full control on it under PPB method.

The integration of participatory approaches into crop genetic resources enables the farmers to access to more genetic resources suitable to them. The farmers are having diverse genetic material to secure the food security. They can use diverse genetic material under certain socioeconomic conditions and environmental variations. The conservation of diverse genetic material and their sustainable utilisation further provide farmers a stable and secured food security for their livelihood. This will also support the researchers, breeders and policymakers in the conservation of PGR and development of suitable crop varieties. More is the access of farmers to crop genetic resources; more is the genetic diversity in the breeding material. This will help in the release of suitable crop varieties under adverse climatic conditions. In situ conservation of PGR for the development of crop varieties provides the job opportunities to the indigenous people in rural areas. On-farm conservation of plant genetic diversity will help the farming communities to meet out their requirements for food in the near future.

12.8 Conclusion

In order to meet the demand for food of the developing world, the researchers and the farmers should work together for conservation and sustainable utilisation of PGR and TK. For effective conservation and sustainable use of PGR and TK, farmers' involvement is necessary to secure the food security. Farmers' participation in developmental programmes will help to bring the change in existing approaches of conservation of PGR and their utilisation. The participatory approaches give the opportunity to the farmers to select the genetic material depending upon their need. The farmers can use their experience for selecting the suitable genotypes. The participatory approaches provide the chance to the farmers to explore and select the suitable genetic material by using their own experiences for diverse agroclimatic conditions. The women have more experience in selecting and maintaining the suitable plant species for household uses. They preserve all the plant species having medicinal and food values along with their wild relatives. This will enhance the genetic diversity among the genotypes to maintain the biodiversity and their sustainable utilisation. The participatory approaches bring both plant breeders and the farmers to select the suitable varieties through the PVS and PPB methods. They share their experiences while selecting the varieties of crops for different environmental conditions. Through participatory approaches, any new agricultural techniques can be diffused. In order to meet the food security, maintenance of PGR and TK diversity is required. Both the public institutions and the farmers are conserving the PGR and TK to enhance the genetic diversity for sustainable utilisation of these resources for food security. These participatory approaches have been developed by the scientists and the policymakers to enhance the participation of farmers for PGR and indigenous/traditional knowledge conservation and their sustainable use.

References

Andersson HC (1992) Gender planning methodology: three papers on incorporating the gender approach in development cooperation programmes. Rapporteur Och Notiser 109. Institution for Kuffurgeografi och ekonomik geografi unid lund Universitet

Anil Kumar N, Arivudai Nambi V, Mampoothiri KUK, Geetha Rani M, King EDIO (2010) MSSRF – biodiversity programme: hindsight and forethought. Chennai, India

Bhag Mal B, Padulosi S, Bala Ravi S (2010) Minor millets in South Asia: learnings from IFAD-NUS project in India and Nepal. Bioversity International, Rome

Boef W, de J, Hardon, Louwaars NP (1997) Integrated organisation of institutional crop development as a system to maintain and stimulate the utilisation of agro-biodiversity at the farm level. Paper presented at the International meeting Managing Plant Genetic Resources in the African Savannah, Bamako Mali

CIAT (1997) New frontiers in participatory research and gender analysis. CGIAR systemwide program on participatory research and gender analysis for Technology Development and Institutional Innovation. CIAT, Cali

Clawson DL (1985) Harvest security and intraspecific diversity in traditional tropical agriculture. Econ Bot 39:56–67

Eyzaguirre P, Iwanaga M (1996) Farmers' contribution to maintaining genetic diversity in crops, and its role within the total genetic resources system. In: Eyzaguirre P, Iwanaga M (eds) Participatory plant breeding. IPGRI, Rome, pp 9–18

Eyzaguirre PB, Raymond R (1995) Rural women: a key to the conservation and sustainable use of agricultural biodiversity. Fourth World Conference on Women-Focus Day on Rural Women. FAO, Beijing

FAO (2012) Food and Agriculture Organization. The State of Food and Agriculture. ftp://ftp.fao.org/docrep/fao/009/a0800e/a0800e.pdf

Horne PM, Stur WW (2003) Developing agricultural solutions with smallholder farmers-how to get started with participatory approaches. Australian Centre for International Agricultural Research (ACIAR) and Centro Internacional de Agricultura Tropical (CIAT). ACIAR Monograph 99, p 119

IDRC (1998) Guidelines for integrating gender analysis into biodiversity research. Sustainable use of biodiversity program initiative. IDRC, Ottawa

Iriarte Lucio, Litza Lazarte, Javier Franco y David Fernández (1999) El Rol del Genero en la Conservación, Localización, y Manejo de la Diversidad Genética de Papa, Tarwi y Maiz. Gender and Genetic Resources Management. IPGRI, FAO, BIOSOMA, Rome

Jarvis D, Hodgkin T, Eyzaguirre P, Ayad G, Sthapit B.R, Guarino L (1998) Farmer selection, natural selection and crop genetic diversity: the need for a basic dataset. In: Jarvis D, Hodgkin T (eds) Strengthening the scientific basis of in situ conservation of agricultural biodiversity on-farm. Options for data collecting and analysis. Proceedings of a workshop to develop tools and procedures for in situ conservation on-farm, Rome, Italy, pp 1–8

King E, Nambi VA, Nagarajan L (2009) Integrated approaches in small millets conservation: a case from Kolli Hills, India. In: Jaenike H, Ganry J, Hoeschle-Zeledon I, Kahane R (eds) Proceeding of the International Symposium on underutilized plant species for food, nutrition, income and sustainable development. Acta Horticulturae (ISHS), Arusha, pp 79–84

Long N, Long A (1992) Battlefields of knowledge: the interlocking of theory and practice in social research and development. Routledge, London

McGuire S, Manicad G, Sperling L (1999) Technical and institutional issues in participatory plant breeding – do from a perspective of farmer plant breeding. A global analysis of issues and current experience. CGIAR Systemwide Program on Participatory Research and Gender Analysis for Technology Development and Institutional Innovation. CIAT, Cali

Naming M, Yu RA, Tu CL (2010) Traditional knowledge conservation and transmission of agrobiodiversity: sharing of experiences by the Penan community in Mulu Sarawak. In: Mirfat AHS, Salma I, Mohd Rani MY, Mohd Norowi H, Mohd Shukri MA, Erny Sabrina MN, Noor Sarinah MN, Siti Noor Aishikin AH, Nor Asiah I (eds) Second National Conference on Agrobiodiversity Conservation and Sustainable Utilization, Agrobiodiversity for Sustainable Economic Development. Malaysian Agricultural Research and Development Institute (MARDI), Malaysia, p 54

Prain G, Hagmann J (2000) Farmers' management of diversity in local systems. In: Almekinders CJM, de Boef WS (eds) Encouraging diversity. The synthesis between crop conservation and development. Intermediate Technology Publications, London, pp 94–100

Quek P, Zhang Z (1997) Documenting indigenous knowledge: the need for an IK journal. IPGRI-APO Newsletter 23

Quek P, Manurung R, Naming M, Kimonyi JM, Dai L, Abraham M, Eyzaguirre P, Morimoto Y, Maundu P, Zhang Z (2009) The traditional knowledge journal (TKJ) methodology empowers communities to document their traditional knowledge. Knowledge share fair for agricultural development and food security. Bioversity International, Rome

Shrestha PK (1998) Gender and generation: role of traditional seed supply systems in the maintenance of agrobiodiversity in Nepal. In: Partap T, Sthapit B (eds) Managing agrobiodiversity: farmers' changing perspectives and institutional responses in the Hindu Kush-Himalayan region. ICIMOD, IPGRI, Kathmandu, pp 143–152

Sperling L, Loevinsohn M (1996) Using diversity: enhancing and maintaining genetic resources on-farm. Proceedings of a workshop held on 19–21 June 1995, New Delhi, India. IDRC, New Delhi

Sthapit BR, Joshi KD, Witcombe JR (1996) Farmer participatory cultivar improvement: a case of high altitude rice from Nepal. Exp Agri 32:479–496

Sthapit BR, Shrestha PK, Upadhyay MP (2006) Good practices: on-farm management of agricultural biodiversity in Nepal. NARC, LI-BIRD, IPGRI and IDRC, Kathmandu

Sthapit BR, Subedi A, Shrestha P, Shrestha PK, Upadhyay MP (2008) Practices supporting community management of farmers' varieties. In: Thijssen MH, Bishaw Z, Beshir A, de Boef WS (eds) Farmer's varieties and seeds. Supporting informal seed supply in Ethiopia. Wageningen International, Wageningen, pp 166–171

Swaminathan MS (2000) Government-Industry-Civil Society: partnerships in integrated gene management. Volvo Environ Ambio 29(2):115–121

Tapia Mario E, De la Torre A (1998) Women farmers and Andean seeds. Gender and genetic resources management. IPGRI/FAO, Rome

Teshome A (1996) Factors maintaining Sorghum Landrace Diversity in north Shewa and south Welo regions of Ethiopia. PhD thesis. Ottawa-Carleton Institute of Biology, Biology Department, Ottawa

Teshome A, Baum B, Fahrig L, Torrance JK, Arnason JT, Lambert JD (1997) Sorghum landrace variation and classification in north Shewa and south Ethiopia. Euphytica 97:255–263

Teshome A, Arnason JT, Torrance JK, Lambert JD, Fahrig L, Baum B (1999a) Traditional farmers' knowledge of sorghum landrace storability in Ethiopia. Econ Bot 53(1):69–78

Teshome A, Fahrig L, Torrance JK, Arnason TJ, Lambert JD, Baum B (1999b) Maintenance of sorghum landrace diversity by farmers' selection in Ethiopia. Econ Bot 53(1):79–88

Uphoff NT, Cohen JM, Goldsmith AA (1979) Feasibility and application of rural development participation: a state-of-the-art-paper. Cornell University, New York

UPWARD (1996) Into action research, partnerships in Asian rootcrop research and development. UPWARDS, Manila

van Veldhuizen L, Waters-Bayer A, Ramirez R, Johnson DA, Thompson J (1997) Farmers' research in practice, ILEIA Readings in Sustainable Agriculture. Intermediate Technology Publications, London

Wilde V, Vainio-Mattila A (1995) Gender analysis and forestry: international training package. FAO, Rome

Witcombe JR, Joshi A, Joshi KD, Sthapit BR (1996) Farmer participatory cultivar improvement. I: varietal selection and breeding methods and their impact on biodiversity. Exp Agri 32:445–460

Bioprospecting: Enhancing the Value of Biodiversity and Intellectual Property

13

Sumer Pal Singh and Anju M. Singh

Abstract

Biodiversity is the fundamental resource for bioprospecting. The three phases of bioprospecting include collection, analysis, and commercialization. Bioprospecting can also include the collection of traditional knowledge relating to the use of genetic resources from local communities. Sharing the benefits of commercialization with the country of genetic resource has to be ensured in bioprospecting. Also where indigenous knowledge is used in development and commercialization of a product, benefit sharing with local communities is imperative. When the resource or traditional knowledge is used for commercialization and benefits not shared with the concerned parties, it leads to biopiracy. Convention on Biological Diversity (CBD), creation of Traditional Knowledge Digital Library (TKDL), intellectual property rights, and bioprospecting policy are some of the ways to ensure that bioprospecting is done in a sustainable and ethical manner and results in fair benefits for the country and local people from whom the genetic resources are prospected. With increasing awareness and formulation of policies by various nations, bioprospecting has become more difficult for private companies, and the bioprospecting research has been scaled down partly because the exchange of genetic materials has become highly regulated and benefit sharing quite complex. Considerable bioprospecting research has shifted to academic and research institutions. It should be ensured by policymakers in all nations that regulation of exchange of genetic resources and traditional knowledge as well as benefit sharing should be such that they should not come in the way of welfare of mankind as a whole.

13.1 Introduction

Biodiversity prospecting, or bioprospecting, is the exploration of biodiversity and biochemicals with the intention of developing and commercializing products. However, not all investigations on biodiversity are considered bioprospecting, e.g., academic research, although these may have

S.P. Singh, Ph.D. (✉) • A.M. Singh, Ph.D.
Division of Genetics, ICAR-Indian Agricultural Research Institute, Pusa, 110012 New Delhi, India
e-mail: sumerpalsingh@yahoo.com; anju_gen@iari.res.in

R.K. Salgotra, B.B. Gupta (eds.), *Plant Genetic Resources and Traditional Knowledge for Food Security*, DOI 10.1007/978-981-10-0060-7_13

commercial applications in the future. Similarly, it does not include the trade in existing ornamental plants, commercial agriculture, or even the local collection and sale of non-timber forest for domestic or subsistence purposes.

Systematic search for the development of new sources of chemical compounds, botanicals, genes, etc. from nature which can be utilized and commercialized by pharmaceutical, agricultural and industrial sectors, etc. is known as bioprospecting (Reid et al. 1993). This includes development of drugs, botanicals, fragrances, cosmetics, and personal care products. Most of the biodiversity is found in the countries of the southern hemisphere, and most commercially exploitable product development takes place in the developed world. For several hundred years, unregulated collection and exploitation of natural genetic resources from the biodiversity-rich countries has taken place, and benefits have been harvested by the developed world companies without recognizing the share of those nations/communities that have been responsible for maintaining that diversity and having prior knowledge of useful properties of those resources (Bhattacharya 2014). Since the 1990s under the auspices of United Nations, the need for setting the terms and conditions for accessing the biodiversity and benefit sharing has been debated, and several regulations and guidelines have come in place. The biodiversity-rich countries have also been sensitized to the need for maintaining and using their biodiversity in a sustainable manner. The present chapter discusses in detail the concept of bioprospecting, Convention on Biological Diversity, and the efforts by Indian Government for the preservation of their biodiversity and traditional knowledge.

13.2 Bioprospecting

Identifying and commercializing biodiversity are bioprospecting. However, not all investigations on biodiversity are considered bioprospecting, e.g., academic research, although these may have commercial applications in the future. Similarly, it does not include the trade in existing ornamental plants, commercial agriculture, or even the local collection and sale of non-timber forest products for domestic or subsistence purposes.

The World Conservation Monitoring Centre recognized 17 mega diverse countries in July 2000 including Australia, Brazil, China, Colombia, Democratic Republic of the Congo (DRC) (formerly Zaire), Ecuador, India, Indonesia, Madagascar, Malaysia, Mexico, Papua New Guinea, Peru, the Philippines, South Africa, the United States of America (USA), and Venezuela. Together, these 17 countries harbor more than 70 % of the earth's species. South Africa is recognized as the third richest center of biodiversity in the world, having well over 20,000 indigenous plant species, apart from animal, marine, and microbial diversity. South African scientists first began looking at the properties of local plants due to farm livestock losses caused by grazing on toxic plants and at indigenous plants eaten by rural communities. This well-developed research capacity combined with the rich biodiversity and traditional knowledge makes South Africa a prime location for bioprospecting. A 1996 review of bioprospecting activities in South Africa showed almost all research institutes in South Africa to be involved in bioprospecting in one form or another. Similarly, India has 81,000 species of fauna and 47,000 of flora, including 15,000 plant varieties unique to the country (UNDP 1999). However, bioprospecting has been at its lowest in our country.

Bioprospecting cuts across a range of different sectors, including the pharmaceutical, agribusiness (biotechnology, seed, crop protection, and horticulture), cosmetic and personal care, fragrance, botanicals, and the food and beverage industries. Bioprospecting can be divided into three phases: collecting, analysis, and commercialization. Samples and indigenous knowledge related to a sample are collected and then undergo analysis – using a variety of different technologies. In the early days, analysis was very time-consuming and costly, but in the late 1980s, techniques were significantly improved, making the screening of natural product molecules simpler, faster, and easier – a factor which affects both the impact and level of bioprospecting. The most difficult phase is the commercialization phase. It can take anywhere between 15 and 20 years for a biomolecule to be tested in trials, taking approval for human consumption or application from statutory bodies for the purpose, only then the product is ready for commercialization. This is not only a time-consuming phase but also can be expensive, e.g., in the USA, upward of US$ 800 million indirect and direct charges are to be spent to bring a drug to the US market (www.pub.ac.za; readers are encouraged to visit this site which has informative videos in addition to documentations on bioprospecting).

13.3 Traditional Knowledge

Local knowledge that is unique to a culture or society is termed as indigenous (traditional) knowledge. It is passed down the generations, usually by word of mouth and cultural rituals or may be documented as in Ayurvedic writings in India such as the Sushruta Samhita. Traditional knowledge is of paramount importance for rural communities and is a deciding factor for their food, health, and agriculture. Local communities have very deep association with biological and natural resources and have a sound knowledge of use and functioning of these resources. According to ASI, there are 4635 ethnic communities in India. In principle each of these communities could be having their own oral medical traditions that have been evolving across time and space. Over 7500 species of plants and several hundred animal species and also metals and minerals are utilized by the folk tradition in India. Similarly, a large proportion (80 %) of the world population especially from undeveloped/developing regions is still dependent on indigenous medicinal knowledge of local plants for their medical needs (Shankar 1997). In fact, most old civilizations of the world such as India and China are rich in traditional knowledge (TK). Frequently used TK in India include knowledge that neem extract is useful as pesticide, turmeric is anticancer, and tulasi leaves/extract can be used to overcome fever, cold, cough, etc., and in China, various balms and rubbing creams having medicinal components are extracted from herbs, various types of teas etc. People are realizing now that traditional knowledge is not old and obsolete, but it is highly evolutionary, adaptive, creative, and even novel.

13.4 Biopiracy and Patents

No country in this world is self-sufficient in terms of biodiversity. Even most biodiversity enriched countries have to go to other parts of the world in order to fulfill their bio-resources need (The Crucible Group 1994). In the past, many staple food crops and cash crops were introduced from other countries, and these crops are playing key role in the economy of recipient countries. Biodiversity-rich country like Brazil is also dependent to take two thirds of calories based on plant species that are found in other countries. Distribution of biodiversity over the world is not in proportion to scientific and technological wealth (Laird and Kerry 2002). Many countries have the advanced technologies for bioprospecting research and development of novel products from this research. They, however, may not be rich in the genetic or natural resources which have to be explored for bioprospecting and are therefore dependent on the source country of biodiversity in that case. Many times, the local knowledge, e.g., of the medicinal properties of a genetic resource exists. Big firms and corporations mostly from the developed countries have been utilizing freely the genetic resources, traditional knowledge, and technologies of developing countries without benefit sharing with the country or community of origin. This unauthorized commercial use of plant genetic resources and traditional knowledge without benefit sharing is biopiracy. For instance, various medicinal plants were discovered by indigenous people and local communities. These plants form the basis of many pharmaceutical industries, and economic benefits (by patenting of these processes/products) earned are largely in the hands of big pharmaceutical companies without giving due recognition and incentives to those who created knowledge about the use of these resources (Kothari and Patel 2006).

As per the WTO/TRIPS agreement, every country has to make available patents for new inventions. A product may be considered patentable when it is slightly modified, discovered, or extracted from nature and, in some jurisdictions, simply because it is described in chemical terms for the first time (Dutfield 1999). In some cases, whole gene sequences or several varieties can be covered by a single patent (Khor 2002). These provisions allowed the big companies to patent the products formed by processing of the genetic resources and earn huge benefits. The Maya ICBG controversy, the rosy periwinkle, the neem tree, and the basmati rice are some famous examples of biopiracy. These are also some successful stories where patents were revoked, but there are many stories of biopiracy which have remained unnoticed/uncontested. India's agriculture being rich in biodiversity has always been an easy prey for big corporations engaging in agribusiness for the purpose of biopiracy (Ministry of Environment and Forests 2010). Different private firms tried to spread genetically modified brinjals in India in the form of Bt brinjals in spite of the fact that India itself is a source of over 2500 different unique varieties of brinjals (Samuels 2012).

13.5 CBD, Bioprospecting Policy, and Access and Benefit Sharing (ABS)

The most significant change and the starting point of international regulation and legislation took place at the Rio Earth Summit in 1992, when participating countries discussed bioprospecting and signed the Convention of Biological Diversity (CBD). The CBD recognizes the need for conservation, sustainable use, and equitable benefit sharing as cornerstones and, for the first time, acknowledged the sovereign rights of countries over their natural resources. It meant that biodiversity could no longer be regarded as common property and that the origin of the material used was recognized. The CBD recognizes the importance of the knowledge, practices, and innovations of indigenous and local communities (Article 8 (j)) and makes provision for prior informed consent to be obtained by any public or private enterprise seeking access to biodiversity resources (Article 15). The CBD is supported by the International Treaty on Plant Genetic Resources for Food and Agriculture (ITPGRFA), which came into existence in 2001. To ensure that bioprospecting is done sustainably, the National Environmental Management: Biodiversity Act, Act 10 of 2004, was promulgated. Subsequently in 2008, bioprospecting and access and benefit sharing regulations were also put in place. One hundred thirty-three countries are party to the CBD, including all Himalayan countries and the European Union with the notable exception of the USA. One hundred sixteen countries are party to the ITPGRFA. If carried out properly, these guidelines can provide economic incentives to the people or community conserving biodiversity, and it will help in local and regional development (Lad and Naresh 2012). Bioprospecting is one part of a package of economic activities that, if implemented in the right perspective, use biodiversity in a sustainable management of natural resources and can help in economic development of a nation. The source country therefore needs to decide on the extent of use of its biodiversity. However, CBD though a legal document cannot be forced on a nation. CBD is based on mutual compliance among the parties. Every country has its right on biological resources and can negotiate for favorable benefit sharing with interested stakeholders (Dutfield 1999). National bioprospecting policy has been developed by several countries to address opportunity and drawbacks of bioprospecting. Issues like intellectual property rights, tenure of land and natural resources, research and development, and conservation and protection of diversity are included in bioprospecting policy.

So far, only about 25 countries have introduced new regulations to facilitate access, and vast majority of countries still have to formulate appropriate laws. However, issues relating to access to biological resources have not been fully resolved. Therefore, the official body of CBD called the Conference of the Parties has been developing and modifying the guidelines to seal the loopholes in the earlier conventions. A new treaty under the CBD was recently adopted in Nagoya in October 2010 called as "Nagoya protocol on Access to Genetic Resources and the Fair and Equitable Sharing of Benefits." This protocol deals more explicitly than previous documents with the use of traditional knowledge associated with genetic resources (Secretariat of the Convention on Biological Diversity 2011). The Biotechnology Industry Organization of the USA has also produced detailed guidelines for its members in engaging in bioprospecting and has designed material transfer agreements to be followed.

Probably the best-known recent examples of bioprospecting and access and benefit sharing are (1) the development by the Council for Scientific and Industrial Research (CSIR), South Africa, and the UK-based company Phytopharm of an anti-obesity drug, based on the San people's traditional knowledge of a Kalahari plant called *Hoodia* (Harvey and Gericke 2011); and (2) scientists at Trivandrum Botanical Garden in India developed two medicines from plants used by tribal people, and royalties from the sale of these drugs now benefit the hill tribes that provided the original leads (Anuradha 2013).

13.6 Traditional Knowledge Digital Library (TKDL): A Tool for Prevention of Misappropriations of Traditional Knowledge

There are several cases where traditional knowledge was used and patented without taking into consideration traditional knowledge holders (Bhattacharya 2014). The misuse of traditional knowledge made it a debatable issue in International Forum, and ways and means of conservation, protection, and benefit sharing were found out in order to avoid the misuse of traditional knowledge. In India also, decision has been taken for documentation and digitization of traditional knowledge in the form of Traditional Knowledge Digital Library (TKDL). It may prove very useful in preventing misappropriation of traditional knowledge by third parties.

TKDL is a collaborative project between the Council of Scientific and Industrial Research (CSIR) and Department of Ayurveda, Yoga, Unani, Siddha, and Homeopathy Systems (AYUSH). The purpose of this project is to document the traditional knowledge of India so that it can be prevented and protected from patenting by other countries. Documentation of our traditional knowledge will help in avoiding the incidence like patenting wound healing properties of turmeric by the United States Patent and Trademark Office (USPTO) and granting patent on antifungal properties of neem by the European Patent Office (EPO). Almost 2000 patents a year are being granted by patent office all over the world despite the fact that these patents are completely based on Indian medicinal system.

Knowledge and information of Ayurveda, Unani, Siddha, and Yoga available in public domain are covered under TKDL. Traditional knowledge existing in local languages like Sanskrit, Urdu, Arabic, Persian, and Tamil is being collected and documented. This knowledge will be made available in five international languages, namely, English, German, Spanish, French, and Japanese. Traditional knowledge on medicinal plants is being digitized using innovative classification system, i.e., Traditional Knowledge Resource Classification (TKRC). The information is being structured under section, class, subclass, group, and subgroup as per the IPC for the convenience of its use by the international patent examiners.

Software associated with Traditional Knowledge Resource Classification (TKRC) converts text available in local language into multiple languages. This software does knowledge-based conversion and not only translation. Traditional terminology like *jwar* will be converted into fever by the software. Therefore, TKDL will act as bridge between information available in local languages and patent examiner at global level (www.tkdl.res.in).

13.7 Conclusion

The CBD has aroused awareness about the use of biodiversity in a sustainable manner. Simultaneously, it has also awakened the biodiversity-rich countries to the need of protecting the indiscriminate and unlawful use of their genetic resources and traditional knowledge. As a result, the legal framework under the auspices of the United Nations is being increasingly implemented by these countries. At the same time, the appetite for bioprospecting by various companies especially the pharmaceutical development companies has significantly diminished since the adoption of CBD by various nations largely because of the complexities relating to access and benefit sharing. More of the research with natural products is shifting to academic institutions. It would not be surprising to see major companies proposing academic research groups to become essential partners in their innovation research and delivery.

References

Anuradha RV (2013) Sharing with the Kanis: a case study from Kerala, India. https://www.cbd.int/financial/bensharing/india-kanis.pdf

Bhattacharya S (2014) Bioprospecting, biopiracy and food security in India: the emerging sides of neoliberalism. Int Lett Soc Humanist Sci 23:49–56

Dutfield G (1999) Sharing the benefits of biodiversity: access regimes and intellectual property rights, science technology and development. Centre for International Development and Belfer Center for Science and International Affairs, Harvard University, Cambridge, MA

Harvey A, Gericke N (2011) Bioprospecting: creating a value for biodiversity. In: Pavlinov IY (ed) Research in biodiversity – models and applications. Published by InTech Janeza Trdine 9, 51000 Rijeka, Croatia. doi:10.5772/1833. ISBN 978-953-307-794-9

Khor M (2002) Intellectual property, biodiversity and sustainable development. Resolving the difficult issues. Zed Books, London

Kothari A, Patel A (2006) Environment and human rights. National Human Rights Commission, New Delhi

Lad MD, Naresh K (2012) Bio-prospecting, IPR, traditional knowledge and its trade secrets. Int Res J Pharm 3(10):72–77

Laird SA, ten Kerry K (2002) Biodiversity prospecting: the commercial use of genetic resources and best practice in benefit-sharing. In: Laird SA (ed) Biodiversity and traditional knowledge: equitable partnerships in practice. Earthscan Publications Limited, London

Ministry of Environment and Forests (2010) National consultations on Bt brinjals: a primer on concerns, issues and prospects. Centre for environment education, India

Reid WV, Laird SA, Meyer CA, Gomez R, Sittenfeld A, Janzen DH, Gollin MA, Juma C (1993) Biodiversity prospecting: using genetic resources for sustainable development. World Resources Institute/Instituto Nacional de Biodiversidad/Rainforest Alliance/African Centre for Technology Studies, Washington, DC/Santo Domingo de Heredia/New York/Nairobi

Samuels J (2012) Genetically engineered Bt brinjal and the implications for plant biodiversity – revisited. Greenpeace, India. Accessed from http://www.greenpeace.org/india/PageFiles/446445/GE-Bt-brinjal-revisited.pdf

Secretariat of the Convention on Biological Diversity (2011) Nagoya protocol on access to genetic resources and the fair and equitable sharing of benefits arising from their utilization to the convention on biological diversity. Secretariat of the Convention on Biological Diversity, Montreal

Shankar D (1997) Traditional medicine and biopiracy. Ancient Sci Life 17(1):67–71

The Crucible Group (1994) People, plants and patents: the impact of intellectual property on trade, plant biodiversity and rural society. IDRC PO Box 8500 Ottawa, Canada K1G 3H9

UNDP (1999) Biopiracy and the patenting of staple food crops. United Nations Development Programme, Human Development Report

14 Role of Women in Conserving Plant Genetic Resources and Related Traditional Knowledge for Food Security

Monika Sood, Veena Gupta, and Awsi Jan

Abstract

Women have been the custodian of plant genetic resources and its related indigenous traditional knowledge from time immemorial. This was the woman who first domesticated the plants, thereby initiating the art of agriculture and science of farming. To ensure an uninterrupted supply of healthy food to her family, her role changed dramatically from a food gatherer to plant selector, from plant domesticator to plant conservator and from food provider to a nutritionist. The role of woman in agriculture production varies from managers of small farm holding to landless labourers, thereby contributing equally to the household income. Not only this, they are the vigilant guards and custodians of our traditional culture, rituals or festivals thereby playing a pivotal role in maintaining and conserving the cultural plant diversity. To enumerate the role of woman in conserving plant genetic resources and their traditional knowledge is a herculean task, but it is the need of the hour to note her contributions in today's techno-savvy world in the scenario of emerging intellectual property regimes.

14.1 Introduction

In most of the developing nations, indigenous people have knowledge, traditional practices and skilfulness to combat agricultural crop production while maintaining the genetic diversity. The Food and Agriculture Organisation (FAO) has recognised the importance of conservation of Plant Genetic Resources (PGR) and Traditional Knowledge (TK) for livelihood and future food security. The on farm/in situ and ex situ conservation of PGR and TK has involved different programmes related to a number of issues such as farmer's rights, livelihood of indigenous people, their health and gender equity (Ramanatha Rao et al. 1999). The FAO is emphasising and provid-

M. Sood, Ph.D. (✉) • A. Jan
Division of Food Science and Technology, Sher-e-Kashmir University of Agricultural Sciences and Technology of Jammu, Chatha, Jammu, Jammu and Kashmir, India
e-mail: monikasoodpht@gmail.com; awsijan@gmail.com

V. Gupta, Ph.D.
Division of Germplasm Conservation, National Bureau of Plant Genetic Resources, New Delhi, India
e-mail: veena@nbpgr.ernet.in

R.K. Salgotra, B.B. Gupta (eds.), *Plant Genetic Resources and Traditional Knowledge for Food Security*, DOI 10.1007/978-981-10-0060-7_14

ing support by using the traditional practices for conservation of PGR and TK on sustainable basis. The projects developed by the FAO involve the preservation of traditional practices, biodiversity and other local cultures of indigenous people.

The food security of the developing and underdeveloped nations is being promoted by FAO along with maintaining the genetic diversity of important PGR for future use. FAO is also taking care of unprecedented happening like impact of climate change, rise in world food prices, disasters, etc. FAO is developing the linkage among the international organisations to conserve the genetic resources and their sustainable utilisation to feed the rising world population and ensure food security in the future (Frankel and Hawkes 1975; Holden and Williams 1984).

Across the globe, and particularly in biodiversity-rich tropical regions, the women have played many roles such as conservator of diverse plants, maintainer of genetic diversity and domesticators of PGR and their sustainable utilisation. The women are the real plant breeders and farmers who maintain the domestication of different plant species and the genetic diversity. Apart from contributing in day-to-day maintenance and conservation of the plant diversity, women have made the ultimate sacrifice to protect plants, and such instances have been the lighthouses for present-day conservationists. Sacrifice of Amrita Devi of Bishnoi community in India is noteworthy in this regard. In the year 1730, the King of Jodhpur ordered his men to cut some *Prosopis cineraria* from Khejarli village of Marwar. Residents of that village are mostly Bishnoi cult having strong regard for plants and environment. Men of that village vigorously opposed this step. On the next day, the King sent his men in the afternoon when all male members of the village were out on their work. Amrita Devi resisted the cutting of trees and sacrificed her life along with her two daughters but did not allow cutting of trees. Thus, the Chipko movement was started. The second milestone was started in different villages of Uttar Pradesh and Uttarakhand where people mostly use this forest to meet their basic needs. Women have played significant role in conserving and maintaining these PGR and TK for their livelihood. The nonviolent protests started at the Dasholi villages by the women to avoid the deforestation. This movement helped to prevent the cutting of various trees and save the region from landslides and floods. The farming women have developed their own systems and gained knowledge of their local ecosystems including the management of pests, the conservation of soils and the development of plant and animal genetic resources (FAO 1999).

In most of the Asian countries like India, women are actively involved in agriculture system. Their involvement in agriculture crop production varies from one region to region. In India, women's role in agriculture is more than 65 % of the total farm labour, and their mode of participation varies from crop production in the farm production to household conservation (Kavya et al. 2009). Although most of the agricultural crop production techniques are being done by the women, the women's contribution is not awarded anywhere in the society. Compared to men, the status of women is low in the society. Their working choices are also restricted due to social issues, rituals and gender issues (Kavya et al. 2009).

In smallholder peasant agriculture, women are generally responsible for selection, adaptation and improvement of plant genetic material. Women generally follow careers in the household in agricultural communities whose responsibilities include family nutrition, cooking, cleaning, health, water collection, animal husbandry and family food crops. It is estimated that the men and women have different knowledge and experiences of crop production (Vernooy 2003). For this reason, women are a resource in and of themselves because of the knowledge they have developed with holistic and diverse values of food cropping and animal welfare systems necessary for supporting their families. Many PGR are an emergent property of the demands of farming households, and while women generally run the household, they are plant breeders generating potentially important crop genetic resources. While agrarian societies today are affected by modernisation of agriculture, the conservation effort will likely come out from those who already have conservative values towards traditional farming (Zimmerer 1991). In situ conservation is not for every farmer and efforts to completely reverse a modern agricultural sector are set to fail; rather strategies must support in situ conservation by farmers already farming with conservation methods. Women are seen as the harbingers of these conservation values because of the encompassing values with which they grow crops and rear animals.

14.2 Importance of PGR and TK in Food Security and Sovereignty

PGR are the basic raw material and closely linked with the development of agriculture and agro-based industries. Climate change and malnutrition are two major crises which threaten the human survival in today's world. The diversity in PGR which has sustained the life on Earth till today will keep on doing so in the future provided that we can exploit the plants and their related TK in a more harmonious manner. The range of PGR is very low, and this can be enhanced by exploring more areas, collection and maintenance of diverse germplasm for future use at national and international levels. Collection and conservation of diverse PGR are the important component of biodiversity, and for the last few decades, more emphasis has been given for maintaining these resources (Frankel and Bennet 1970).

On the other hand, TK is knowledge which is experienced over the years through traditional practices by the indigenous communities. TK is the unique one which is developed by the local people and is linked with their rituals and cultures in a particular region of a country (Grenier 1998). The indigenous knowledge is being transferred from one generation to another generation. The knowledge is mostly transferred to the next generation orally, and no documentation has been made (Sen 1981). The indigenous knowledge has been developed by the indigenous people without involving any modern scientific methods. Moreover, most of this knowledge remains in the public domain without sharing of benefits. The indigenous people are using these valuable resources for their daily need and to secure their livelihood. These people have made important role in maintaining the genetic diversity among the plant species in an informal manner and to conserve these resources.

Sincere efforts have been made to improve the methods of in situ conservation of PGR and TK resources (Jarvis 1999; Sthapit and Jarvis 1999). The emphasis has also been given for the effective maintenance of wild relatives of crop species. Although women are collecting and conserving the diverse plant species, the in situ and ex situ conservation methods are complementary to it. On farm conservation basically a part of in situ conservation offers multidisciplinary approach to conserve the local land races/ diversity including wild species and their relatives. Recent past had witnessed the development of high-yielding varieties which possess very narrow and highly uniform genetic base. These varieties are posing threat to our landraces or farmer's varieties or old native diversity which still possess wider unexploited genetic base to face the emerging problems of climate changes and environmental crisis.

14.3 Multiple Roles of Women in Conservation of PGR and TK

The studies have shown that most of the plant species are collected and conserved by the women for household consumptions. These plant species are later on widely grown as landrace varieties, and some of the species have been used in breeding programmes for the development of new varieties (Posey 1999). Although women are maintaining the important germplasm, no one is acknowledging the contribution of women. The household tasks of women include the growing of different plant species in the kitchen garden, maintenance of these germplasm, processing of these genetic resources and conservation for the next season (Hamrick 1993). The women are using their experiences and traditional practices for maintenance of plant species. These conserved plant species have huge genetic diversity. The policy-makers of national and international organisations should promote the moral values of these women who are involved in the conservation and maintenance of these important germplasms. Some incentives should be provided to women for further strengthening of these resources. Till date women are not conserving and maintaining these resources only for their livelihood and other household uses. The government organisations and NGOs should provide some incentives and training to these women for systematic conservation and sustainable utilisation of these resources (ADB 2013).

14.3.1 Role of Women in Food Production

Most of the world food has been produced by the women. They are mostly producing the staple foods directly or indirectly. Women's contribution in world food production is more than 60 % in most of the developing countries (FAO 1995). The role of women in food production varies from country to country, and women share in food production is much higher in underdeveloped and developing countries (FAO 1995). In African countries most of the agriculture production tasks are carried out by the women. They are responsible for sowing of crop, transplanting, harvesting and postharvest management of all crops (Figs. 14.1 and 14.2). Besides this women are also involved in rearing cows, buffaloes, goats, pigs, rabbits and sheep and raising of poultry.

All the household jobs like cooking of food, collection of firewood and rearing of animals are carried out by the women. Women play significant role in conservation of plant species and their sustainable maintenance. In rural areas most women are using woods for fuels collected from the forests. Most of the rural women are collecting firewood to meet out daily need to sustain their livelihood and other necessary household items.

14.3.2 Women as Preservers of Biological Diversity

For food security, conservation of genetic diversity in PGR is very imperative for providing food for future generations. Women are proving the food and health care to the families by using PGR and TK in a balanced manners. They are engaged in conservation of plant species for future utilisation and to meet the need of the family in crop failure or any unprecedented incidence. Women do experiments with their PGR materials for better utilisation (Karl 1996). They cultivate different plant species of different varieties used for traditional culture, food and medicines. Women are growing seeds of diverse plant species and exchange these seeds among themselves. The exchange of seed among women enhances the genetic diversity. In some countries, women associations have been formed for production of seeds of different varieties, their conservation and distribution among them. These associations are mostly dealing with cash crops like vegetables (Alordo 2008). These practices increase the genetic diversity and preservation of PGR and TK. Unfortunately, the women's role in crop production, germplasm conservation and maintenance of genetic diversity is overlooked. Women provides the basic household needs in the form of foods, fuels and medicines from plant species and help to preserve these resources on sustainable basis for future use (ISDR 2008).

14.3.3 Women Act as the Gatherer of Plant Species

The plant species are the important and major component of the rural people, and their survivability depends on these species. Women are mainly linked with the collection of different plant species. In most of the developing countries, the livelihood of rural people is totally dependent on the use of PGR and TK. In a study, it is estimated that more than 79 % of the total plant species collected for food are done by women (Barry and Schlegel 1982).

Fig. 14.1 Traditional method of planting paddy crop in the Himalayas

Fig. 14.2 Traditional method of sowing wheat crop paddy

The wild plants are the source of important genetic material to be used for crop improvement. The wild species also provide important food material at the time of famine, scarcity of food and outbreak of diseases and insect-pests. The plant species collected by the men and women vary from their preferences and responsibilities while collecting these resources. Compared to men, women have more experience and knowledge of collection of plant species. Women's perception of collection of plant species is also different from the men (Flickenger 1997). Women's preference for collection of plant species are related to household uses such as plant- or herb-related food, fibre, medicine and fuel, whereas men mostly prefer to gather the plant related to enhance agriculture production and fodder for animals.

14.3.4 Woman's Role in Gardening

Home gardens or backyard gardens are the growing of different plant species of different crops that preserve not only the plant diversity but also the knowledge about their uses through women. Women provide the secondary source of income in the family, and they provide food, medicines for health, spices, fodder for animals and fuel. The women have more genetic diversity in plant species in the backyard compared to cultivated ones in the field. Therefore, women have generated an important repository of plant diversity of a particular crop (Soemarwoto et al. 1976).

The backyard gardens are the 'genetic backstop' which preserve the important plant species with great genetic diversity (Niriez 1987). But the backyard plant genetic diversity conserved by the women is not well recognised (Watson and Eyzaguirre 2001). In the backyard garden, maximum and continuous evolution of crop plants takes place due to in situ conservation of these important species. It has been studied that before the plantation of new plant species on a large scale in a field, the new cultivars are first tried at home garden for testing (Okigbo 1985). For example, introduction of potato worldwide for cultivation occurred through backyard cultivation practices. Similarly, the maize crop has been diffused all over the world by settling down of women in new areas (Niriez 1987). The knowledge for cultivation of plant varieties are also transmitted from one generation to another through backyard production technologies by women (Keys 1999).

Women have more skill, knowledge and responsibilities for cultivation of crop species than the men which varies from one culture to another and from one region to another. Women's cultivation practices in the home gardening are mostly taken very lightly in a disparaged manner. The work and contribution of women are supplementary to the production of food crops, but their work is never recognised and documented. The low visibility and perception of women's work lead to devaluation of their contribution towards conservation of important PGR (Lambrou and Piana 2006). The backyard cultivation of plant species by women is providing supplementary food for the society in most of the underdeveloped and developing countries. There is a need to give some rewards and recognitions to the women who have made the contribution in the conservation and maintenance of plant genetic diversity.

14.3.5 Women Act as the Herbalist

The estimate shows that more than 80 % of the population is using medicines derived from plant species (Farnsworth et al. 1985). There is a need to focus on the research on medicinal plants and folk medicines (McClain 1989). In folk medicines, shamans have great magical power, whereas the herbalists are the men who are usually treating the ill persons through herbs. However, midwives also cure the ill persons by using the herb plants. *Ghar ka vaid* (*home doctor*) or *Dadi Ma's* (*grandmother's*) *prescription* is always a welcome first aid even today in rural India. The knowledge of women about the use of plant species for curing the ill persons are not too much explored (McClain 1989; Good 1987; Kothari 2003). It is the woman who has skilfully developed the standard procedure for the methodology of usage of plants to make a drug. The knowledge of women for treating ill person is often learned by other women or by girls (Howard-Borjas and Cuijpers 2002).

The men and women have different knowledge for the use of medicinal plants for the treatment of ill persons. They don't have different knowledge, but their knowledge is also structured differently for treating the persons by using medicinal plants. In most countries the knowledge of women about the use of plant species for the treatment of ill persons are generally neglected (Kothari 2003). But the contribution made by the women with regard to conserving the biodiversity and the use of medicinal plants is remarkable.

14.3.6 Role of Women in Crop Evolution

In the introduction and domestication of plant species, women play a significant role. Women have different roles in crop evolution. These are the small indigenous farmers who traditionally acted as plant breeder of most of the crops (Sperling and Berkowitz 1994). The men and women are having their different visions and criteria for the selection of crop plants. Generally, women select the plants keeping in view the household utilities. In the selection of any crop plant, women have the selection criteria such as medicinal tonic values and food values of the plants. Women's selection criteria have been ignored by the earlier plant breeders. Moreover, women use the crop variety by different ways. For example, they use rice not only for food purposes but also straw for fodder and hatching purposes (Jiggins 1986). Women's selection criteria for crop species are more precise than the men, as women are more concerned with the nutritional value, cooking quality and postharvest management. The studies showed that the men have taken the selection criteria of women while selecting any crop variety. But women's selection criteria are mostly ignored by the researchers and policymakers because they are only concerned with the suitability of a variety in different zones (Boster 1985).

14.3.7 Women as Pioneers in Postharvest Technology

The postharvest management of a crop variety is more important, and the farmers get the real price of a crop commodity after proper postharvest management. The value of foods is not judged by its quantity. The nutritional values of the foods are more imperative. In some societies, foods are the main components of indigenous peoples to exchange among themselves to show their hospitality in festivals and cultures. The indigenous knowledge regarding cooking of food items is being transmitted from generation to generation mostly by mother to her daughter.

The postharvest management of food is more complex and dynamic and women have specific knowledge for this. Traditional skills of women for postharvest management of food have been correlated with scientific innovations. From time immemorial, women have engaged themselves in postharvest aspects like preparation of dehydrated products, preserves, prickles, jams, juices and squashes from fruit and vegetable genetic resources. Other important technologies like grading, packaging, marketing and storage of products are also performed by women with little credit. Women are using traditional techniques for the fermentation of plants to increase the nutritional values and reduce the spoilage. After harvesting the field crops, women have used different techniques to control the incidence of insect-pests and diseases (Norton 1985; Madge 1994).

It is imperative to understand the role of men and women while framing the methods and procedure for effective diffusion of knowledge to the farming communities. The differentiation in skills and knowledge of men and women is significantly effecting the conservation of diverse PGR and TK for future food security. This differentiation affects prototypes of control, participation and sustainable utilisation of these resources.

The traditional ways of storage and preservation of food were associated with women. The issues of food processing such as winnowing, seed selecting, threshing, shelling, pounding, drying and cooking (preserving) were more of a woman's job. Men were more concerned with construction work. Women have played a significant role for future food security, even though they are challenged by their physical and decision-making powers within patriarchal societies (Matsa and Manuku 2013). Through women's TK of food storage and preservation, they have contributed enormously to food security in many nations (World Bank 2003).

14.4 Gender Bias in Rights to PGR and TK

Most of the ethnobotanists ignore the contributions made by the women. The gender biasness is there for the protection of rights to PGR and TK. The researchers have the opinion and assume that the men are the main representatives of society. Moreover, most of the policies are made considering that men are the main representatives of the societies. Women's role which is the main operational of the traditional practices in the developing countries is ignored (Duke and Vasquez 1994).

The debates are being held at the international forum for Intellectual Property (IP) protection of PGR and TK. The emphasis is given to the fact that local farmers' rights should be protected for the PGR and TK which they have developed. They should be given due share of benefits for these resources. The indigenous peoples' concepts and knowledge are mostly ignored by the IP protection international organisations. Moreover, little is known about the rights of indigenous people, and research priorities should be given in this direction (Cleveland and Murray 1997; Mgbeoji 2002). Policies should be framed to provide compensation and rights to local people particularly the women (Howard 2003).

14.5 Challenges for Women to Achieve Food Security

Women have played significant role to achieve present food security. But the roles of women are ever ignored. They are facing a number of challenges in the society due to gender biasness. Their representation is very less in the developing countries compared to men. There are certain challenges which are affecting the roles of women to ensure future food security.

14.5.1 Ignorance of Women's Contributions in Conservation of PGR

The role of women has been ignored in most of the developmental programmes. Women are actively involved in conservation of important PGR, but their roles have never been documented anywhere. Women's work has never been recognised, and their experience and knowledge remain unpublished. Women are involved in their daily household activities like maintenance of home garden diversity and conservation of PGR and TK, but their activities have never been counted and acknowledged. The female gender is not involved in participating in any social as well as developmental activities. Men are actively involved and considered in all the developmental activities, but their role in traditional practices is very less.

The gender blindness is there for the role of men and women in the societies. This is due to the lack of awareness and less defined roles of women and men in agricultural production techniques. The policymakers, researchers, planners and extensionists develop various programmes with lack of knowledge of the role of women in the developmental processes. These programmes are mainly executed by rural women in the society, and men are only participating in public meeting and other activities. The women are the real workers in the field for the production of agricultural crops and conservation of PGR.

The policymakers, researchers and planners should focus on the role of women in conserving important PGR and TK for food security. Data on the role of women in traditional practices should be collected and documented. The compensation should be provided to farm women for conservation and protection of germplasm. This will help to further increase genetic diversity of these genetic resources for food security.

14.5.2 Lack of Developmental Policies for Women

Although planners and the policymakers are aware about the role of women in farming communities for the production of agricultural crops, they have never documented and mentioned it in any developmental programmes. The policymakers know the crucial roles of women in agriculture to ensure food security. There is no specific policy developed for specifying the exact role of women in agricultural production technologies. Rarely, the roles of women have been addressed, and where these have been addressed, these are not in practice. Therefore, the farm women remain debarred of any recognition and rewards. Women and men have different roles in crop production, but the recognitions have been given to men job not to women. Generally women tasks have been ignored and neglected in most of the agricultural developmental programmes (Plucknett et al. 1987).

The government is giving due consideration for framing the policies for production of export quality crops. Moreover, the government is earning huge foreign exchange, but no attention has been given to food crops for household food security which are being cultivated by women. The markets for the sale of local household products are overlooked by the government.

The main cause of recognition of women's roles in agricultural crop production is due to the lack of collection of data from farming women communities. Women's role and knowledge in food security have never been acknowledged. Moreover, the representation at national and international levels is very less. Very few women representatives are participating in decision and policymaking programmes. Women are very less in number in the management and decision-making positions (Karl 1996).

14.5.3 Rights to Use the Resources for Crop Production

For the production of agricultural crops, the use of various inputs is needed by both men and women. Generally, women have less access to land use, credit from bank, agricultural inputs, production technology, training and other extension activities. The productivity of crop variety depends upon the availability of these resources for both men and women. Compared to men, women are having less loans from banks, agricultural inputs, training and other facilities related to crop production. Moreover, the participation of women in developmental and extension activities is very less. This is due to gender biasness, discriminations with women, cultural attitudes and weak rules and regulations related to the welfare of women. Women are not involved in any decision-making policies and framing the new laws. In developing countries, women are only involved in household activities such as cultivation of crop in kitchen garden, postharvest management practices, cooking and rearing children. The females are not allowed to be part in any social and modern developmental activities. In a study it has appeared that if women are allowed to access all the resources to be used in agricultural production, there is a significant increase in productivity and development of the society (Saito 1994).

14.5.4 Ownership Rights of Land

New agricultural production technologies are not being used effectively due to smallholding of farmers' land in the developing countries. Land fragmentation is taking place due to nuclear families in many regions of the world. This has created pressure on land degradation, use of land for other purposes, environmental degradation and erosion of genetic resources. The smallholdings of farmers' land also discourage the farmers for taking credits from banks and use of other resources. Moreover, farmers have less access to improved technologies related for crop production and other services. As the farmers have less access to agricultural input, they are less engaged in the developmental activities related to sustainable agriculture.

Mostly men have the rights on the land and women are debarred of these rights. All the credits and beneficiary schemes of government are on the name of men only. Women do not have any rights over land and membership of government-funded schemes. Therefore, most of the input schemes go into the hands of men who are not real users of these inputs. More than 60 % of agricultural activities are carried out by the women in developing countries, but they are debarred from these benefits.

Women access to land is also restricted due to social and cultural reasons. Although some laws have given the rights to women for land, there are certain customary practices that restrict the women to access all the beneficiaries' inputs. Moreover, the legislations have given the ownership title of land to the male head of the family which also restrict the females to have rights on land to access the benefits.

14.5.5 Participation in Rural Developmental Programmes

Most of the developmental organisations have membership drive before implementing any scheme. Women are having very less access to the membership drives. There are certain schemes exclusively for women but their social and cultural reasons restrict them to get the benefits. Moreover, they are not aware of most of the developmental programmes due to less participation. Women should be aware about all the schemes and rural developmental programmes so that they can get equal opportunity to access the benefits.

Special membership drive for woman should be launched through local extension workers so that more women can be added in developmental programmes. Women should be involved in policy and planning programmes, and their views and experiences should be included. If women are involved in the planning and policymaking decisions, there implementation is easy and effective. The land reform should be made to give titles to women also in the society for sustainable development (FAO 1994).

14.5.6 Less Access to Agricultural Production Technology

To increase the productivity of crop varieties, access to modern agricultural techniques is important. To access these modern techniques, it is important that the farmers should have right on land. In most cases, the owner of the land has distributed the land for cultivation to landless farmers, and the owner is not interested in membership drives or they do not have time for this. The landless farmers cultivate the land without accessing any inputs. They grow the crops by using maximum natural resources without adding any modern technologies, and they are entitled to get credits for input. Women are mostly engaged in household activities such as cooking and postharvest management practices and lose their rights on land and modern production techniques. Women are involved in laborious tasks related to food and backyard activities. There is a need that appropriate measure should be taken for proper distribution of tasks to men and women to different developmental works related to agriculture to ensure food security.

14.5.7 Lack of Proper Trainings to Women

The lack of women access to membership limits their access to a number of rural developmental programmes. Most of the organisations and rural developmental agencies are providing trainings to persons having their memberships. Moreover, these developmental organisations are providing memberships to only candidates who have their rights on land. In most of the developing countries, the legislations are framed in such a way that the rights on land are given to male head of the family in a society. Women do not have rights on land, and so they are excluded from all the benefits given by the organisations. In agriculture production system, women play an important role, but the trainings provided to men are not translated to women. Another factor that limits the access of women to trainings is the lack of female trainers because some of the societies do not allow the women to work with male persons. Moreover, male trainer does not understand the problems of farming women. The female trainers are only providing the trainings to farm women related to household activities only. They are least concerned with the training required for conservation and utilisation of important PGR and TK related to food security.

14.5.8 Less Involvement of Women in Marketing

The inception of globalisation leads the liberalisation of trade throughout the world. For globalisation of trade, the membership is required. In most of developing countries, women are not having any membership because of lack of access to land. The marketing cooperatives have fixed members to sell their products. Another reason for less involvement of women in marketing is due to poor development of rural infrastructure, and women are not accessing the markets regularly. Due to these difficulties, women are not producing surplus farm products. They only produce the food for their household consumption.

14.6 Recognising the Participation of Women in Conservation of PGR and TK

To meet the ever-increasing food demand for growing population, it is imperative to conserve these genetic resources. PGR and TK should be utilised on sustainable basis. The pressure on land has to be reduced by avoiding over-exploitation of these resources. To protect the over-exploitation and sustainable utilisation of these resources, there is a need to implement national and international programmes effectively. In most of the developing countries, more than 60 % agriculture work is being carried out by women. The role of women varies from region to region and country to country. Besides field work, women are involved in household activities. But women's role is neither documented nor cited anywhere. Women are fully involved in the conservation and maintenance of genetic diversity in PGR and TK for their sustainable utilisation to ensure food security. The governments should frame the policies and legislation to recognise the role of women. More programmes should be launched for conservation and sustainable uses of these resources (Watanabe et al. 1998).

Both the governmental and the non-governmental organisations operating at national and international level should include in their programmes activities that include supporting and strengthening the role of women in the farming and conservation of PGR, so that they may introduce innovations in the way they intervene in the farming, conservation and use of resources. There is a need to provide more recognition to farm women and to determine the responsibility of women in the process of conservation and sustainable utilisation of PGR and TK. To do this, it is important to carry out a quantitative study at different levels and how these relate to the roles of the men and women farmers in deciding which genetic materials are used and conserved (Hamrick et al. 1993).

Culinary is another area dominated by women, and their role in conserving PGR is larger than life. The examples and case studies are endless. Every area of developing and developed countries is full of women conservationists. It is high time to highlight the role of such unsung women champions of biodiversity conservation. Such success stories have high motivational impact. Proper encouragement is very important for continuing and propagating such schemes and programmes. Proper regard from the society and the governmental organisations is urgently needed. There are some governmental schemes for working women, and the women from unorganised sectors have contributed significantly in the biodiversity conservation and the environmental management. Supporting such workers is equally important. It will not be an exaggeration to say that until unsung women are not credited properly and are involved in environmental management, desired result will not be achieved.

The men and women have different experiences for selection, conservation and sustainable uses of plant materials. The difference in knowledge of men and women is not only affecting the conservation and sustainable utilisation of plant species but also has implications on food, health and other developmental activities for food security. There is urgent need to take some of the following positive steps to recognise the contribution made by women in the conservation of biodiversity in the world.

1. Emphasis should be given for documenting the role of women for selection of genotypes, maintenance of genetic diversity and preservation of PGR and TK.
2. Recognition and documentation should be done with the traditional knowledge held by women in agriculture production techniques.
3. The traditional technical knowledge held by the women must be translated into scientific

methods and transferred to other areas of PGR and TK management.

4. Women's view should be taken while planning a developmental programme related to conservation of PGR and TK.
5. Women's right on technical traditional knowledge should be protected, and they should be involved in decision-making process.
6. Assessment of women to land should be provided through strong legislations.
7. Integrated agricultural extension services for sustainable agriculture to ensure food security should be developed and improved.
8. More awareness programmes should be launched to aware the policymakers and planners about the roles of women in preserving PGR for food security.
9. Capacity-building programmes should be launched to promote women's internal strength to take effective decisions.
10. Development of women household-friendly integrated agriculture programmes.

14.6.1 Development Programmes for Women

In the developing countries, the policymakers and planners should develop the programmes especially for women to strengthen their participation and decision-making capacity. Women's role should be highlighted in conserving the important PGR. The traditional practices of women used in farming system and household activities should be well supported in the developmental programmes and making the policies. Women rights on important PGR and TK should be well protected before implementing any developmental programmes.

14.6.2 Legislation

PGR and TK are important components of food security, and to protect these resources, the legislations should be framed. To avoid over-exploitation of PGR and TK due to privatisation and commercialization of these resources, some strong laws should be framed. No doubt some laws have been formulated, but this legislation is being inappropriately operated. With the inception of globalisation of trade, the private companies are over-exploiting these resources for their benefits without sharing their benefits with the farmers. Due to this PGR and TK are under the threat to ensure food security in the future. The commercial companies are taking these resources to other parts of the nation for further utilisation and contributing nothing to local area where these resources are being maintained (McNeely 1988). The rural women are continuously promoting genetic diversity in PGR and TK and sustaining the world food security. The governments should frame strong legislations to protect the interests of women in rural area for further improvement of PGR and TK.

14.6.3 Representation in Decision-Making Processes

Women are most of the time excluded from the decision-making at farm level, society level, regional level and national level. Women's decisions are more often ignored at different levels. In some cases women's decisions are also not seriously considered at household activities. The gender biasness is there in the decision-making of most of the programmes. The policymakers and planners at the time of implementation of projects consider men as the main representatives of a society. Women should also be given equal opportunity while framing and implementing the developmental projects. Their decisions should be given due consideration for effective conservation of PGR and TK for food security.

14.6.4 Women Access to Agricultural Inputs and Other Resources

The government is providing the subsidies on various agricultural inputs such as fertilisers, seeds, irrigation and plant protection chemicals in developmental programmes. All these benefits go to men in the society because men have rights on land. Women are losing their interest in the conservation and improvement of PGR and TK because they do not have their rights on farm land. Women are feeling insecurity due to the lack of access to land. They are not using their full potential in conserving the PGR and TK and their sustainable utilisation to ensure food security. Moreover, in rural areas males show their dominance due to customary and traditional cultures, and the women are excluded from decision-making. The knowledge and experiences of women are not utilised for farm activities and other household activities in the society.

Women are mostly cultivating their own selected crop varieties, and they have every access to modern released crop varieties because of the lack of knowledge to grow these varieties. Men have accessibility to the crop varieties, but they fail to fully translate to women their practices. Moreover, women are not involved in any participatory approach for selection of suitable genetic materials. The males of society are only entitled to take part in the participatory approaches and selection of new cultivar. Therefore, the governments should launch such programmes in which farm women should be encouraged to take active part in the developmental programmes. This will not only improve the awareness and status of rural women in decision-making but also help in enhancing the genetic diversity in PGR and TK.

14.6.5 Formation of Farm Women Groups

Women are engaged in household activities and conservation of important PGR and TK. But they are not receiving any compensation for it to meet their livelihood security. Moreover, they are unable to access any resources such as land, credits and other agricultural inputs. In most of the formal functions, men are taking part and women are excluded from these programmes. To encourage the participation of women in all these activities, small self-help groups should be formed (Fig. 14.3). These groups will help the farm women to access credits and other crop production inputs. The women will also be trained in decision-making in household and other agricultural practices. This will help the women to take their decisions in the conservation and management of resources.

14.7 Role of Women in PGR and TK Conservation Case Studies from India

In Asian countries like India, women are involved in a number of household and agriculture production activities. The role of women varies from one part of the country to another. Among the developing nations, India is rich in genetic diversity and a hot spot of biodiversity. Women have played a significant role in the collection, conservation and sustainable utilisation of diverse plant species (Fig. 14.4). They are the main conservator of different plant species with enhanced genetic diversity to be used in crop improvement. Women have been custodian of plant genetic diversity and indigenous knowledge from time immemorial. They have adequate knowledge regarding importance of plants as food, medicine, clothing and raw materials for various household purposes. Plants are the backbones of religious events and rituals. Developing and least developed countries are full of unsung women champions of genetic resource conservation. It is difficult to mention all of them, but few important among them are described below:

14.7.1 Rural Education and Development Services

Women of Dharmapuri village of Tamil Nadu in India formed an organisation called rural education of development services (REDS) which took up the task of conserving PGR and TK diversity of their area. The role of ex situ conservation took shape, and now 35 medicinal gardens have been established which conserve and cultivate more than 158 medicinal plants. The

Fig. 14.3 Awareness about the formation small self-help groups

Fig. 14.4 Collection of diverse plant species from the forest

approach of these women soon started paying dividend, and apart from conservation of the medicinal plants, the attempt led to socio-economic growth of the area.

14.7.2 Herbs for All, Health for All

Women from Panangofe, Pampuchathamana and Bharathnoor villages in Kerala in India formed an organisation and got training for scientific conservation of medicinal plants. The women got training about collecting, maintaining, conserving and propagating medicinal plants. The women then started training other village women, and now the area is strongly committed to conservation of PGR and TK diversity.

14.7.3 Raksha Bandhan in Garhwal Himalayas

As mentioned earlier, women play an important role in maintaining the rituals of a particular area. Bachchni Devi of Garhwal Himalayas in India successfully exploited this for the conservation and protection of trees and other PGR. She organised a group of women of her locality and encouraged them to tie sacred thread to trees. In this way Raksha Bandhan of tree was done. This event became popular in Garhwal and helped in the conservation of trees and other genetic resources.

14.7.4 Kamala Pujar of Orissa

Koraput district of Orissa in India is abode of over 1,800 landraces of rice. It is unfortunate that presently the cultivation of only 15 high-yielding varieties is done. Landraces are better adapted to the local environment and are often not as susceptible to diseases as the high-yielding varieties. Normally the landraces are tolerant to climatic changes and also hence are of great value as a tool for the climate change adaptability. Landraces are also a treasure of various valuable genes which may prove to be an important tool for crop improvement. Too much emphasis on high-yielding verities has accelerated erosion of the landraces. Kamala Pujari took responsibility of conserving landraces of rice with the help of some other women of her district. Now these landraces are also being cultivated in many areas of Orissa along with the high-yielding varieties. Contribution of Mrs. Pujari was acclaimed at international forum also, and she was awarded for this at the World Summit at Johannesburg.

14.7.5 Maa Ghodadei Mahila Samiti

Illegal falling of trees is a serious problem and it has got a far reaching negative effect on the environment. Women of Nayagarh district of Orissa in India took this problem seriously and accepted the challenge of the forest protection. Women organised themselves and formed a group known as the Maa Ghodadei Mahila Samiti (MGMS) specially aimed at checking illegal falling of tree from forest. Members of the Samiti used to patrol forest area even during the night and catch the culprits. The women force has created such an impact that dreaded criminals now do not dare to enter forest for the illegal activities. Thus, contribution of women has helped in protecting the forest.

14.8 Conclusion

In almost all parts of the world, women are the main creature of things which are used for daily household consumption. They are the first collector of PGR and experiment on them in the backyard near to home. Women select the plant species for food, medicines and nutritional values. Women collect the plant species along with their wild relatives to maintain genetic diversity. They identify and multiply the best land race for wider cultivation in the field. Women are supplying the diverse genetic materials to plant breeders for the development of new crop variety. Women's selection criteria for PGR are different from men. They have more vision, knowledge and skills for the selection of plant species which help to maintain the genetic diversity in selected plant species. Besides household activities women also contribute more than 60 % in agricultural crop production in the field. They are involved in sowing the seeds, weeding, harvesting and threshing of crops and postharvest management practices. Although women are involved in household activities and other production of field crops, their role has never been appreciated and documented.

Women have less or lack of accessibility to land, credits and other agricultural services which adversely affect the interests of women to conserve the PGR and TK. Women's role has been ignored in most of the developing countries. Most of the developed nations are commercialising the PGR and TK conserved by the women, but the rural women are not being provided with any compensation. The moral values of women in rural areas are decreasing day by day. The policymakers and planners should frame the legislation related to the welfare of women. They should involve women in decision-making and sharing of benefits out of utilisation of PGR and TK conserved by the women. The governments should make strong laws to provide equal participation of women in all the developmental programmes related to agriculture. Small self-help groups should be formed in the rural areas which help the rural women to access the land, credits and other agricultural services. Emphasis should be given to provide capacity-building trainings to women that will boost their morale and decision-making capabilities. This will help in the selection of the best genetic materials through participatory approaches and development of modern crop varieties. During selection of plant species, women always select and conserve the diverse genotypes for future food security of the world.

References

ADB (2013) Gender equality and food security-women's empowerment as a tool against hunger. Asian Development Bank (ADB), Mandaluyong City, pp 19–42

Alordo R (2008) Saving the roots of biodiversity. Latin America Press, Lima

Barry H, Schlegel A (1982) Cross-cultural codes on contributions by women to subsistence. Ethnology 21(2):165–188

Boster JS (1985) Selection for perceptual distinctiveness: evidence from Aguaruna cultivars of *Manihot esculenta*. Econ Bot 39(3):310–325

Cleveland DA, Murray SC (1997) The world's crop genetic resources and the rights of indigenous farmers. Curr Anthropol 38:477–515

Duke J, Vasquez R (1994) Amazonian ethnobotanical dictionary. CRC Press, Boca Raton

FAO (1994) Women, agriculture and rural development: a synthesis report of the Africa region. FAO, Rome

FAO (1995) A fairer future for rural women. FAO, Rome

FAO (1999) Women – users, preservers and managers of agrobiodiversity. Food and Agriculture Organisation, FAO, Rome

Farnsworth NR, Akerele O, Bingel AS, Soejarto DD, Guo Z (1985) Medicinal plants in therapy. Bull World Health Organ 63:965–981

Flickenger D (1997) Rehabilitation of degraded tropical forests in India's Western Ghats: silvicultural and socio-economic implications of multiple species plantations. PhD thesis, University of Florida. University Microfilms International, Ann Arbor

Frankel OH, Bennett E (1970) Genetic resources in plants – their exploration and conservation, vol 11, IBP Handbook. Blackwell, Oxford

Frankel OH, Hawkes JG (1975) Crop genetic resources for today and tomorrow. Cambridge University Press, Cambridge, p 492

Good CM (1987) Ethnomedical systems in Africa. Guilford Press, Hove

Grenier L (1998) Working with indigenous knowledge: a guide for researchers. IDRC, Ottawa

Hamrick JL (1993) Genetic diversity and conservation of tropical trees. In: Drysdale RM, John SET, Yapa AC (eds) Proceedings an international symposium on genetic conservation and production of tropical tree seed, Chiang Mai, Thailand. ASEAN-Canada Forest Tree Centre, Thailand, pp 1–10

Hamrick JL, Murawski DA, Nason JD (1993) The influence of seed dispersal mechanisms on the genetic structure of tropical tree populations. Vegetation 107(108):281–297

Holden JHW, Williams JT (1984) Crop genetic resources: conservation and evaluation. George Allen and Unwin, London

Howard P (2003) Women and the plant world: an exploration. In: Howard P (ed) Women and plants. Gender relations in biodiversity management and conservation. Zed Press/Palgrave-Macmillan, London/New York, pp 1–48

Howard-Borjas P, Cuijpers W (2002) Gender relations in local plant genetic resource management and conservation. In: Doelle HW, DaSilva E (ed) Biotechnology, in encyclopedia for life support systems. EOLSS, Cambridge

International Strategy for Disaster Reduction (ISDR) (2008) Gender perspectives: integrating disaster risk reduction into climate change adaptation good practices and lessons learned. http://www.unisdr.org/eng/about_isdr/isdr-publication

Jarvis DI (1999) Strengthening the scientific basis of in situ conservation of agricultural biodiversity on farm. Bot Lith Suppl 2:79–90

Jiggins J (1986) Gender-related impacts and the work of the international agricultural research centers., Consultative group on international agricultural research. World Bank, Washington, DC, p 17

Karl M (1996) Inseparable: the crucial role of women in food security. Isis International, Quezon City

Kavya D, Anjali G, Veena G (2009) The hand that rocks the cradle, feeds the nation. Curr Sci 96(8):1016

Keys E (1999) Kaqchikel gardens: women, children, and multiple roles of gardens among the Maya of Highland Guatemala. Yearb Conf Lat Am Geogr 25:89–100

Kothari B (2003) The invisible queen in the plant kingdom: gender perspectives in medical ethnobotany. In: Howard P (ed) Women and plants. Gender relations in biodiversity management and conservation. Zed Press/Palgrave-Macmillan, London/New York, pp 150–164

Lambrou Y, Piana G (2006) Gender: the missing component of the response to climate change. Gender and Population Division, Sustainable Development Department, Food and Agriculture Organization of the United States, New York

Madge C (1994) Collected food and domestic knowledge in The Gambia, West Africa. Geogr J 160(3):280–294

Matsa W, Manuku M (2013) Traditional science of seed and crop yield preservation: exploring the contribution of women to indigenous knowledge-systems in Zimbabwe. Int J Soc Sci Humanit 3(4):234–245

McClain CS (1989) Women as healers: cross-cultural perspectives. Rutgers University Press, New Brunswick

McNeely JA (1988) Economics and biological diversity. IUCN, Gland

Mgbeoji ICM (2002) Patents and plants: rethinking the role of international law in relation to the appropriation of traditional knowledge of the uses of plants (TKUP). JSD, Dalhousie University/University Microfilms International, Halifax/Ann Arbor

Niriez VK (1987) Household gardens: theoretical and policy considerations. Agric Syst 23:167–186

Norton H (1985) Women and resources of the Northwest Coast: documentation from the 18th and early 19th centuries. PhD thesis, University of Washington, University Microfilms International, Ann Arbor

Okigbo BN (1985) Homegardens in tropical Africa. In: Landauer K, Brazil M (eds) Tropical home gardens. United Nations University Press, Tokyo

Plucknett DL, Smith NJH, Williams JT, Anishetty NM (1987) Genebanks and world's food. Princeton University Press, Princeton

Posey D (1999) Cultural and spiritual values of biodiversity: a complementary contribution to the global biodiversity assessment. United Nations Environment Programme and Intermediate Technology Publications, London

Ramanatha Rao V, Quek P, Mal B, Ming-De Z (1999) Role of IPGRI in promoting research on PGR conservation and use, and GPA implementation, with a focus on Asia and the Pacific. In: Ming-De Z, Zongwen Z, Ramanatha Rao V (eds) Proceedings of a national workshop on conservation and utilization of plant genetic resources in China. IPGRI Office for East Asia, Beijing

Saito KA (1994) Raising the productivity of women farmers in Sub-Saharan Africa, World Bank Discussion. World Bank, Washington, DC, p 230

Sen AK (1981) Poverty and famines: an essay on entitlement and deprivation. Clarendon Press, Oxford

Soemarwoto O, Soemarwoto I, Karyono, Soekartadiredja, EM, Ramlan A (1976) The Javanese home gardens as an integrated ecosystem: science for better environment. In: Proceedings of the International Congress Human Environment, Science Council of Japan, Tokyo

Sperling L, Berkowitz P (1994) Partners in selection: bean breeders and women bean experts in Rwanda. CGIAR, Washington, DC

Sthapit BR, Jarvis D (1999) On farm conservation of crop genetic resources through use. In: Mal B, Mathur PN, Ramanatha Rao V (eds) South Asia Network on Plant Genetic Resources (SANPGR). IPGRI South Asia Office, New Delhi, pp 151–166

Vernooy R (2003) Seeds that give: participatory plant breeding. International Development Research Centre, Ottawa

Watanabe KN, Ramanatha Rao V, Iwanaga M (1998) International trends on the conservation and use of plant genetic resources. Plant Biotechnol 15(3):115–122

Watson JW, Eyzaguirre PB (2001) Proceedings of the second international home gardens, Federal Republic of Germany. International Plant Genetic Resources Institute, Rome

World Bank (2003) Gender equality and the millennium development goals. World Bank, Washington, DC

Zimmerer KS (1991) Seeds of peasant subsistence: agrarian structure, crop ecology and Quechua agriculture in reference to the loss of biological biodiversity in the southern Peruvian Andes. PhD thesis, University of California, Berkeley. University Microfilms International, Ann Arbor

15 Therapeutic Potential of Plant Genetic Resources and Traditional Knowledge: A Panoramic View of the Flora Indigenous to North West Himalayas

Muslima Nazir, Roohi Mushtaq, Showkat Ahmad Zargar, Aijaz Ahmad Wani, and Sajad Majeed Zargar

Abstract

In the recent years, the uses of traditional medicinal plants for curing the diseases have been increased worldwide. Plant genetic resources and traditional knowledge provide the basic needs to secure the health of human. The medicinal plants have therapeutic properties for the control of common diseases. In most of the developing countries, indigenous people are using the herbal plant species for treatment of diseases. The medicinal plants have antibacterial properties which can be used to inhibit the growth of microorganism of diseases. The medicinal plant species along with TK of indigenous people help to control a number of diseases. Indigenous people are conserving the traditional PGR and TK over years without any incentives. The pharmaceutical companies in the developing countries are over exploiting these valuable resources without sharing of benefits. There is need to protect these valuable resources under intellectual property protection laws. Indigenous people should be provided with equal sharing of benefits arising out of these resources. Moreover, the developed nations prefer to use herbal plant-derived medicines over the allopathic medicines. Considering the importance of PGR-derived products in human health, the

M. Nazir • S.A. Zargar • A.A. Wani, Ph.D.
Department of Botany, University of Kashmir,
Srinagar, Jammu and Kashmir, India
e-mail: muslima023@gmail.com;
showkatzargar1@gmail.com; aijazbotku@gmail.com

R. Mushtaq, Ph.D.
Department of Biotechnology and Bioinformatics,
S.P. College, Srinagar, Jammu and Kashmir, India
e-mail: rohikhan400@gmail.com

S.M. Zargar, Ph.D. (✉)
Centre for Plant Biotechnology, Division of
Biotechnology, SKUAST-K,
Shalimar, Srinagar, Jammu and Kashmir, India
e-mail: smzargar@gmail.com

R.K. Salgotra, B.B. Gupta (eds.), *Plant Genetic Resources and Traditional Knowledge for Food Security*, DOI 10.1007/978-981-10-0060-7_15

medicinal plant species should be conserved. Both in situ and ex situ/on farm conservation methods can be used for the preservation of these valuable resources. International and national regulations should be framed for conservation and sustainable utilization of PGR and TK.

15.1 Introduction

Traditional medicines are attracting the world communities, particularly the developed nations. Today, more than 80 % of the world population is depending on the traditional medicines. The plant-derived drugs are the primary source of medicines to be used for health care. More than 25 % of the medicines are derived from the plant herbs, and others are synthetic analogues based on prototype compounds isolated from plants.

The demand for traditional medicines is increasing all over the world. To meet the demand of traditional medicines, area under herbal plants is increasing at faster rate in the developing and developed nations. For the development of traditional medicines, the herbal materials are still taken from the wild plant species naturally grown in the forests. Very small portion of herbal plants for medicines has been taken from the cultivated ones. The commercialization of wild herbal plants is under great threat and is likely to be extinct.

It has been observed that the local and indigenous people are not equally benefitted from the development of traditional medicines. Although the indigenous farmers are preserving these wild herbal plants, they are debarred of sharing of benefits. After the commercialization, the herbal plants become extinct due to overexploitation. The herbal plants should be used judiciously and on sustainable basis. Most of the pharmaceuticals companies are overexploiting these plant resources. These resources should be conserved and used on sustainable basis for future food security.

The relation of human and plant herbs is very old and inseparable. The shamans have deep knowledge about the herbal plants, and they are considered as the first professional in botany. No doubt in the developed nations the communities have switched to allopathic drugs, but the plant herbs still have their importance in the developing countries. In most of the under developed and developing countries, the therapies of local and indigenous people are done with plant herbs along with chants, spiritual ceremonies, and dance. These therapies are done to remove the illness of persons in the environment. The shamans in the developing countries have knowledge about the treatment of a number of diseases (Schultes 1979). They are the skilled persons and have knowledge about the importance of different plant herbs used for the treatment of diseases. The knowledge of shamans about the use of different plant herbs for the treatment of diseases are not documented yet. They are transferring the knowledge of treatment with the use of plant herbs orally from father to son or daughter.

Most of the hospitals are dependent on allopathic medicines for the treatment of diseases. No doubt allopathic medicines are integrated with traditional medicines in most of the hospitals, but people prefer to use the allopathic medicines. In the rural areas, people are still treating their common diseases by using the herbal plants in most of the developing countries. Now most of the countries have independent hospitals for the treatment of diseases by using traditional medicines. There is great demand for these medicines in these countries. They are collecting all these plant materials from the rural forests, and some of the countries have started to cultivate the plant herbs. In future, the importance of these herbal plants may increase due to more awareness about the use of traditional medicines for the treatment of common diseases. In this regard, World Health Organization (WHO) has launched a campaign for preservation and sustainable utilization of herbal plants for the treatment of common diseases (Akerele 1991). This campaign has initiated the process of effective use of herbal plants for the treatment of common diseases (Rabineau 1991).

With the inception of modern advanced technologies, the use of chemical drugs has increased significantly. This development is due to immediate and specific effect of allopathic medicines for the treatment of diseases. People have started to switch from the traditional medicines to allopathic ones for the treatment of diseases. Over 25 % of drugs in the Organization for Economic Co-operation and Development (OECD) countries have proved that these drugs are slightly altered from the higher plant drugs (The Lancet 1994).

Plants have anchored mother earth long before human life came into existence. It is said that plants were endowed with materials for survival of the dependent man and animals and these self-sustaining creatures are by and large the sustainers of this planet. The plant resource of a geographical zone is the backbone of all the life present there and represents the socioeconomic status of that place. In addition to their basic function that is photosynthesis, plants per se are little factories of antioxidants, polyphenols, vitamins, and many other useful compounds. Nature offers a huge diversity of plant resource to be explored and utilized for various human benefits. From the very earliest days of civilization, plants have been used for healing, and today 80 % of the population of the world is using traditional medicine (Singh and Verma 2008). The WHO has estimated that the present demand of medicinal plants is worth more than US$ 14 billion per year (Kala et al. 2006). Moreover, green diversity provides a reservoir of raw material having tremendous economic importance that improves the quality of human life to a great extent.

15.2 Importance of PGR and TK and Their Uses

The importance of Traditional Knowledge (TK) is a way of enlarging our understanding of local indigenous practices and environments. The importance of TK and treatment of common diseases with the use of herbal plants has grown at faster rate in recent years. Today, most of the medicinal plants are being used for various pharmaceutical purposes (Fig. 15.1). Beginning moderately with the local classification of plants (Conklin 1957), firewood (Metzger and Williams

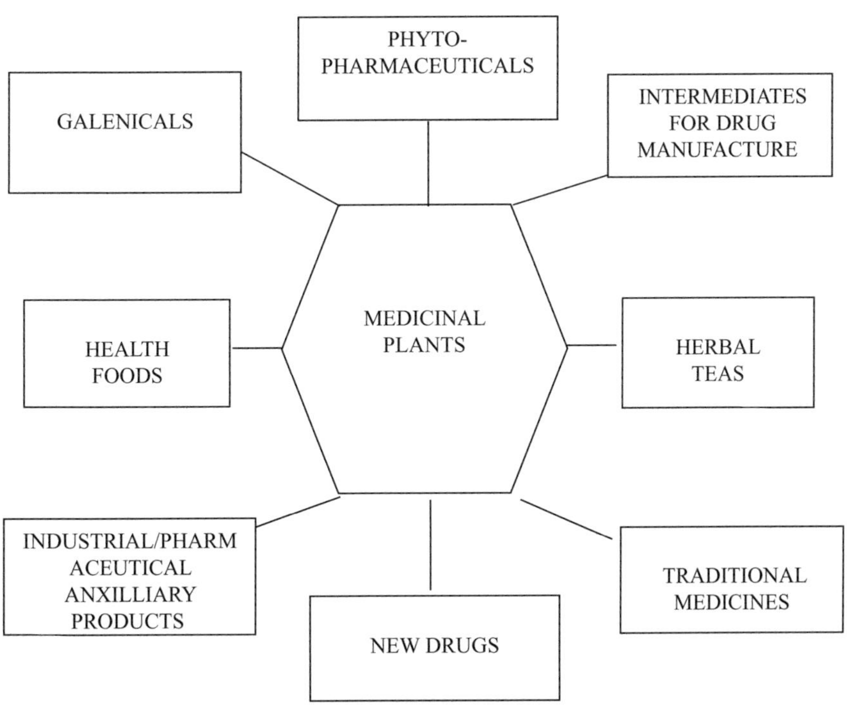

Fig. 15.1 Medicinal plants with various pharmaceutical applications

1966), colors (Berlin and Kay 1969), and other domains, this field addresses complex issues (Colfer et al. 1989; Joshi 1997; Sinclair and Walker 1999).

TK has already contributed significantly for treating the diseases of people. The shamans have used their indigenous knowledge for the treatment of numerous diseases. Anthropologists study the interaction between the herbal plant species and the human behavior and the ecology, subsistence patterns, division of labor, seasonal variation in income, etc. (Colchester 1981; Posey 1983; Bird-David 1992; Balee 1993; Colfer et al. 2000). The indigenous people are contributing significantly to allopathic medicines by using their knowledge of medicines, fibers, wood, food, and wildlife (Clay 1988; Colfer et al. 1997). The following is the number of issues related to examine TK: (1) Intellectual Property Rights (IPRs), (2) immunity differentiation, (3) different standards related to indigenous knowledge, (4) different use of plant species, and (5) TK and their association with the livelihoods of the indigenous people.

In traditional medicine systems, the indigenous people look on the whole habitat, whereas new systems of medicines are focusing on single plant habitat or single crop species of PGR (Scott 1998; Sivaramakrishnan 2000). There are multiple uses of PGR in traditional medicine systems. Scientists have worked thoroughly on the indigenous knowledge and the natural habitats (Joshi et al. 2004a). Indigenous people's knowledge on PGR and their uses for the treatment of common diseases have attracted the attentions of the international communities. Indigenous knowledge of the rural communities on PGR plays a significant role for sustainable utilization of these resources for the welfare of local and farming communities. They are conserving these PGRs for a long time and are using these in their daily needs. The local and indigenous people have better understanding about the agroecological system (Sinclair and Walker 1999; Sinclair and Joshi 2000). Moreover, the use of TK in humankinds can be separated from their livelihoods. The rural communities are totally dependent on the TK and traditional practices.

TK of indigenous people are not in static in nature and changes over time with experiences of the people and improvement in traditional practices. Indeed, new plant species and crops have been introduced in a country for cultivation along with its package and practices. The farmers are integrating these new crop species with their indigenous PGR. The practices applied to the cultivation of introduced crop species are also applied to their indigenous PGRs which have significant effect on their traditional practices. The small and local farmers are adopting the new innovations and practices more easily which will increase their farm produce (Dove 2003).

It has been observed that the indigenous people's knowledge is not culture and traditional bound and it changes with new innovations and techniques (Richards 1994; Sinclair and Walker 1999). The new developmental programs for the management of PGR are also affecting the TK of the indigenous people for conservation of plant herbs to be used in traditional medicines, e.g., the fodder management in hills (Joshi 1997), soil management (Waliszewski et al. 2003; Shrestha 2000), and coffee-based farming (Joshi et al. 2004b). The overall aim is to enable the indigenous people to enhance their capabilities for the better management of PGR.

15.3 PGR and Biomedical Knowledge

The medicines manufactured from PGR have been used to treat the common diseases of human. The men and women shamans have some special TK for the manufacturing of medicines from the herbal plants and their use for the treatment of diseases. The indigenous knowledge for the development of medicines from plants and their use are passed down to the next generation. The researchers have identified the various bioactive plant metabolites and their uses for the treatment of diseases. PGRs have been used for the development of important drugs for commercial purposes, e.g., diosgenin from *Dioscorea nipponica* (Kang et al. 2011) and analgesic aspirin from willow bark (*Salix* spp.) (Miner and Hoffhines 2007). The researchers have developed more interest in the discovery of important and potential herbal plant species for developing more drugs (Li and Vederas 2011). However, plant-based discovery and development of drugs has number of challenges. Moreover, the knowledge and available information on plant species to be used for traditional medicines is very meager. The use of plant species for development of traditional medicines poses a number of challenges in pharmacology. The integration of indigenous knowledge on herbal plant species with biomedical resources can provide the foundation for essential prioritization strategies (Sharma and Sarka 2013).

India and China are using huge quantity of traditional medicine for the treatment of common diseases. In these countries, research is also going on at priority and at larger scale. India and China are investing more in Research and Development (R&D) industries. These industries use natural products for the development of traditional medicines. The demand for traditional medicines has been significantly increased in the world markets. The globalization of trade makes the importance of traditional medicines for the cure of common diseases. Moreover, the people of developed nations are using more traditional medicines compared to the allopathic medicines for the treatment of diseases.

Although traditional medicines are expensive and slow curing, the treatment of disease is very reliable. The slow process of treatment can be improved with the help of modern techniques. The computational approaches are used for drug designing and discovery of new medicines (Sharma and Sarkar 2012). The conventional methods of improving the discovery and development process of traditional medicines may delay quality control and clinical testing. The scientific names and the different local names given to the herbal plant species may pose difficulty and challenges in the development of new medicines (Zhao et al. 2006). Moreover, the natural products are so mixed that it is very difficult to purify and develop pure traditional medicines. Agronomical practices, purification, and active gradients are the great challenges for the development of traditional medicines. The active gradient may be affected by weather conditions and processing methods. The lack of knowledge of toxicity of herbal plant species produces bottlenecks for the development of traditional medicines, and the rectifications of these bottlenecks are the priorities of modern science (Sharma and Sarka 2013).

15.4 Ownership of PGR and TK

PGRs are generally owned by the local and indigenous communities who are the owners of the land. The CBD has recognized the rights of nations over the PGR. Benefits must be shared among nations who are directly and indirectly linked with the conservation and sustainable utilization of PGR. However, it is very difficult to identify the owners in case of the TK. In the case of TK, indigenous communities are the real owners of the TK, and the benefit sharing must go to these people. The legal owner of the TK may be the individual, the local people, the communities, the farmers, the governments, and NGOs (Koo et al. 2004). Moreover, the communities are mostly defined by the geographical boundaries or regions, and the sharing of benefits with the real recipients is also a difficult process.

When dealing with TK and PGR, it has been observed that the values of PGR and TK cannot be accessed with equal yard stick. TK is mostly a public domain property that belongs to a community. TK does not meet the standard laws for IP protection, whereas PGR can be protected under IP protection laws. There are certain laws for the protection of PGR, and these laws vary from one country to another (Koo et al. 2004). Therefore, the international organizations must have collaborations on IP protection laws related to PGR and TK. These laws must address the benefit-sharing issues in the form of royalties and capacity building by providing trainings.

15.5 Challenges in Development of Therapeutic Products from PGR

The development of traditional medicines from the natural products is very difficult. PGRs are used for the development of commercial products. The PGR materials are mostly collected from the developing countries for further processing. To collect the materials from the developing countries is very difficult, and there are a number of hurdles. In order to utilize the PGR materials from the source counties, the conservation, protection of endangered plant species, sustainable utilization, and equitable sharing of benefits are the important issues that need to be resolved. The more complex issues are related to sharing of benefits arising out of utilization of TK and related knowledge. It is difficult to share the benefits of common natural products. The utilization of TK and related indigenous knowledge and indigenous PGR and traditional practices are the main issues that need to be resolved with regard to sustainable utilization, sharing of benefits, IP protection, etc. These resources are being over exploited by colonial time in the developing countries. The IP protection and equal sharing of benefits out of utilization of PGR and TK are linked with local politics, national and international politics, and international laws for the protection of these resources. The sharing of benefits of TK and related knowledge and PGR is a challenge involved with controversies over years in the developing countries which need to be addressed under guidelines of CBD and the Agreement on Trade-Related Aspects of Intellectual Property Rights (TRIPS) of the World Trade Organization (WTO) (Sampath 2005). The main controversies are related to appropriate valuation of TK and related indigenous knowledge and PGR, equal sharing of benefits, and identification of ownerships of IP.

Overexploitation of TK and PGR by the pharmaceutical companies of the developed nations is a national and international concern. Moreover, the global decline in PGR has damaged several important medicine plant species (Hamilton 2004). The CBD, the Convention on International Trade in Endangered Species (CITES), the Forest Service, and the National Center for the Preservation of Medicinal Herbs (NCPMH) are keeping track to protect these endangered medicinal plant species worldwide (Lui et al. 2005). The overexploitation of natural habitats and PGR is the major challenge for the conservation and sustainable utilization of these resources possessing medicinal values (Hamilton 2004). Most of the developing countries' forests are affected by the overexploitation and patenting of natural resources by private companies. These factors are also affecting the morals of local and indigenous communities for conservation and sustainable utilization of their indigenous knowledge and technical know-how related to medicinal plant species (Roberson 2008). The CBD and CITES have taken initiatives for conservation of herbal plant species and their sustainable utilization. The CBD has ensured to provide equitable benefit sharing to the indigenous people and to develop the local, national, and international laws for preservation and protection of TK.

15.6 Strategies to Conserve Plant Diversity of Medicinal Plants

Until now the available national and international PGR-related agreements recognize the sovereignty of countries where these resources occur within their borders. However, the onus to conserve (using both ex situ and in situ approaches) and use them rests with countries, and current agreements stress the importance of equable sharing for their utilization. The Article 8 of the CBD requires that countries should develop their own guidelines of selected areas for in situ conservation, protection of areas, and sustainable utilization of PGR related to medicines. Countries also require promoting the developments related to environment, rehabilitating degraded lands and ecosystems, and controlling or eradicating exotic species that threaten the native species. Countries also should ensure compatibility between conservation of PGR and sustainable use. They should also respect and preserve the TK, innovations, and traditional practices of indigenous communities.

Additionally, there should be regulatory mechanism for the protection of medicinal plant species in the developing countries like India. In these countries, there is need to provide guidelines, financial support, and other supports for the conservation of PGR, particularly in situ conservation. In situ conservation and on farm conservation involve the management of land race varieties by the indigenous farmers within traditional agricultural systems (Frankel et al. 1995; Altieri and Merrick 1988; Brush 1991). In recent years, the in situ conservation and on farm conservations are important approaches which are being used by farmers over years.

The richness of biological diversity in Himalayas particularly in Jammu and Kashmir, Uttarakhand, and Himachal Pradesh of India is evident; it is mainly due to varied climatic, altitudinal, and ecological habitats. We have mountains that fall in hot to cold regions, and hence the number of plant species is enormous. There have been increasing rates of threats of depletion to these biological resources including medicinal plants (an important bioresource of Kashmir valley) due to immense biotic and abiotic stresses. A total of 560 plant species of India have been included in the International Union for Conservation of Nature and Natural Resources (IUCN) Red List of Threatened species, out of which 247 species are in the threatened category (Kapai et al. 2010). The changing climatic conditions may play a crucial role in depletion of this biological diversity. Various groups across the country are involved in identifying threatened species, conservation, and analysis of their metabolic profiles, and identification of a number of useful biomolecules having therapeutic potential from these species has also been done. However, there is still a great potential within these species yet to be explored. These species can serve as a rich source of germplasm for identification of novel genes, proteins, and metabolites. Various high-throughput techniques like genomics, transcriptomics, proteomics, and metabolomics can help us in the accomplishment of this goal. The conservation of the threatened plant species is a very important task. Conservation is possible by both in situ and ex situ methods. In situ methods have been followed to conserve the biological diversity by defining biosphere reserves and protected areas. Although protecting plant species in their natural habitats is the best method of conservation, but managing these protected areas becomes a major concern.

Ex situ conservation has its origin in setting up botanical gardens. Globally, ex situ conservation now plays a significant role in biodiversity conservation. Seed banks/gene banks, tissue culture, captive breeding, aquaria, and forest nurseries have recently been added to ex situ conservation networks and have proven to be much useful, quick, and effective. As such biotechnology provides us solution to conserve these medicinal plants that have great potential. The need is to develop gene banks and various other efficient in vitro systems to avoid loss of these important bio-resources.

15.7 Therapeutic Potential of Indigenous Flora in North-West Himalaya

Located in the north-western part of India, Jammu and Kashmir – a predominantly Himalayan state with its three geographically distinct divisions, viz., Jammu, Kashmir, and Ladakh – is immensely rich in biological and cultural diversity. This State possesses great altitudinal variation, diverse geological formation, and different climatic zones resulting in the diversity in its flora and fauna. The most prominent feature of our planet is the existence of life, and the most prominent feature of life is diversity. Himalaya harbors a rich diversity of medicinal plants, owing to its topographic variations spanning from valley floor through terraced table lands and dense forests up to alpine peaks.

Plant wealth of the Indian Himalayan region is known for its unique, natural, and socioeconomic values. The Jammu and Kashmir State of Himalayas is often considered as paradise on earth, rich in biodiversity. This is one of the Himalayan biodiversity hot spot. This region of Himalayan represents a rich source of biodiversity and promising scientific and economic benefits. The Jammu and Kashmir State is a treasure house of medicinal plants.

Most of the drugs have been derived from medicinal plant species which are the important source of these drugs. The use of herbal plant species for drugs is safe and economical (Kumar et al. 2011). In olden days, people are using the medicines extracted from plant species to cure a number of common diseases. In developing countries, like India, indigenous people are using the medicinal plant species as antimicrobials for the treatment of various diseases. These antimicrobials are extracted from medicinal plants and used for the treatment of various diseases caused by microorganisms (Jain 1994). The antimicrobial agents inhibit the growth of disease-causing microorganisms and help in curing the diseases (Craig 1998). The antimicrobial agents derived from the medicinal plants invade the disease-causing microorganisms through the biochemical mode of action and inhibit the further growth of microorganisms (Pellecure et al. 1976).

Saussurea lappa
Kashmiri name: Kuth
Trade name: Costus
Family: Asteraceae
Therapeutic potential: Costus (Kuth) has been in use in the Indian medical system since ancient times. This herb grows particularly along the higher belts of Tibet, Karnah. It is among the 214 endangered medicinal and aromatic plants of Himalayas (Nayar and Shastry 1990). The oil extracted from costus contains essential oil (1.5 %), resinoids (6 %), insulin (18 %), and saussurine and other alkaloids (0.05 %). Biological actions of Costus oil are hyposensitive, bronchiodialatory, and antiseptic against *Streptococcus* and *Streptophylococcus*. Since time immemorial, the remedial actions of Costus oil in leprosy have been observed and roots are used for curing asthma and cold. Costus root when pulverized and muddled with sessanum oil is applied to a rheumatic limb. Costus possesses various bioactivities such as antifungal (Barrero et al. 2000), antidiabetic (Upadhyay et al. 1996), anthelmintic (Seki et al. 1991), and antitumor (Ko et al. 2005). It is also believed that one part of granulated root scrambled with three parts of sugar can be used to heal stomach ulcers.

Artemisia absinthium
Kashmiri name: Tethwen
Trade name: Artemesia
Family: Asteraceae
Therapeutic potential: Artemesia (Tethwen) grows naturally on uncultivated, arid ground, on rocky slopes, and at the edge of footpaths and fields and is a white hoary shrub. Most of the Artemisia plant species have been categorized as endangered, but among the critically endangered is the *Artemisia amygdalina* endemic to valley. Plant scientists have found its cultivation difficult in the plains of valley, because of being a plant of high altitude pastures and forests. It is an important part of Unani and Ayurvedic systems of medicine, having antimicrobial, antiseptic, antispasmodic, carminative, chologogue, febrifuge, and antihelminthic uses. Tethwen is used as a vermifuge in indigenous systems of medicine for the cure of chronic fevers and for the inflammation of the liver, as an antispasmodic and antiseptic (Koul 1997). Essential oil extracted from Artemisia has antimicrobial (Juteau et al. 2003) and antifungal activity (Saban et al. 2005). Artemisia was scientifically claimed antihelminthic action against gastrointestinal nematode infections in sheep (Khurshid and Mudasir 2012).

Amaranthus caudatus
Kashmiri name: Leesa
Trade name: Amaranths
Family: Amaranthaceae
Therapeutic potential: The extract of leaves and inflorescence is used in expectorants and to treat fever. Amaranths are rich in nutritional value. Amaranths also contain high content of micronutrients such as b-carotene, iron, calcium, vitamin C, and folic acid (Priya et al. 2007).

Berberis aristata
Kashmiri name: Kawe dach
Trade name: Indian Barberry (Daruhaldi)
Family: Berberidaceae
Therapeutic potential: The bark of the root is dried and crushed to make powder; the pow-

der is taken orally and is useful in Jaundices, back pain, weakness, and fractures.

Cichorium intybus

Kashmiri name: Jungli hand

Trade name: Chicory

Family: Asteraceae

Therapeutic potential: In this case, root extract is mixed with water and sugar to cure thyroid. Besides, this Chicory is used for the treatment of various ailments. Uterus and tumor cancers can be treated by taking the juice of Chicory (Judzentiene and Budiene 2008). In some parts of the developing countries, Chicory plant's leaves, roots, and stems can be used for treatment of jaundice. Chicory plant syrup is used as tonic and medicines for the children (van Wyk et al. 1997). Chicory is used as ointments for healing wounds (Sezik et al. 2001).

Dioscorea deltoidea

Kashmiri name: Krech

Trade name: Yam

Family: Dioscoreaceae

Therapeutic potential: Yam root tuber juice is taken as juice for the treatment of roundworm. It is also used to alleviate constipation and to cure diseases such as asthma and arthritis.

Euphorbia wallichii

Kashmiri name: Dub

Trade name: Wallich spurge

Family name: Euphorbiaceae

Therapeutic potential: It is used for the treatment of nerve troubles and dropsy. Wallich is used for the treatments of warts and skin infections.

Mentha arvensis

Kashmiri name: Pudina/pudni

Trade name: Mint

Family: Lamiaceae

Therapeutic potential: Pudina, through decoction or paste of the plant, can be used for the treatment of asthma, cough, rheumatism, indigestion, and diarrhea. Gastroenteritis is also solved by using the leaves as tea.

Nymphaea stellata

Kashmiri name: Bum posh

Trade name: Blue water lily

Family: Nymphaeaceae

Therapeutic potential: This herbal plant is used for the treatment of indigestion and is also anti-hepatotoxic and antidiabetic by using the flower extract.

Podophyllum hexandrum

Kashmiri name: Wanwangun

Trade name: Himalayan may apple

Family: Podophyllaceae

Therapeutic potential: The juice extract from the fruit and seed is used for the treatment of diarrhea, tumor, heart abnormalities, and chronic constipation.

Rheum emodi

Kashmiri name: Pumba chalan

Trade name: Rhubarb

Family: Polygonaceae

Therapeutic potential: Rhubarb is used for the treatment of rheumatic, pain, and wound.

Rumex acetosa

Kashmiri name: Abij

Trade name: Sorrel

Family: Polygonaceae

Therapeutic potential: The vegetable from the plant is used in young phase. The medicinal plant is used for the treatment of chest problems, astringent, hardness of muscles, asthma, and skin diseases. The dried plant is powdered and mixed with oil and ghee to make the paste.

Taraxacum offcinale

Kashmiri name: Hand

Trade name: Common dandelion

Family: Asteraceae

Therapeutic potential: Common dandelion is used for the treatment of chronic cough, internal ulcers, asthma, infection, abdominal swelling, stomach cramps, acidity, and urine irritation.

Urtica dioica

Kashmiri name: Soi

Trade name: Nettle

Family: Urticaceae

Therapeutic potential: Nettle is used for curing diseases such as fever, fractures, stomach pain, wounds, dandruff, skin infection, paralyzed limbs, and nose bleeding. In this case, the plant is crushed and turned into paste which is used for treatment.

15.8 Conclusion

PGR and TK are indispensable components of life which play significant roles in humankind. These valuable resources have been used for a long time. The herbal plant species have been used in the treatment of a number of diseases of indigenous people. The shamans are using their indigenous knowledge for curing the common diseases of local and indigenous people by using the medicinal plants. The extract of herbal plant species has antibacterial property which is used to inhibit the growth of microorganisms of common diseases. Moreover, the new synthetic medicines developed by modern technologies have plant-based driven molecules. The medicinal plants are diverse in nature and have huge potential values which need to be investigated. These valuable resources are being overexploited in the developing countries by the developed nations. The strong laws should be framed for conservation, sustainable utilization, and equitable sharing of benefits arising out of use of PGR and TK. TK and indigenous knowledge should be protected under IP protection. PGRs have therapeutic values, and their crude extracted is potentially used for the treatment of various diseases. The extract of the herbal plant species has antidiabetic, antioxidant, antistress, antihyperlipidemic, and antibacterial properties. Most of the medicinal plants are owned by the local and indigenous people, and there is need to conserve (in situ and ex situ/on farm) these PGR to meet the future needs of the world communities. Indigenous people should be given some inceptives for further management of these resources. Moreover, these people should be provided with equal sharing of benefits arising out of utilization of PGR and TK for future management of resources.

References

Akerele O (1991) Medicinal plants: policies and priorities. In: Akerele O, Heywood V, Synge H (eds) Conservation of medicinal plants. Cambridge University Press, Cambridge, pp 3–11

Altieri MA, Merrick LC (1988) Agroecology and in-situ conservation of native crop diversity in the third world. In: Wilson E, Peter F (eds) Biodiversity. National Academy of Sciences, Washington, DC, pp 361–369

Balee W (1993) Footprints of the forest: Ka'apor ethnobotany – the historical ecology of plant utilization by an Amazonian people. Columbia University Press, New York, p 416

Barrero AF, Oltra JE, Alvarez M, Raslan DS, Saude DA, Akssira M (2000) New sources and antifungal activity of sesquiterpene lactones. Fitoterapia 71:60–64

Berlin B, Kay P (1969) Basic color terms: their universality and evolution. University of California Press, Berkeley, p 178

Bird-David N (1992) Beyond the 'hunting and gathering mode of subsistence: culture-sensitive observations on the Nayaka and other modern hunter-gatherers. Man (NS) 27:19–44

Brush SB (1991) A farmer-based approach to conserving crop germplasm. J Econ Bot 45(2):153–165

Clay JW (1988) Indigenous peoples and tropical forests: models of land use and management from Latin America, Cultural survival report 27. Cultural Survival Inc, Cambridge, MA, p 116

Colchester M (1981) Ecological modeling and indigenous systems of resource use: some examples from the Amazon of South Venezuela. Antropologica 55:51–72

Colfer CJP, Yost R, Agus F, Evensen S (1989) Expert systems: a possible link from field work to policy in farming systems. AI Appl Nat Resour Manag 3:31–40

Colfer CJP, Peluso NL, Chin SC (1997) Beyond slash and burn: building on indigenous management of Borneo's tropical rain forests. , Advances in economic botany, vol 11. New York Botanical Garden, Bronx, p 236

Colfer CJP, Wadley RL, Salim A, Dudley RG (2000) Understanding patterns of resource use and consumption: a prelude to co-management. Borneo Res Bull 31:29–88

Conklin HC (1957) Hanunóo agriculture: a report on an integral system of shifting cultivation in the Philippines. FAO forestry paper 20. FAO, Rome

Craig WA (1998) Pharmacokinetics/pharmacodynamic parameters: rationale for antibacterial dosing of mice and men. Clin Infect Dis 26:1–12

Dove M (2003) So far from power, so near to the forest: a structural analysis of gain and blame in tropical forest development. In: Padoch C, Peluso NL (eds) Borneo in transition: people, forests, conservation and development. Oxford University Press, Selangor, pp 50–70

Frankel O, Brown ADH, Burdon JJ (1995) The conservation of plant biodiversity. Cambridge University Press, Cambridge, p 299

Hamilton AC (2004) Medicinal plants, conservation and livelihoods. Biodivers Conserv 13(8):1477–1517

Jain SK (1994) Ethnobotany and research on medicinal plants in India. Ciba Found Symp 185:153–164

Joshi L (1997) Incorporating farmers' knowledge in the planning of interdisciplinary research and extension. PhD thesis, University of Wales, Bangor, p 272

Joshi L, Shrestha P, Moss C, Sinclair FL (2004a) Locally derived knowledge of soil fertility and its emerging role in integrated natural resource management. In: van Noordwijk M, Cadisch G, Ong C (eds) Belowground interactions in tropical agroecosystems: concepts and models with multiple plant components, chapter 2. CABI International, Wallingford, pp 17–39

Joshi L, Schalenbourg W, Johansson L, Khasanah N, Stefanus E, Fagerstrom M, van Noordwijk M (2004b) Soil and water movement: combining local and modellers' ecological knowledge in scaling up from plot to landscape. In: van Noordwijk M, Cadisch G, Ong C (eds) Belowground interactions in tropical agroecosystems: concepts and models with multiple plant components. CABI International, Wallingford, pp 349–364

Judzentiene A, Budiene J (2008) Volatile constituents from aerial parts and roots of *Cichorium intybus* L. (chicory) grown in Lithuania. Chemija 19:25–28

Juteau F, Jerkovic I, Masotti V, Milos M, Mastelic J, Bessiere JM, Viano J (2003) Composition and antimicrobial activity of essential oil of *Artemisia absinthium* from Croatia and France. Plant Med 69:158–161

Kala CP, Dhyani PP, Sajwan BS (2006) Developing the medicinal plants sector in northern India: challenges and opportunities. J Ethnobiol Ethnomed 3:32–11

Kang TH, Moon E, Hong BN (2011) Diosgenin from Dioscorea nipponica ameliorates diabetic neuropathy by inducing nerve growth factor. Biol Pharm Bull 34:1493–1498

Kapai VY, Kapoor P, Rao IU (2010) In Vitro propagation for conservation of rare and threatened plants of India – a review. Intl J Biol Technol 1(2):1–14

Khurshid AT, Mudasir AT (2012) Preliminary studies on plants with anthelmintic properties in Kashmir—the north-west temperate Himalayan Region of India. Chin Med 3:106–112

Ko SG, Kim HP, Jin DH, Bae HS, Kim SH, Park GH (2005) *Saussurea lappa* induces G2-growth arrest and apoptosis in AGS gastric cancer cells. Cancer Lett 220:11–19

Koo B, Nottenburg C, Pardey PG (2004) Plants and intellectual property: an international appraisal. Science 306:1295

Koul MK (1997) Medicinal plants of Kashmir and Ladakh, temperate and cold arid Himalaya. Indus Publishing Company, New Delhi

Kumar V, Andola HC, Lohani H, Chauhan N (2011) Pharmacological review on Ocimum sanctum Linnaeus: a queen of herbs. J Pharm Res 4:366–368

Li JW, Vederas JC (2011) Drug discovery and natural products: end of era or an endless frontier? Biomed Khim 57:148–160

Lui S, Wei M, Moore R (2005) RxNorm: prescription for electronic drug information exchange. IT Prof 7:17–23

Miner J, Hoffhines A (2007) The discovery of aspirin's antithrombotic effects. Tex Heart Inst J 34:179–186

Nayar MP, Shastry ARK (1990) Red data book of Indian plants. Botanical Survey of India, Kolkata

Pellecure S, Allegrini J, Buochberg S (1976) *Huiles essenticlles* bactericides et fongicides. Rev Inst Pasteur Lyon 9:135–159

Posey D (1983) Indigenous ecological knowledge and development of the Amazon. In: Moran E (ed) The dilemma of Amazonian development. Westview Press, Boulder, pp 225–257

Priya VP, Celine VA, Gokulapalan C, Rajamony L (2007) Screening amaranth genotypes (*Amaranthus spp.*) for yield and resistance to leaf blight caused by Rhizoctonia solani Kuhn. Plant Genet Res Newsl 147:1–4

Rabineau I (1991) Towards a Caribbean pharmacopoeia. Endacaribe, Santo Domingo

Richards P (1994) Local knowledge formation and validation: the case of rice in Sierra Leone. In: Scoones I, Thompson J (eds) Beyond farmer first. Intermediate Technology Publications, London, pp 165–170

Roberson E (2008) Medicinal plants at risk-nature's pharmacy, our treasure chest: why we must conserve our natural heritage. http://www.biologicaldiversityorg/publications/papers/Medicinal_Plants_042008_lorespdf

Saban K, Recep K, Ahmet M, Ahmet CAA, Ali Y (2005) Determination of the chemical composition and antioxidant activity of the essential oil of *Artemisia dracunculus* and of the antifungal and antibacterial activities of Turkish *Artemisiaabsinthium Artemisia dracunculus*, *Artemisiasantonicum*, and *Artemisiaspicigera* essential oils. J Agric Food Chem 53:9452–9458

Sampath PG (2005) Regulating bioprospecting: institutions for drug research access and benefit-sharing. The United Nations University Publisher, Tokyo/New York/Paris

Schultes RE (1979) The Amazonia as a source of new economic plants. Econ Bot 33:259–266

Scott JC (1998) Seeing like a state: how certain schemes to improve the human condition have failed. Yale University Press, New Haven, p 445

Seki K, Hashimoto A, Kobayashi H, Kawahara Y, Yamahara J (1991) Motility inhibitory effect on Anchusan and Jintan and its active components in Anisakis type larvae. Yakuri to Chiryo 19:265–289

Sezik EE, Yesilada G, Honda Y, Takaishi Y, Takeda TT (2001) Traditional medicine in Turkey X. Folk medicine in Central Anatolia. J Ethnopharmacol 75:95–115

Sharma V, Sarkar IN (2012) Bioinformatics opportunities for identification and study of medicinal plants. Brief Bioinform. doi:10.1093/bib/bbs021

Sharma V, Sarka IN (2013) Leveraging biodiversity knowledge for potential phyto-therapeutic applications. J Am Med Inform Assoc 20:668–679

Shrestha PK (2000) Synthesis of the participatory rural appraisal and farmers' knowledge survey at Landruk, Bandipur and Nayatola. LI-BIRD, Pokhara, p 76

Sinclair FL, Joshi L (2000) Taking local knowledge about trees seriously. In: Lawrence A (ed) Forestry, forest users and research: new ways of learning. ETFRN, Wageningen, pp 41–61

Sinclair FL, Walker DH (1999) A utilitarian approach to the incorporation of local knowledge in agroforestry research and extension. In: Buck LE, Lassoie JP, Fernandes ECM (eds) Agroforestry in sustainable agricultural systems. CRC Press LLC, Boca Raton, pp 245–275

Singh SP, Verma S (2008) Current and future status of herbal medicines. Vet World 1(11):347–350

Sivaramakrishnan K (2000) State sciences and development histories: encoding local forestry knowledge in Bengal. Dev Chang 31(1):61–90

The Lancet (1994) Pharmaceuticals from plants: great potential, few funds. Lancet 343:1513–1515

Upadhyay OP, Singh RH, Dutta SK (1996) Studies on antidiabetic medicinal plants used in Indian folklore. Aryavaidyan 9:159–167

Van Wyk AE, Van Oudtshoorn B, Gericke N (1997) Medicinal plants of South Africa. Briza Publications, Pretoria

Waliszewski WS, Mabote R, Sinclair FL (2003) Local knowledge about rangeland agroecology in the highlands of Lesotho. A guide to using the agroecological knowledge toolkit (AKT). School of Agricultural and Forest Sciences, University of Wales, Bangor

Zhao Z, Hu Y, Liang Z (2006) Authentication is fundamental for standardization of Chinese medicines. Planta Med 72:865–874

MIX
Papier aus verantwortungsvollen Quellen
Paper from responsible sources
FSC® C105338

If you have any concerns about our products,
you can contact us on
ProductSafety@springernature.com

In case Publisher is established outside the EU,
the EU authorized representative is:
Springer Nature Customer Service Center GmbH
Europaplatz 3, 69115 Heidelberg, Germany

Printed by Libri Plureos GmbH
in Hamburg, Germany